Textile Material Research—Statistical Tools of Experimental Data Analysis

Jiří Militký · Dana Křemenáková ·
Mohanapriya Venkataraman

Textile Material Research—Statistical Tools of Experimental Data Analysis

Springer

Jiří Militký
Department of Material Engineering
Faculty of Textile Engineering
Technical University of Liberec (TUL)
Liberec, Czech Republic

Dana Křemenáková
Department of Material Engineering
Faculty of Textile Engineering
Technical University of Liberec (TUL)
Liberec, Czech Republic

Mohanapriya Venkataraman
Department of Material Engineering
Faculty of Textile Engineering
Technical University of Liberec (TUL)
Liberec, Czech Republic

ISBN 978-981-97-6606-2 ISBN 978-981-97-6607-9 (eBook)
https://doi.org/10.1007/978-981-97-6607-9

This Springer imprint is published by the registered company Springer Nature Singapore Pte Ltd.
The registered company address is: 152 Beach Road, #21-01/04 Gateway East, Singapore 189721, Singapore

Contents

Chapter 1
Basics of Probability and Statistics

You have to dream before your dreams can come true [Abdul Kalam].

Abstract The essential concepts of probability and mathematical statistics are presented in a consistent form without detailed derivations. Some parts are also oriented to the description of special characteristics of random variables used in the textile metrology and in the engineering statistics. The practical limitations of standard methods based on the assumptions of normality and independence are shown. This chapter is written as a collection of topics supporting the easier readability of subsequent chapters.

1.1 Introduction

In this chapter, basic concepts and connections from the field of probability theory and mathematical statistics are discussed in a condensed form. Some connections are also given with the processing of metrological information and so-called engineering statistics. More detailed information on these and related problems, including full derivations, can be found in a number of mathematical statistics textbooks. Some problems of engineering statistics are discussed in monographs. (Meloun & Militký, 2011; Pham, 2023).

1.2 Concepts Related to Probability

The classic definition of probability is based on the concept of relative frequency. Let some phenomenon A (e.g. a head appears on top after a coin toss) has been observed in m trials (coin tosses). This phenomenon has occurred a total of m_A times. The relative frequency of occurrence of phenomenon A is given by the relation

J. Militký et al., *Textile Material Research—Statistical Tools of Experimental Data Analysis*, https://doi.org/10.1007/978-981-97-6607-9_1

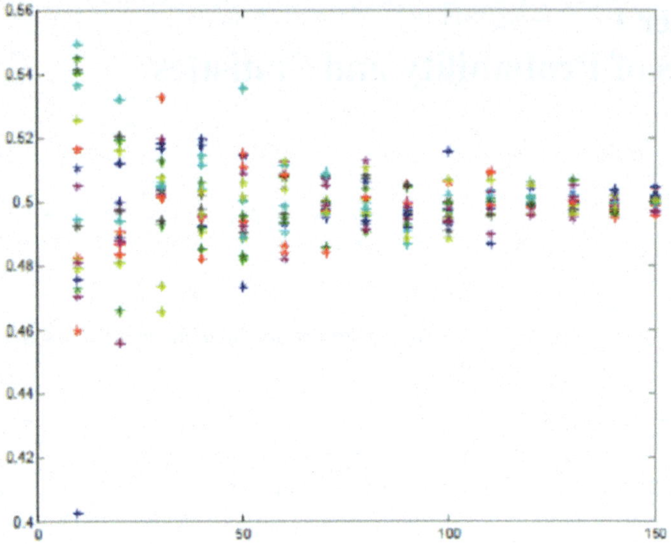

Fig. 1.1 Frequency f_A (on the *y*-axis) of appearing heads when rolling the dice for different numbers of throws (on the *x*-axis)

$$f_A = \frac{m_A}{m} \tag{1.1}$$

If the *m* trials are repeated, different values of f_A are obtained. Usually, the higher the value of *m*, the lower the spread of the values of f_A (see Fig. 1.1).

Phenomena that have this property are called statistically stable. The classic definition of probability is:

"for $m \to \infty$ $f_A = P(A)$, where $P(A)$ is the probability of event *A*".

This definition applies to statistically stable phenomena, when f_A converges to the constant $P(A)$ for $m \to \infty$. In the following, the designation *P*(statement) will mean the probability with which the statement given in parentheses is true. Thus, for example, $P(X < 5)$, indicates the probability with which the random variable *X* acquires values smaller than number 5. Phenomena that have this property are called statistically stable.

Symbolism and interrelationships between phenomena

- **Definite phenomenon** *S* (always occurs) $P(S) = 1$.
- **Impossible phenomenon** Ø (never occurs), then $P(\emptyset) = 0$.
- **Complementary phenomenon** *C* to the phenomenon *A* is $C = S - A$ and therefore $P(C) = 1 - P(A)$.
- **Union of phenomena** *A* and *B* (*A* or *B* or both occur simultaneously) $U = A \cup B$. This means that *U* is a phenomenon when at least one of the phenomena *A*, *B* occurs.

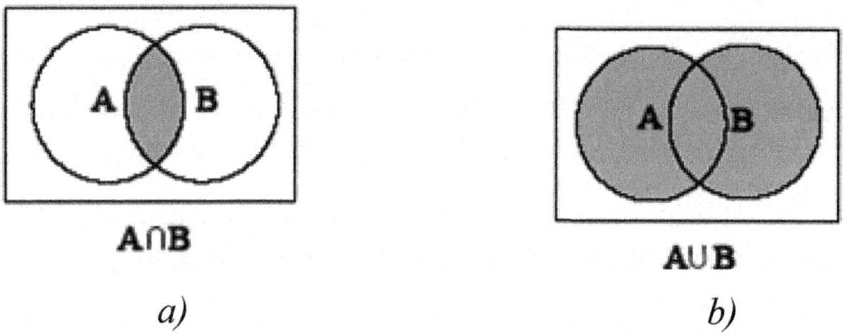

Fig. 1.2 Intersection (**a**) and union (**b**) of two random phenomena

- **Intersection of phenomena** A and B (both phenomena occur simultaneously) $I = A \cap B$. This means that I is a phenomenon when exactly both phenomena A, B occur (Fig. 1.2).
- **Incompatible (mutually exclusive or disjoint)** phenomena A and B (cannot occur simultaneously) $A \cap B = \emptyset$ (Fig. 1.2).
- **Elementary phenomenon** E (it cannot be expressed by union of other phenomena—it is not divisible). $0 \leq P(E) \leq 1$.
- The probability of the union of two phenomena A and B is generally given by the relation

$$P(A \cup B) = P(A) + P(B) - P(A \cap B)$$

For the mutually exclusive phenomena $A \cap B = \emptyset$ it holds that

$$P(A \cup B) = P(A) + P(B)$$

- **Rule of addition of probabilities**: The probability that at least one of the mutually exclusive events A_i will occur is equal to the sum of the probabilities $P(A_i)$.
- **Conditional probability** is the probability $P(A/B)$ that event A will occur given that event B occurred

$$P(A/B) = P(A \cap B)/P(B)$$

- **Probability of simultaneous occurrence** of events A and B

$$P(A \cap B) = P(A/B)P(B) = P(B/A)P(A)$$

- **Independent phenomena** A and B (the occurrence of A is not affected by the occurrence of phenomenon B) so $P(A/B) = P(A)$ and then

$$P(A \cap B) = P(A)P(B)$$

Rule of multiplication of probabilities: the probability of simultaneous occurrence of independent phenomena. A_i is equal to the product of probabilities $P(A_i)$.

Bayes relation $P(B/A) = P(A/B)P(B)/P(A)$.

$P(B)$ is a priori information, $P(B/A)$ is a posteriori information after the end of the experiment, $P(A/B)$ is information from the data. It can therefore be seen that the a priori information supplemented with information from the data leads to the refinement of a posteriori information after the end of the experiment.

1.3 Random Variables

A random variable is generally defined as a quantity that occurs in a chosen interval with a certain probability. Information about the behavior of a random variable X can be obtained from its realizations x (numerical results of experiments). All possible realizations of $\{x\}$ form a statistical ensemble. Its behavior is studied on a representative random sample (x_i) $i = 1 \ldots N$. **Probability theory** deals with the mathematical description of the behavior of the random variable X. **Mathematical statistics** are used to estimate the behavior of X from realizations x_i.

Discrete random variable: acquires only certain discrete levels (values) $x_i = 0, 1, 2, \ldots K$. To fully describe this random variable, the **probability function** $p(x_i) = P(X = x_i)$ is used, which gives the probability with which X acquires the value of precisely the level xi. Let n_i be the number of realizations at level "x_i" and N be the total number of realizations, then

$$p(x_i) = P(X = x_i) \approx n_i/N$$

From the definition of probability, it must hold that $\sum_{i=0}^{K} p(x_i) = 1$. Another function fully describing the behavior of a discrete random variable is the **cumulative probability**, or **distribution function** $F(x_i)$. This function gives the probability with which a discrete random variable acquires levels less than or equal to x_i, i.e.

$$F(x_i) = P(X \le x_i) = \sum_{j=0}^{i} p(x_i)$$

The relationship between the functions $p(x_i)$ and $F(x_i)$ is shown in Fig. 1.3.

Since the measurement results are always rounded (limited by the resolution of the measuring device, the finite range of numbers in computers, etc.), the measurement results are actually discrete random variables. If the number of levels is large enough (as in the case of rounding), the idea of a **continuous random variable** is chosen to describe the data, which takes on an arbitrary value x from the definition interval. For the probability that a continuous random variable lies in an infinitesimally small interval, it holds that

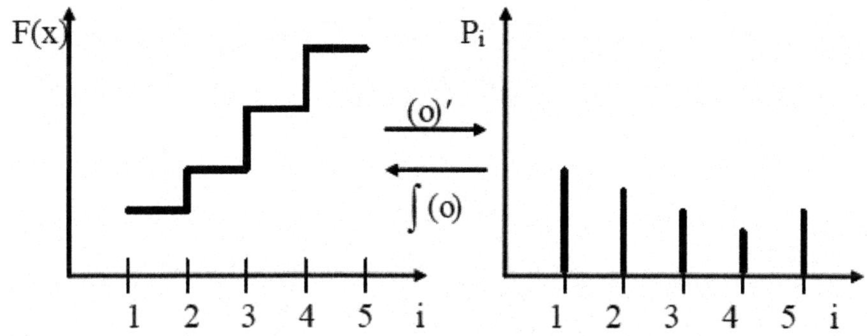

Fig. 1.3 Distribution and probability functions of a discrete distribution

$$P(x \leq X \leq x + \mathrm{d}x) \approx f(x)\mathrm{d}x \tag{1.2}$$

where $f(x)$ is the probability density function. For continuous random variables, from Eq. (1.2) it follows that regardless of the size of x, $P(X = x) = 0$. This seemingly paradoxical result is actually a consequence of the fact that real numbers are defined to an infinite number of decimal places. The relationship between the distribution function and the probability density function of continuous variable is shown in Fig. 1.4.

It is clear that the probability density function $f(x)$ is not a probability, but a relation holds

$$f(x) = \frac{P(x \leq X \leq x + \mathrm{d}x)}{\mathrm{d}x}$$

Figure 1.5 shows how the probability density is related to the probability with which a continuous random variable is found in a given interval.

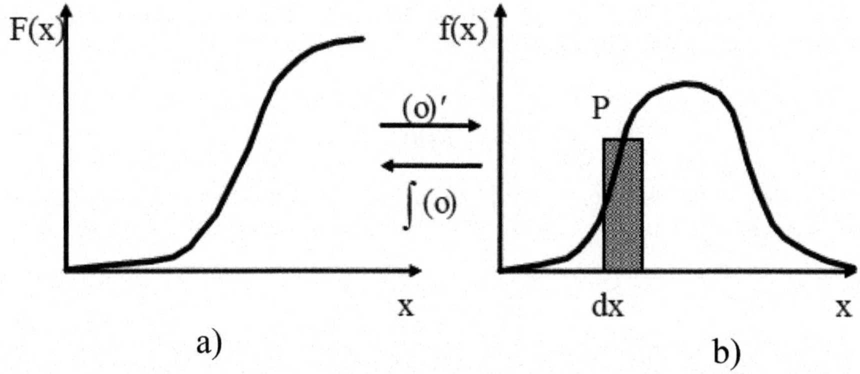

Fig. 1.4 **a** Distribution function, **b** probability density function of a continuous random variable

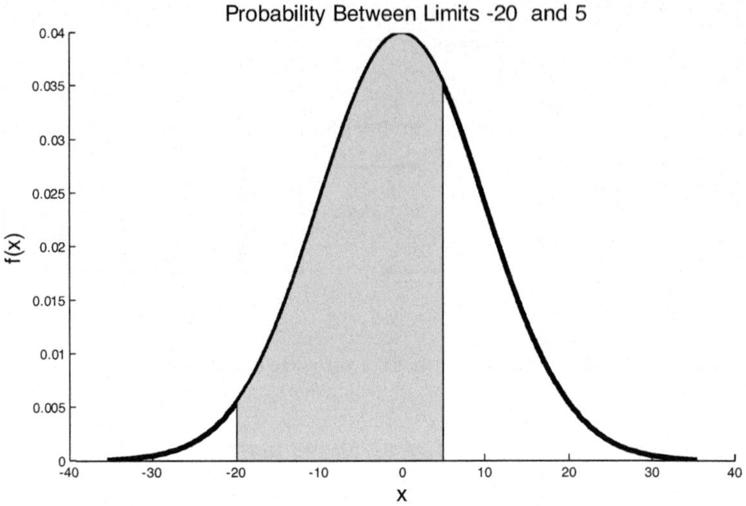

Fig. 1.5 Relationship between probability density $f(x)$ and probability—gray area

Because probability is defined in the interval from zero to one, it is clear that the total area under the probability density function must be equal to one (certainty) and therefore probability density has dimension reciprocal to dimension of random variable x. This is important in metrology because by changing of random variable x unit size (e.g. degree in Celsius to Kelvins or mm to cm) is changing the shape of probability density function.

It is commonly assumed that a continuous random variable is defined on the entire real axis. In many practical cases, the range of values of a random variable is limited by, e.g. measuring equipment, measurement conditions, physical limitations, etc. Let X be a continuous random variable from the interval $[-\infty, \infty]$, but the measurement results X^* are from the limited interval $[A, B]$. Then a **truncated distribution** can be used. If the probability density of the random variable X is $f(x)$, then the probability density of the variable X^* is equal to

$$f^*(x) = \frac{f(x)}{F(B) - F(A)}$$

The distribution function of X is $F(x)$, but the distribution function of the quantity X^* is equal

$$F^*(x) = \frac{F(x) - F(A)}{F(B) - F(A)}$$

I. **Probability density function** $f(x)$—denoted as pdf.

The probability density function $f(x)$ is the first derivative of the distribution function

$$f(x) = \mathrm{d}F(x)/\mathrm{d}x.$$

Properties of $f(x)$:

1. positive $f(x) \geq 0$
2. normalized $\int f(x)\mathrm{d}x = 1$
3. For two random variables x and y with densities $f(x)$ and $g(y)$, the random variable $z = x + y$ has the distribution $h(z) = \int f(x)g(z - x)\mathrm{d}x = \int f(z - y)g(y)\mathrm{d}y$.
4. The units of $f(x)$ are [1/(random variable units)].

If x and y have the same uniform (rectangular) distribution, their sum has a triangular distribution. In general, the maximum value of the probability density is related to the units of measurement (the normalization condition (2) must be maintained). If the probability density is zero in some interval, the probability of occurrence of a random variable in this interval is also zero.

II. **Distribution function F(x)**—abbreviation cdf (cumulative density function).

The dimensionless $F(x)$ corresponds to the probability with which a random variable x has values smaller than x. Thus

$$F(x) = P(X \leq x) \text{ or } F(x) = \int_{-\infty}^{x} f(x)\mathrm{d}x$$

Properties of F(x):

1. Bounded from below by $F(-\infty) = 0$, and from above by $F(-\infty) = 1$
2. Non-decreasing, i.e. $F(x + \mathrm{d}x) \geq F(x)$ (probability is non-negative)
3. It is valid that $P(a \leq X < b) = F(b) - F(a)$ i.e. graphically as in Fig. 1.6:

The distribution function is simply related to the **reliability function** $R(x)$. If the random variable X is the time until failure t (of material, equipment, etc.) from the interval $0 \leq t < \infty$, the reliability function $R(t)$ expresses the probability with which failure will not occur until time t, i.e. $R(t) = 1 - F(t)$.

Properties of $R(t)$:

1. Bounded from below by $R(0) = 1$, and from above by $R(\infty) = 0$
2. Non-increasing, i.e. $R(t + \mathrm{d}t) \leq R(t)$.

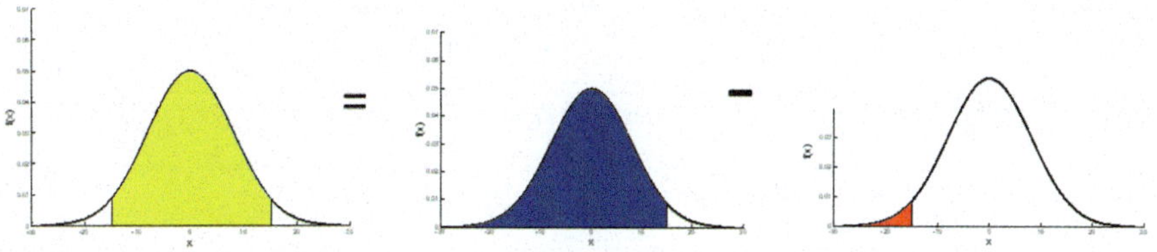

Fig. 1.6 Properties of $F(x)$

The probability density corresponding to the reliability function is given by the relation

$$f_R(t) = -\frac{\mathrm{d}R(t)}{\mathrm{d}t}$$

This function is also often referred to as the **fault density**. The ratio of the fault density to the reliability function is referred to as the **fault intensity** $\lambda(t)$, which is defined by the relation

$$\lambda(t) = \frac{f_R(t)}{R(t)}$$

The interrelationships between these functions are shown in Table 1.1.

Functions $f(t)$ and $\lambda(t)$ have dimension [1/time]. The intensity of faults expresses approximately the probability that a fault will occur in the interval $(t + \Delta t)$ for Δt approaching to zero.

III. **Quantile function** $Q(u)$ for $0 \leq u < 1$.

The quantile function $Q(u)$ is the inverse function of the distribution function $Q(u) = F^{-1}(x)$ The following relationship is also valid

$$F(x) = \int_{-\infty}^{Q(u)} f(x)\mathrm{d}x$$

The notation α-quantile $\tilde{x}_\alpha$ is used for $Q(\alpha)$. The α-quantile is value of random variable x for with is probability that x is below $\tilde{x}_\alpha$ just α.

$$P(x \leq \tilde{x}_\alpha) = \alpha \tag{1.3}$$

Table 1.1 Relationships between functions describing the behavior of the random variable t

	$R(t)$	$F(t)$	$f(t)$	$\lambda(t)$
$R(t)$	$R(t)$	$1 - F(t)$	$1 - \int_0^t f(x)\mathrm{d}x$	$\exp\left[-\int_0^t \lambda(x)\mathrm{d}x\right]$
$F(t)$	$1 - R(t)$	$F(t)$	$\int_0^t f(x)\mathrm{d}x$	$1 - \exp\left[-\int_0^t \lambda(x)\mathrm{d}x\right]$
$f(t)$	$-\frac{\mathrm{d}R(t)}{\mathrm{d}t}$	$\frac{\mathrm{d}F(t)}{\mathrm{d}t}$	$f(t)$	$\lambda(t)\exp\left[-\int_0^t \lambda(x)\mathrm{d}x\right]$
$\lambda(t)$	$-\frac{\mathrm{d}(\ln R(t))}{\mathrm{d}t}$	$\frac{\mathrm{d}F(t)}{\mathrm{d}t}/(1 - F(t))$	$\frac{f(t)}{1-\int_0^t f(x)\mathrm{d}x}$	$\lambda(t)$

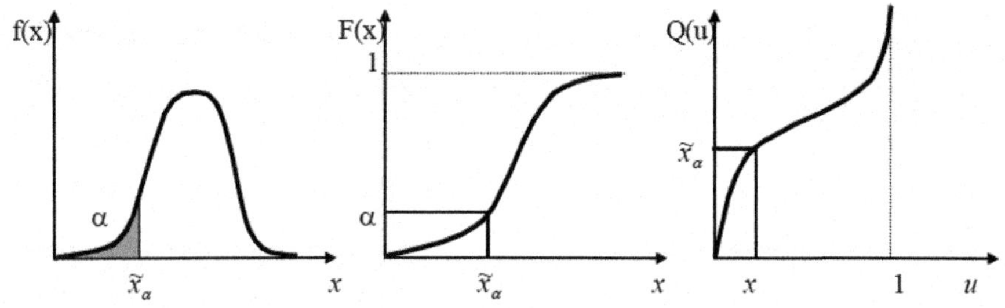

Fig. 1.7 Relationships between probability density, distribution function and quantile function

Properties of Q(u):

1. Mirroring the distribution: $Q(1 - u)$ is the mirror image of the distribution $Q(u)$ around the origin.
2. If $Q_1(u)$ and $Q_2(u)$ are quantile functions, their weighted sum with weights w and $1 - w$ is $Q(u) = wQ_1(u) + (1 - w)Q_2(u)$ and the quantile function is lying between these two functions
3. The quantile function for reciprocal variable $1/x$ is $1/Q(1 - u)$
4. If variable $z = T(x)$ (for suitable function $T(.)$) is a non-decreasing function of x, then $T(Q(u))$ is also a quantile function (Fig. 1.7).

Quantile frequency function $q(u) = \left[\mathrm{d}Q(u)\right]/\mathrm{d}u$.

Frequency quantile function (eff - cue function) $f[Q(u)]$.

It is valid, that $f\left[Q(u)\right]q(u) = 1$.

In textile metrology, a function called **staple diagram** $S(u)$ for $0 \leq u \leq 1$is used to describe the distribution of fiber lengths l in an interval $0 \leq l \leq l_m$. This function is the inverse of the so-called diagram $d(l)$, defined in the same way as the reliability function $d(l) = R(l)$, i.e.

$$S(u) = d^{-1}(l) = (1 - F(l))^{-1} \text{ or } F(l) = 1 - \int_{S(u)}^{l_m} f(l)\mathrm{d}l \tag{1.4}$$

In Fig. 1.8 is schematically shown the diagram $R(l)$ and the staple diagram $S(u)$.

The definition of a staple diagram is motivated by its preparation. It is based on a fibrous tuft containing a total of N fibers. These fibers are gradually (according to size from the longest to the shortest) placed on the velvet mat. The resulting dependence has the length of fibers l on the y-axis and the number of longer fibers or equal, of size l on the x-axis, respectively. The probability of the occurrence of fibers longer than l is then estimated by the relation $u \approx n_l/N$.

Depending on the measurement method, the fiber length distribution can be described by the so-called mass probability density $f_w(l)$, which expresses the mass of fibers $w(l)$ whose length lies in the interval $[l - \mathrm{d}l, l + \mathrm{d}l]$, or by the frequency

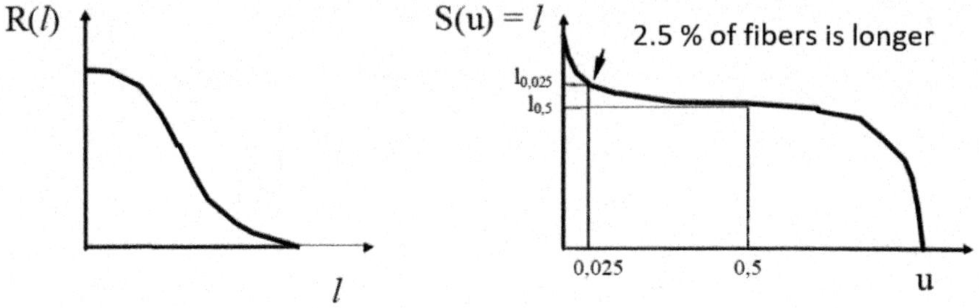

Fig. 1.8 Diagram $R(l)$ and staple diagram $S(u)$

probability density $f_n(l)$, which expresses the number of fibers $n(l)$, the length of which lies in the interval $[l - \mathrm{d}l, l + \mathrm{d}l]$. Frequency and mass diagrams are then defined by relationships

$$d_w(l) = \int_l^{l_m} f_w(l)\mathrm{d}l \quad d_n(l) = \int_l^{l_m} f_n(l)\mathrm{d}l \tag{1.5}$$

Especially when measuring the fibrous fringe, i.e. the protruding lengths of fibers clamped in one place, the so-called **fibrogram G(*l*)** is obtained (Azzouz et al., 2008). Frequency and mass fibrograms are defined by relationships

$$G_w(l) = \frac{1}{E_w(l)} \int_l^{l_m} d_w(l)\mathrm{d}l \quad G_n(l) = \frac{1}{E_n(l)} \int_l^{l_m} f_n(l)\mathrm{d}l \tag{1.6}$$

Probability density functions can have different shapes characterizing the behavior of a random variable (see Fig. 1.9).

In metrology, measurement errors are commonly assumed to have a unimodal distribution with one maximum (mode) on the probability density function (see Fig. 1.10a, b). Bimodality (Fig. 1.9c) or polymodality is commonly considered to be an example of a mixture measurements (e.g. measuring lengths of a cotton/polyester blend without separating the individual components). When processing this type of data, the task is to separate the individual components and analyze the individual components separately. However, a bimodal distribution was also found in the experimental error distribution of some measuring devices (Novitskii & Zograf, 1985).

Symmetric distributions (around the "*S*" value) have the following properties:

$$f(x) = f(2S - x) \quad F(x) = 1 - F(2S - x) \quad Q(1 - u) = 2S - Q(u)$$

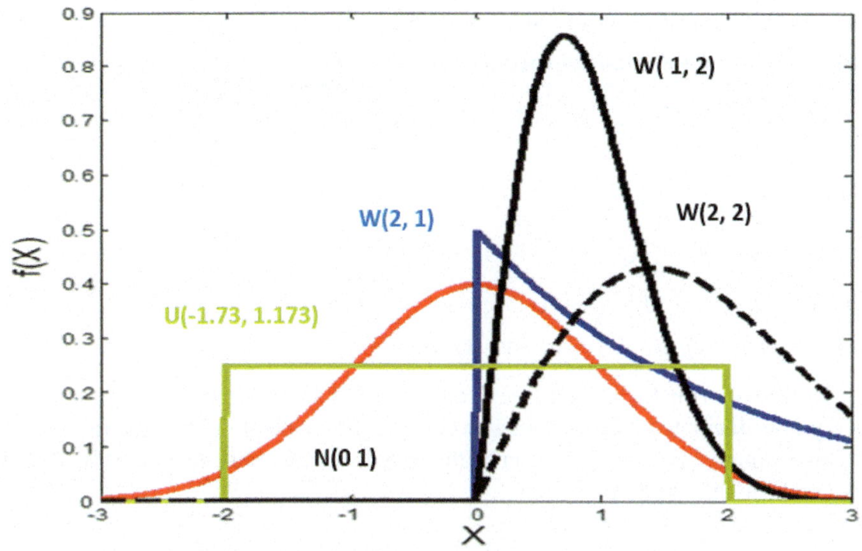

Fig. 1.9 Selected probability density function shapes (normal distribution N, uniform distribution U, and various Weibull distributions W)

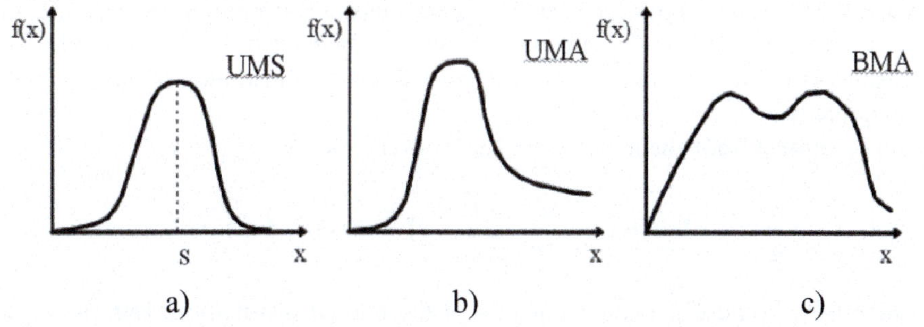

Fig. 1.10 Different forms of probability density function: **a** symmetric unimodal, **b** asymmetric unimodal, **c** bimodal

Distribution of a transformed random variable

In many cases, the values obtained by measurement need to be suitably transformed in order to obtain the desired result. For example, the cross-sectional area of circular fibers is calculated from the measurement of their radii. Another common motivation for a transformation is to improve the distribution of the original random variable (suppressing asymmetry, approaching to normality, etc.) In general, a random variable X with pdf $f(x)$ is available and a particular "monotonic" transformation $y = \mathrm{g}(x)$ is known. We need to find the distribution of the random variable Y, i.e. $h(y)$:

(a) for **discrete random variables** is valid $h(y) = f\left(g^{-1}(y)\right)$
(b) for **continuous random variables** is valid $h(y) = \frac{\mathrm{d}g^{-1}(y)}{\mathrm{d}y} f\left(g^{-1}(y)\right)$

where $\mathrm{d}g^{-1}(y)/\mathrm{d}y$ is a normalization factor referred to as the Jacobian of the transformation.

Examples

(a) $y = ax + b,\ g^{-1}(y) = \frac{y-b}{a}\ h(y) = \frac{1}{|a|} f\left(\frac{y-b}{a}\right)$
(b) $y = ax^2,\ g^{-1}(y) = \sqrt{\frac{y}{a}}\ h(y) = \frac{1}{2\sqrt{ay}} f\left(\sqrt{\frac{y}{a}}\right)$
(c) $y = aexp(x),\ g^{-1}(y) = \ln\left(\frac{y}{a}\right)\ h(y) = \frac{1}{|y|} f\left(\ln\left(\frac{y}{a}\right)\right)$

The **linear transformation** (see a) only changes the position and dispersion, but every **non-linear transformation** also changes the shape of the distribution.

Interestingly, the transformation by the own distribution functions $y = F(x)$ leads to a uniform distribution

$$f(y) = f\left[F^{-1}(y)\right]\left[\mathrm{d}F^{-1}(y)\right]/\mathrm{d}y = f\left[Q(y)\right]q(y) = 1$$

In metrological applications, there is often a situation where the resulting random variable (e.g. measurement error) is the sum of random variables with some distributions. The distribution of the resulting sum is then a convolution (see property ad 3) of the probability density functions. How quickly the distribution of the sum of equally distributed random variables changes is a good illustration of the approach to normality.

The **central limit theorem** states that for every $x < y$ is

$$\lim_{n\to\infty} P[x \le \frac{S_n - n\mu}{\sigma\sqrt{n}} \le y] = \Phi(y) - \Phi(x)$$

where $\Phi(x)$ is the distribution function of the normal distribution. Here is $S_n = X_1 + X_2 + \cdots + X_n$ the sum of n independent, equally distributed, random variables with mean μ and standard deviation σ.

1.4 Characteristics of Random Variables

Numerical characteristics known as moments are commonly used to characterize the shape peculiarities of the distribution. **General moments** M_k and their estimates from a random sample $\{x_i\}$ $i = 1, \ldots N$ are defined by the relations

$$M_K = \int_{-\infty}^{\infty} x^K f(x)\mathrm{d}x \quad \hat{M}_K = \frac{1}{N}\sum_{i=1}^{N} x_i^K$$

It is obvious that $\hat{M}_1$ is actually the arithmetic mean.

The **central moments** C_k and their estimates $\hat{C}_k$ from a random sample $\{x_i\}$ $i = 1, \ldots N$ are defined by the relations

$$C_K = \int_{-\infty}^{\infty} (x - M_1)^K f(x)\mathrm{d}x \quad \hat{C}_K = \frac{1}{N}\sum_{i=1}^{N} (x_i - \hat{M}_1)^K$$

If the mean value operator $E(x) = M_1$ is defined, it can be central moments written as $C_K(x) = E[x - E(x)]^K$.

So, the central moment is actually the mean value of the K-th power of the deviations from the mean value. It can be easily shown that the first central moment must be equal to zero, i.e. $C_1 = 0$. This means that the sum of the deviations of the individual elements of the sample from the arithmetic mean is always equal to zero.

The mean value (mathematical expectation) $M_1 = E(x) = \mu$ is commonly used to characterize the **location** of a random variable with distribution $f(x)$, for which

$$E(x) = \int_{-\infty}^{\infty} xf(x)\mathrm{d}x$$

In Fig. 1.11a two unimodal symmetric probability distributions differing in mean value are shown. It is noticeable that in this case the mean value corresponds to the position of the maximum probability density on the x-axis, and the distributions differing only in the mean value are only shifted along the x-axis.

Properties of the mean value: $E(x) = M1 = \mu$.

The units of the mean value are the same as the units of the realization of the random variable X.

1. Constants A, B $E(Ax \pm B) = AE(x) \pm B$ This fact is very important from a metrological point of view, because it shows that if different types of units are used, which can be expressed as a linear transformation $Y = A X + B$, the mean

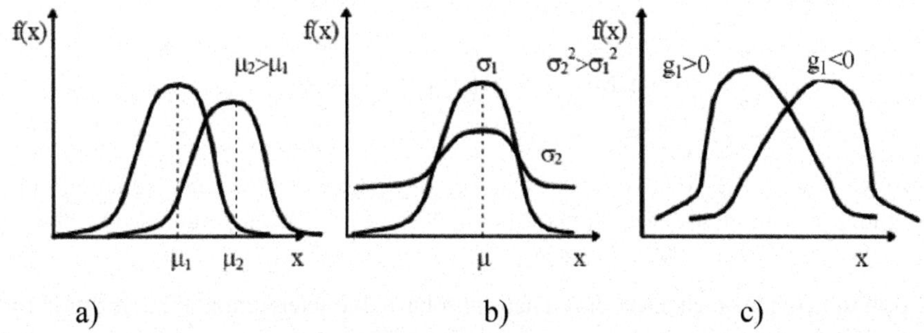

Fig. 1.11 Different characteristics of the distribution of random variables

value $E(y)$ can be easily determined as $A\ E(x) + B$ without the need to convert individual measurements to another system of units.

2. Independent random variables $(X,\ Y)$ $E(x \pm y) = E(x) \pm E(y) or E(xy) = E(x)E(y)$.
3. correlated random variables $(X,\ Y)$

The degree of dependence is expressed by covariance $\operatorname{cov}(x, y) = E(xy) - E(x)E(y)$ and then

$$E(xy) = E(x)E(y) + \operatorname{cov}(x, y)$$

Centering: removing dependence on the mean value $X^0 = X - E(x) \Rightarrow E\left(x^0\right) = 0$.

4. The mean value from the quantile function $E(x) = \int_0^1 Q(u)\mathrm{d}u$

In textile metrology, a number of other location characteristics are used specifically for the case of measuring fiber lengths. The **mean frequency and weight length** of the fibers are defined relations

$$E_w(l) = \int\limits_0^{l_m} lf_w(l)\mathrm{d}l \text{ and } E_n(l) = \int\limits_0^{l_m} lf_n(l)\mathrm{d}l$$

Upper Half Mean Fiber Length (UHML) is the average length of fibers longer than the median length (it is the upper half of the fiber lengths). For the frequency distribution of lengths, the UHML is expressed in the form (Azzouz et al., 2008)

$$\text{UHML} = \frac{1}{d_n(\hat{l}_{0.5})} \int\limits_{\hat{l}_{0.5}}^{l_m} lf_n(l)\mathrm{d}l$$

If the probability density of fiber lengths $f_n(l)$ can be approximated by the probability density of a normal distribution with mean value $E_n(l)$ and variance $D_n(l)$, then (Azzouz et al., 2008)

$$\text{UHML} = E_n(l)\left(1 + 0.8\frac{\sqrt{D_n(l)}}{E_n(l)}\right)$$

Mode: Mo is the relative maximum on the probability density function. For symmetric unimodal distributions, $E(x) = \text{Mo}$, i.e. the mean value corresponds to the maximum on the function $f(x)$.

Median is the 50% quantile. It is a natural robust characterization of the location for which it is

$$P(X \leq \tilde{x}_{0.5}) = 0.5 \quad \text{t.j.} \quad \tilde{x}_{0.5} = Q(0.5)$$

Upper Quartile Length (UQL) is defined as the length exceeded by only 25% of the longest fibers. For the frequency distribution of lengths, the UQL is expressed in the form (Azzouz et al., 2008)

$$\int_{\mathrm{UQL}}^{l_m} f_n(l)\mathrm{d}l = d_n(\mathrm{UQL}) = 0.25$$

If the probability density function of fiber lengths $f_n(l)$ can be approximated by the probability density of a normal distribution with mean value $E_n(l)$ and variance $D_n(l)$, then (Azzouz et al., 2008)

$$\mathrm{UQL} = E_n(l)\left(1 + 0.67\frac{\sqrt{D_n(l)}}{E_n(l)}\right)$$

For symmetric unimodal distributions, the equalities $E(x) = \mathrm{Mo} = \tilde{x}_{0.5}$ is valid. The relationships between these characteristics for different types of unimodal distributions are shown in Fig. 1.12.

For distributions skewed to higher values is $E(x) \gg \mathrm{Mo}$ and for distributions skewed to lower values, $\mathrm{Mo} \gg E(x)$. It can be seen that the mean value is always farthest from the mode value.

For positive x, it is valid that $\tilde{x}_\alpha = E(x)/(1-\alpha)$ and therefore at most $\tilde{x}_{0.5} = 2E(x)$.

To characterize of spread or the **variability** (concentration around the mean value) the second central moment $C_2 = D(x) = \sigma^2$ referred to as variance is used.

$$D(x) = \int_{-\infty}^{\infty} (x - E(x))^2 f(x)\mathrm{d}x = E[x - E(x)]^2$$

The square root of the variance $\sigma = \sqrt{D(x)}$ is called the standard deviation.

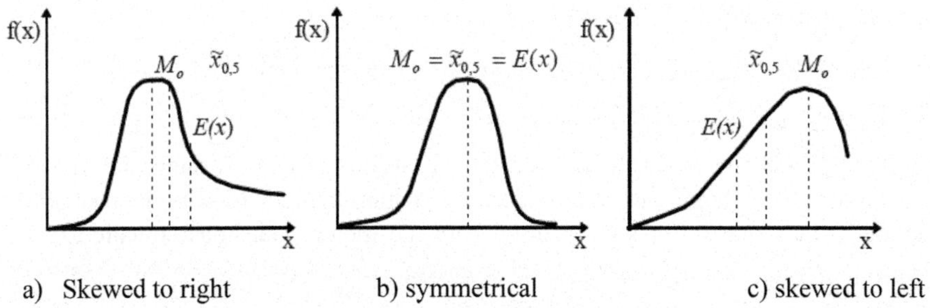

Fig. 1.12 Location characteristics for the unimodal distributions

Properties of variance: $\mathbf{D(x) = C_2 = \sigma^2}$.

The units of the variance are the same as the squared units of the realization of the random variable X, which can lead to problems in expressing them.

Variance is defined as:

$$D(x) = E[x - E(x)]^2 = E\left(x^2 - 2xE(x) + E(x)^2\right) = E(x^2) - E(x)^2$$

So, the variance is the mean of the square of a random variable minus the square of its mean. The last relationship, which is sometimes referred to as a "calculator form", is advantageous in that it is sufficient to load successively individual values (variable P_1), their squares (variable P_2), and their number (variable P_0) for arbitrarily large sample without storing individual elements. After storage is complete, an estimate of the mean is determined as P_1/P_0 and an estimate of the variance as P_2/P_0 - $(P_1/P_0)^2$. The disadvantage of this simple procedure is that, due to rounding errors on computers, a negative variance may be found (of course, it should be at least zero).

1. Constants A, B $D(Ax \pm B) = A^2 D(x)$

This fact is again very important from a metrological point of view, because it shows that if different types of units are used, which can be expressed as a linear transformation $Y = A\,X + B$, the variance $D(y)$ can be determined easily as $A^2\,D(x)$ without the need to convert individual measurements to another system of units.

2. Independent random variables (X, Y) $D(x \pm y) = D(x) + D(y)$. So the variance does not decrease.

For centered X and Y is valid that $D(xy) = D(x)D(y) + E(x)^2 D(y) + E(y)^2 D(x)$.

3. Correlated random variables (X, Y)

$$D(x \pm y) = D(x) + D(y) + 2\,\mathrm{cov}(x, y)$$
$$D(xy) = D(x)D(y) + E(x)^2 D(y) + E(y)^2 D(x) + 2E(x)\,E(y)\mathrm{cov}(x, y)$$

4. For arbitrary A is $E\left[(x - A)^2\right] \geq D(X)$. Thus, the variance minimizes the root mean square deviation from the mean value.
5. Variance is always not negative $\sigma^2 \geq 0$.
6. Variance from a quantile function $D(x) = \int_0^1 [Q(u) - E(x)]^2 \mathrm{d}u$.
7. Following relation is valid $|\tilde{x}_{0.5} - \mu| \leq \sigma$.

In Fig. 1.11b two probability density functions that differ in variance are shown. It can be seen that a larger variance leads to a distribution that with a higher probability admits values that differ significantly from the mean value (has so-called longer tails).

Chebyshev's inequality is used to estimate the probability with which the random variable X appears in the selected interval around the mean value:

$$P(E(x) - \varepsilon\sqrt{D(x)} \leq X \leq E(x) + \varepsilon\sqrt{D(x)}) \geq 1 - c/\varepsilon^2$$

The constant c is generally: $c = 1$, for the unimodal distribution it is: $c = 0.57$ and for the symmetric distribution it is: $c = 0.44$. From this inequality it is possible to determine, for example, the probability $P(X \in E(x) \pm 2\sigma)$ that the random variable X lies in the interval $E(x) \pm 2\sigma$:

For general distribution is $P(X \in E(x) \pm 2\sigma) = 0.75$,
For symmetric distribution is $P(X \in E(x) \pm 2\sigma) = 0.889$,
For unimodal distribution is $P(X \in E(x) \pm 2\sigma) = 0.859$,
For normal distribution is $P(X \in E(x) \pm 2\sigma) = 0.954$.

It is therefore evident that the interval $E(x) \pm 2\sigma$ contains at least 75% of all possible realizations of a continuous random variable.

In textile metrology, special variance characteristics are used specifically for the case of measuring fiber lengths. The **variances** of the mean frequency and weight length of the fibers are defined by the relations

$$D_w(l) = \int_0^{l_m} (l - E_w(l))^2 f_w(l)\mathrm{d}l \quad E_n(l) = \int_0^{l_m} (l - E_n(l))^2 f_n(l)\mathrm{d}l$$

Standardization (scoring) is often performed in practical statistics to remove location and variance dependence.

$$U = \frac{X^0}{\sqrt{D(x)}} = \frac{X - E(x)}{\sqrt{D(x)}}$$

It is valid, that:

1. $E(U) = 0$ and $D(U) = 1$, the quantity U is therefore dimensionless and does not depend on position and scale (expressed by standard deviation). This allows easy comparison of samples differing in location and variance. However, on the other hand, the noise component is highlighted, and if the initial selection is an insignificant "noise", a significant signal will result after scoring.
2. $f(x) = (1/s)f(u)$,
3. $F(x) = F(u)$,
4. $Q(z) = m + sQ_0(z)$ for $\leq z < 1$.

Among the other scale characteristics, the standard deviation is particularly important, because its units are the same as the units of the realization of the random variable X. Multiples of the standard deviation are also used in determining measurement uncertainties.

Standard deviation $\sigma = \sqrt{D(x)}$, beware, it is a non-linear transformation.

1. Independent random variables: X and Y $\sigma(x + y) = \sqrt{\sigma_x^2 + \sigma_y^2}$,

2. Fully correlated random variables X and Y with covariance $\mathrm{cov}(x, y) = \sqrt{\sigma_x^2 \sigma_y^2}$: $\sigma(x + y) = \sigma_x^2 + \sigma_y^2$.

Average absolute deviation $d = E[|x - E(x)|]$, for a better approximation to the standard deviation, in the case of a normal distribution, $d = 1.2533\, E[|x - E(x)|]$ is chosen.

In general, the mean absolute deviation is less than or equal to the standard deviation, that is, $d \leq \sqrt{D(x)}$. For a normal distribution, the ratio $d^2/\sigma^2 = 2/\pi$. Note that the mean absolute deviation is not robust, even if it uses an absolute value (Rousseeuw & Croux, 1993).

Interquartile deviation: $q = 0.5\,(\tilde{x}_{0.75} - \tilde{x}_{0.25})$.

In general, $q \leq \sqrt{2D(x)}$. For a normal distribution is $q \approx 0.675\sigma$.

A special characteristic of variability is the concentration K defined as

$$K = \frac{1}{2\sigma\sqrt{3}} \approx \frac{0.2887}{\sigma}$$

Concentration is closely related to the approximation of the probability density function of a given distribution by the probability density of a uniform distribution (rectangle in the interval from a to b with height)$1/(b-a)$, which has the same mean and variance (see Fig. 1.13).

For a uniform distribution, it holds that $\mu = E(x) = (a + b)/2 \;\; \sigma^2 = (b - a)^2/12$. From these relations it can be determined that $a = \mu - \sigma\sqrt{3} \;\; -b = \mu + \sigma\sqrt{3}$. The difference $b - a = 2\sigma\sqrt{3}$ is called the concentration interval, and the concentration is its reciprocal.

A **relative measure of variability** is the coefficient of variation V, which is often used in metrology as a relative measure of measurement accuracy.

$$V = \frac{\sigma}{E(x)}$$

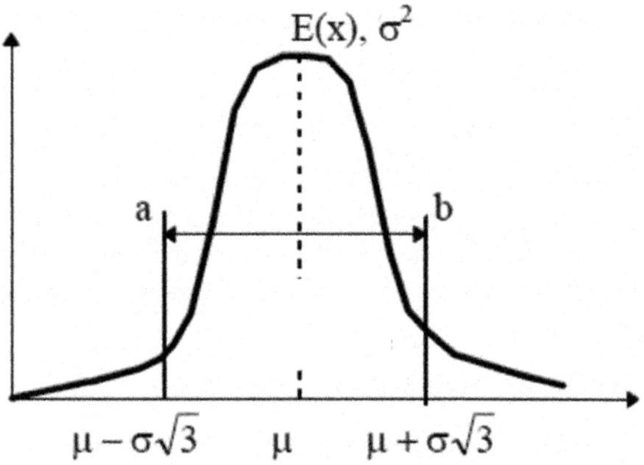

Fig. 1.13 Approximation of the probability density function (curve) by a rectangular distribution

It can be seen that the coefficient of variation shows what proportion of the mean value is the standard deviation. Measurements with a coefficient of variation below 0.01 are referred to as very accurate, with a coefficient of variation up to 0.05 as accurate and above 0.05 as not very accurate. However, this is true if the variance is primarily due to the variance of the measuring device. In the event that the dispersion is mainly caused by fluctuations in the characteristics of the measured material, this division is meaningless.

Properties of the coefficient of variation

1. $V(A) = 0$,
2. $V(Ax) = V(x)$,
3. $V(x + y) = \frac{\sqrt{V^2(x)\,E^2(x)+V^2(y)\,E^2(y)}}{E(x)+E(y)}$.

If $V(x) = V(y)$ and $E(x) = E(y)$, then $V(x + y) = V(x)/\sqrt{2}$. For the case of n identical strands, the resulting coefficient of variation is equal to $V = V(x)/\sqrt{n}$. This relationship shows the resulting non-uniformity created by the association of n identical slivers during drafting.

Shape peculiarities of probability density functions can be characterized by skewness and kurtosis (see Fig. 1.11c). The third normalized central moment is used to characterize the **skewness** of the distribution

$$g_1(X) = \frac{C_3(x)}{\sigma^2\,\sigma}$$

It is true that for symmetric distributions all odd central moments are zero and thus the skewness is also zero. For a distribution skewed to the right (i.e. to higher values) the skewness is positive and for a distribution skewed to the left (i.e. to lower values) the skewness is negative.

Properties of the skewness

1. $g_1(A + Bx) = g_1(x)$,
2. For $z_n = (\overline{x} - \mu)\sqrt{n}/\sigma$ is $g_1(z_n) = g_1(\overline{x}) = g_1(x)/\sqrt{n}$,
3. $P(z_n \le x) \approx \Phi(x) - g_1(x)\,(x^2 - 1)\,\phi(x)/6\sqrt{n}$, where $\Phi(x)$ is the distribution function of the standardized normal distribution and $\phi(x)$ is the corresponding probability density function.

The fourth normalized central moment is used to characterize the **kurtosis** of the distribution

$$g_2(x) = \frac{C_4(x)}{\sigma^2\,\sigma^2}$$

Properties of the kurtosis

1. $g_2(A + Bx) = g_2(x)$,
2. For $z_n = (\overline{x} - \mu)\sqrt{n}/\sigma$ is $g_2(z_n) = g_2(\overline{x}) = 3 + (g_2(x) - 3)/n$
3. $g_2 \ge g_1^2 + 1$.

Probability densities with different skewness and kurtosis are shown in Fig. 1.14. The kurtosis (excess) g_2 is suitable only for the characterization of symmetric distributions with different "tail lengths". For a normal distribution is $g_2 = 3$. For a distribution with "short" tails is $g_2 < 3$. The limiting is the rectangular distribution, for which $g_2 = 1.8$. For distributions with "long" tails is $g_2 > 3$. The limit is the two-sided exponential (Laplace) distribution, for which $g_2 = 6$. For asymmetric distributions, the kurtosis can be considerably higher. For the exponential distribution is $g_2 = 9$.

In some cases, in metrology, the shape of the probability density function $f(x)$ is also characterized by the **entropy coefficient** E, for which $E = 0.5\ \exp(H)/\sigma$, where H is the Shannon measure of entropy defined by

$$H = \int_{-\infty}^{\infty} f(x)\ln(f(x))\mathrm{d}x$$

By default, H is used to express the distribution of measurement errors, where the mean value is $E(x) = 0$.

Maximum value $E = 2.066$ has a normal distribution. It is usually true that $1.11 < E < 2.055$. It is obvious that for a known distribution of a random variable defined, for example, by the probability density $f(x)$, all basic statistical characteristics can be determined directly from the above expressions. The problem is the estimation of moments from the data, when for example the coefficient of variation and skewness estimates are strongly biased even for large samples from non-symmetric distributions ($N > 1000$) (Vogel & Fennessey, 1993).

Hosking (Hosking, 1990) introduced L-moments as analogs of classical moments, expressed as a linear combination of order statistics $x_{(i)}$ [for definitions, see Sect. 6. 5, Eq. (6.67)]. These are actually probability weighted moments (PMW) defined by the relation

$$M_{p,r,s} = E\left(x^p[F(x)]^r[1 - F(x)]^s\right)$$

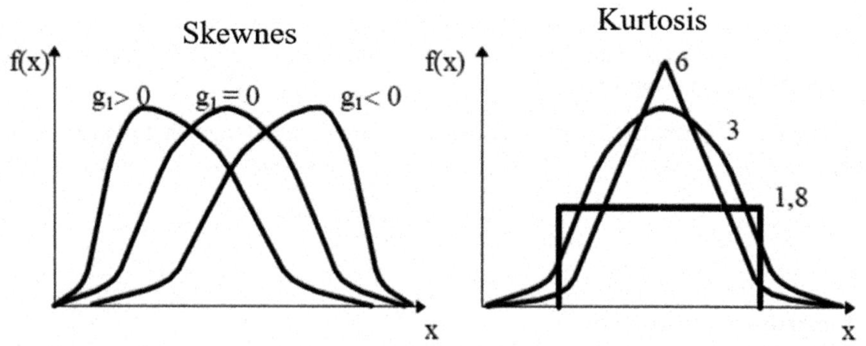

Fig. 1.14 Probability density functions for different skewness and kurtosis

where $F(x)$ is distribution function x and p, r, s are real numbers. The moments $M_{p,0,0}$ for $p = 1, 2, 3...$ are standard general moments of random variable x. Moments $M_{1,r,0} = \beta$, can be defined as $\beta_r = E\left(x[F(x)]^r\right)$. The unbiased moment β_r estimate has the form

$$b_r = \frac{1}{N}\frac{\sum_{i=1}^{N}\binom{i-1}{r}x_{(i)}}{\binom{N-1}{r}} \quad r = 0, 1, \ldots (N-1)$$

PWMs are generally suitable for distribution parameter estimates that can be easily expressed analytically using a quantile function (Gumbel, exponential).

The first four L-moments are determined by the relations

$$L_1 = \beta_0, L_2 = 2\beta_1 - \beta_0, L_3 = 6\beta_2 - 6\beta_1 + \beta_0, L_4 = 20\beta_3 - 30\beta_2 + 12\beta_1 + \beta_0$$

Estimates of these moments $\left(\hat{L}_1, \hat{L}_2, \hat{L}_3, \hat{L}_4\right)$ are obtained by substituting moments β_r by their estimates b_r.

Moment L_1 is the L-mean and L_2 is the L-standard deviation. L-moment ratios can be used to define skewness and kurtosis. The L-skewness τ_2 is defined by $\tau_2 = L_3/L_2$ and the L-kurtosis τ_3 is given by $\tau_3 = L_4/L_2$. The coefficient of variation, L-CV τ_1 is given by the relation $\tau_1 = L_2/L_1$. For positive data, τ_1 lies in the interval (0,1) and is identical to the Gini coefficient.

For the standardized normal distribution is $L_1 = 0, L_2 = 1/\sqrt{\pi}, L_3 = 0, L_4 \approx 0.123$. The meaning of L-moments for selected distributions can be seen from Table 1.2.

L-moments are inherently robust and have a wider range of applications than classic moments. The problem is that for some known distributions (e.g. lognormal,

Table 1.2 Significance of L moments for selected distributions [adopted from (Wyckoff et al., 1980)]

Distribution	Parameters	L_1	L_2	τ_2	τ_3
Rectangular	a, b	$(a+b)/2$	$(b–a)/6$	0	0
Logistic	μ, s	μ	s	0	1/6
Normal	μ, σ^2	μ	$\sigma/\sqrt{\pi}$	0	0.1226
Laplace	μ, b	μ	$3b/4$	0	$1/(3\sqrt{2})$
Student 2 df	$v = 2$	0	$\pi/2^{3/2}$	0	3/8
Student4 df	$v = 4$	0	$15\pi/64$	0	111/512
Exponential	λ	$1/\lambda$	$1/(2\lambda)$	1/3	1/6
Gumbell	μ, β	$\mu + \gamma\beta^*$	$\beta \log 2$	0.1699	0.1504

$^*\gamma$ is Euler–Mascheroni constant 0.57721

gamma, etc.), relations for direct determining of the parameters using L-moments are not available and Monte Carlo simulations must be used. In work (Karvanen, 2006), a distribution system is proposed, for which their parameters can be easily determined by using L-moments.

1.5 Discrete Distributions

An overview of the basic types of discrete distributions is given in the book (Meloun & Militký, 2011). Only the binomial geometric and Poisson distributions are given here. However, it is shown how to estimate which distribution is appropriate to describe particular samples.

Binomial distribution B (N, p)

The binomial distribution is one of the most used in practical problems. It is directly related to the categorization of data into intervals. It is often used in calculations related to quality and reliability assessment. It is related to the processing of data from the ordinal scale to cardinal one and to the estimation of probabilities through relative frequencies. The binomial distribution has a random variable X expressing the number of occurrences of phenomenon A (a favorable outcome) in N independent trials. The probability of event A (favorable outcome) occurring in one trial is p and that of event $\overline{A}$ (unfavorable outcome) is $q = 1 - p$.

Probability function $P(x = i) = \binom{N}{x} * p^X * (1 - p)^{N-x}$

Where $\binom{N}{x} = \frac{N!}{x!\,(N-x)!}$. Especially it is $0! = 1$ and also $\binom{N}{N} = 1$.

Distribution function: $F(x) = \sum_{i=0}^{x} \binom{N}{i} p^i\,(1 - p)^{N-i}$.

Reliability function: $R(x) = \sum_{i=x}^{N} \binom{N}{i} p^i\,(1 - p)^{N-i}$.

Mean value: $E(x) = N\,p$.

Variance: $D(x) = Np(1 - p)$.

Mode: $p(N + 1) - 1 \leq M_o \leq p(N + 1)$.

Skewness: $g_1 = \frac{1-2p}{\sqrt{Np(1-p)}}$.

Kurtosis: $g_2 = \frac{1-6p(1-p)}{Np(1-p)} + 3$.

Estimator of parameter: $\hat{p} = \frac{x_p}{N}$, where x_p is the number of favorable events in N independent trials.

Example

Based on a long-term analysis of clothing production, it was determined that 99% of clothing belongs to the group that meets the prescribed quality criteria. The purpose is to determine the probability that in an order of 30 randomly selected clothes, all satisfactory items will be found. The solution can be directly obtained using the reliability function, which determines the probability with which a random variable takes on values equal to, or greater than the specified value. By direct using of reliability function is found.

$$R(30) = \sum_{i=30}^{30} \binom{N}{i} 0.99^i \, (1 - 0.99)^{N-i} = 0.99^{30} = 0.7397$$

The conclusion is that in 74% of cases will the entire order be satisfactory.

Geometric distribution G(p)

The geometric distribution has a random variable X expressing the number of independent trials until the first favorable result is reached. The probability of a favorable result occurring in one trial is p.

Probability function: $P(x = i) = p(1 - p)^{x-1}$.

Distribution function: $F(x) = 1 - (1 - p)^x$.

Mean value: $E(x) = 1/p$.

Variance: $D(x) = (1 - p)/p^2$.

Mode: $M_o = 1$.

Skewness: $g_1 = \frac{2-p}{\sqrt{(1-p)}}$.

Kurtosis: $g_2 = 9 + \frac{p^2}{1-p}$.

Estimator of parameter: $\hat{p} = x_p/N$, where x_p is the number of independent trials until the first favorable outcome.

Poisson distribution Po (λ)

The Poisson distribution is used instead of the binomial distribution in cases where the number of trials is unknown (Pham, 2023). The Poisson distribution has a random variable X that is equal to the number of events in a given time or space interval.

Probability function: $P(x = i) = \lambda^x \, e^{-\lambda/x!}$.

Distribution function: $F(x) = \sum_{i=0}^{x} \lambda^i \, e^{-\lambda/i!}$.

Mean value: $E(x) = \lambda$.

Variance: $D(x) = \lambda$.

Skewness: $g_1 = \frac{1}{\sqrt{\lambda}}$.

Kurtosis: $g_2 = 3 + \frac{1}{\lambda}$.

Estimator of parameter: $\hat{\lambda} = \overline{x}$, where $\overline{x}$ is the arithmetic mean of the number of phenomena in a given time or space interval for which it is

$$\hat{\lambda} = \frac{\sum_{x=0}^{k} x n_x}{\sum_{x=0}^{k} n_x} = \frac{\sum_{x=0}^{k} x n_x}{N}$$

Here $k + 1$ is the number of levels, **N** is the total experimentally determined number of phenomena (size of sample), and n_x is the number of phenomena at the xth level. For example, when evaluating the number of defects on the textile, there are levels: no defects ($x = 0$), one to 5 defects ($x = 1$), from 5 to 15 defects ($x = 2$) and more than 15 defects ($x = 3$).

In reliability calculations, the parameter λ is referred to as the constant failure rate (Pham, 2023). Then the reliability function $R(t)$ expresses the probability of the occurrence of k or fewer failures by the selected time t is defined by the relation

$$R(t) = \sum_{i=0}^{k} \frac{(\lambda t)^i e^{-\lambda t}}{i!}$$

It is valid, that Poisson distribution is limit case of Binomial distribution for huge number of trials N.

$$(\lim B(N, p)) \quad \text{for} \quad N \to \infty = \frac{(Np)^x \exp(-Np)}{x!} = \text{Po}(Np)$$

This means that for large N and small p, the Binomial distribution turns into a Poisson distribution and $\lambda = Np$. This is clearly visible in Fig. 1.15.

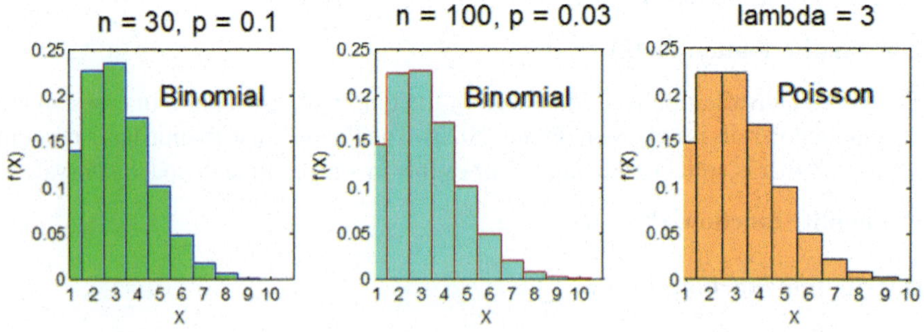

Fig. 1.15 Differences between binomial and Poisson distributions for large N

For the characterization of different discrete distributions based on the knowledge of the number of events n_x at all levels of x (from 0 to K) and the corresponding probabilities, a *graph of the frequency ratio* is suitable, which is based on the fact that a number of discrete distributions are valid

$$\frac{xp(x)}{p(x-1)} = C_o + C_1x$$

where C_o and C_1 characterize the given distribution (see Table 1.3).

The graph of the frequency ratio is the dependence $y = \frac{xp(x)}{p(x-1)}$ on x.

If the points on this graph lie approximately on a straight line, the type of discrete distribution and its parameters can be estimated from the slope and intercept of this straight line. Figure 1.16 shows a graph of the frequency ratio for data generated from the Poisson distribution with parameter $\lambda = 2$ for $K = 8$. A total of $N = 400$ values was generated (Meloun & Militký, 2011).

It can be seen that despite the relatively high variance, there is a regression line with an approximately zero slope and an intercept close to 2.

Table 1.3 Slope C_1 and intercept C_0 of the straight line in the frequency ratio graph

Distribution	$p(x)$	Slope C_1	Intercept C_0
Poisson	$e^{-\lambda}\lambda^x/x!$	0	λ
Binomial	$\binom{n}{p}p^x(1-p)^{(n-x)}$	$-p/(1-p)$	$p(n+1)/(1-p)$
Negative binomial	$\binom{n+x-1}{x}p^n(1-p)^x$	$(1-p)$	$(n-1)(1-p)$
Geometric	$p(1-p)^{(x-1)}$	$(1-p)$	0

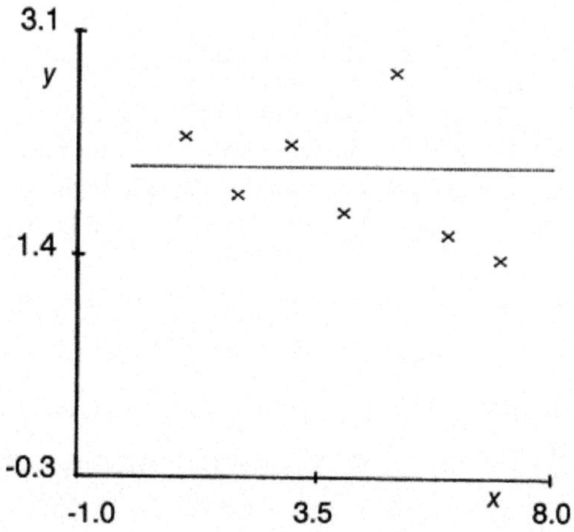

Fig. 1.16 Frequency ratio plot for data generated from Poisson distribution (Meloun & Militký, 2011)

1.6 Continuous Distributions

An overview of the basic types of continuous distributions is given in the book (Meloun & Militký, 2011). Here, only the basic classes of distributions that are common in measuring instrument error analysis, the flexible Weibull distribution and the normal distribution (Novitskii & Zograf, 1985) are presented.

A. Class of trapezoidal distributions

It arises from the convolution of two rectangular distributions with the same mean value and ranges ($x_{max} - x_{min}$) equal to A and B. According to the ratio B/A (it is considered that A is always greater than B), different distributions differ in the length of the tails. An overview of situations in which uniformly distributed errors occur is presented in the work (Novitskii & Zograf, 1985).

B. Class of arcsine distributions

Arcsine distribution occurs wherever the values of a periodic (sine) signal (e.g. alternating voltage) are determined at constant intervals. It is defined in the interval ($-$ A, A) and its probability density is

$$f(x) = \frac{1}{\pi\sqrt{A^2 - x^2}}$$

Other distributions from this class are compositions of two different arcsine distributions (with parameters $A1$, $A2$, which correspond to sampling a signal composed of two periodic components differing in frequency). According to the $A1/A2$ ratio, different distributions of this system are then resulted. If the sampled signal is composed of two components of significantly different frequencies, a bimodal distribution result. This can be the case, for example, when a material whose properties change periodically is measured on a deterministic device (thickness measurement).

C. Class of symmetric distributions of exponential form

This group of symmetric distributions is very popular both in theoretical considerations and in practical applications because it includes the uniform, normal, and Laplace distributions as special cases. It also occurs most often during measurement. The probability density of this class can be expressed in the form

$$f(x) = \frac{\alpha}{2\lambda\Gamma(1/\alpha)}\exp\left(-\left|\frac{x-\mu}{\lambda\sigma}\right|^{\alpha}\right)$$

where μ is the mean value, σ is the standard deviation, $\Gamma(x)$ is the gamma function and

$$\lambda = \sqrt{\frac{\Gamma(1/\alpha)}{\Gamma(3/\alpha)}}$$

The parameter α is a shape factor, determining both the shape around the mean value and the length of the tails. For $\alpha < 1$ it is a distribution close to **Cauchy**, for $\alpha = 1$ it is **Laplace**, for $\alpha = 2$ it is **normal**, for $\alpha > 2$ it is approximately **trapezoidal**, and for $\alpha \to \infty$ it is a **rectangular** distribution. For this class of distribution, the kurtosis g_2 can be expressed in the form

$$g_2 = \Gamma(1/\alpha)\Gamma(5/\alpha)/[\Gamma(3/\alpha)]^2$$

and the entropic coefficient in the form

$$E = 1/\alpha \exp(1/\alpha)\sqrt{\Gamma(1/\alpha)/\Gamma(3/\alpha)}$$

For example: for $\alpha = 0.25$ is $g_2 = 458$, $E = 0.085$, for $\alpha = 1$ is $g_2 = 6$, $E = 1.92$ and for $\alpha = 2$ is $g_2 = 3$, $E = 2.066$.

Attention, it often happens that this class of distribution is considered universal for the description of experimental errors ε, which is far from reality. Convolution of the uniform distribution (with standard deviation σ_R) and the exponential distribution (with standard deviation σ_E) results in a class of **flattened distributions**. It is flatter in the vicinity of mean value and has long tails. The shape of this distribution class is determined by the parameter $C = \sigma_R/\sigma_E$. It is valid for them that at the same value of kurtosis g_2 as for the exponential class of distribution, they have a significantly lower entropy coefficient E.

D. Class of Student distribution

Student distribution belongs to sampling distributions (it characterizes the distribution of the mean value of the sample from the normal distribution). For the case of using this distribution in the description of measurement errors, it is used in the simplest (normalized) form, where the probability density is given by the relation.

$$f(x) = \frac{\Gamma((\alpha+1)/2)}{\sqrt{2\pi}\,\Gamma(\alpha/2)(1+x^2/\alpha)^{\frac{\alpha+1}{\alpha}}}$$

Here the shape factor α corresponds to the degrees of freedom ($\alpha = N - 1$). For $\alpha > 4$, it is possible to determine the standard deviation σ according to the relationship $\sigma = \sqrt{\alpha/(\alpha-2)}$, kurtosis according to the formula

$$g_2 = (3\alpha - 2)/(\alpha - 4)$$

and entropic coefficient from expression

$$E = \frac{\Gamma(\alpha/2)\sqrt{\pi(\alpha-2)}}{2\Gamma((\alpha+1)/2)} \exp[\beta(\alpha)(\alpha+1)]$$

where $\beta(1) = \ln(2)$, $\beta(2) = 1 - \ln(2)$ and

$$\beta(\alpha) = \left[1 - \frac{1}{2} + \frac{1}{3} + \frac{(-1)^{\alpha-1}}{\alpha - 1} - \ln(2)\right](-1)^{\alpha}$$

An interesting peculiarity of this class of distribution is that for $\alpha = 4$ is $g_2 = \infty$ but the entropic coefficient is $E = 1.903$. Moreover, for $\alpha = 6$, the kurtosis is equal to $g_2 = 6$ as for the Laplace distribution, but the entropic coefficient $E = 2.0053$ is considerably higher. For $\alpha \to \infty$, this distribution turns into a normal distribution. This class of distribution is an example that the kurtosis g_2 does not uniquely describe the shape of symmetric unimodal distributions. Various types of other analytical distributions (including skewed ones) are discussed in detail in (Novitskii & Zograf, 1985).

E. Weibull distribution

The Weibull distribution is named after the Swedish physicist Wallodi Weibull, who used it in 1939 to describe the breaking strength of materials. However, this distribution was introduced by Fisher and Tippett in 1928 as the limit distribution (type III) of the smallest order statistic from various distributions. The Weibull distribution has been successfully used in the field of reliability, quality control, and materials engineering, especially to describe the fatigue of materials (Rinne, 2008). It is very rich in shape and is suitable for describing a variety of experimental data. The probability density of the Weibull distribution is defined in range $A < x < \infty$ has the form

$$f(x) = \beta\alpha^{-\beta}(x - A)^{\beta-1}\exp\{-\alpha^{-\beta}(x - A)^{\beta}\}$$

The notation $W(A, \alpha, \beta)$ is used for the Weibull distribution. The parameter A is the lowest value (threshold), $\alpha > 0$ is the scale parameter, and $\beta > 0$ is the shape parameter. Changes in A cause only a shift in the probability density on the real axis. Changes in α are similar as changes of the standard deviation of normal distribution, i.e. they change the spread of the random variable x (Figs. 1.17 and 1.18).

The shape factor β causes changes in the shape of the probability density function (Fig. 1.18).

The distribution function of the Weibull distribution has the form $F(x) = 1 - \exp\{-\alpha^{-\beta}(x - A)^{\beta}\}$.

and the fault intensity is expressed as

$$\lambda(x) = \beta\alpha^{-\beta}(x - A)^{\beta-1}$$

For $\beta > 1$, the Weibull distribution is unimodal, and for $\beta < 1$, the probability density decreases continuously with increasing x (exponential distribution). By standardization

$$z = \left(\frac{x - A}{\alpha}\right)^{\beta} = U^{\beta}$$

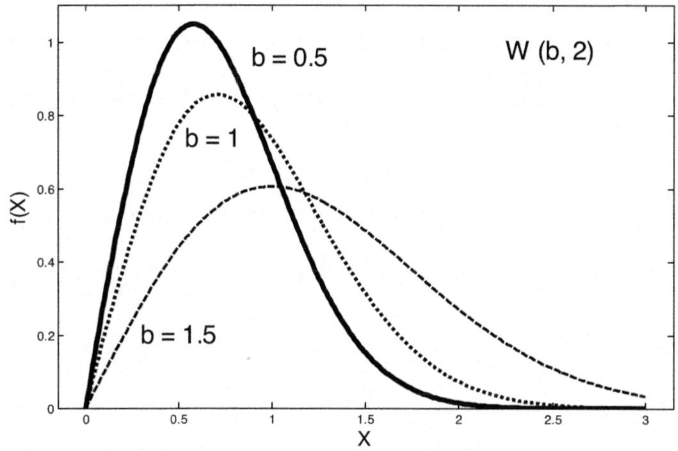

Fig. 1.17 Effect of scale parameter on the course of the Weibull distribution

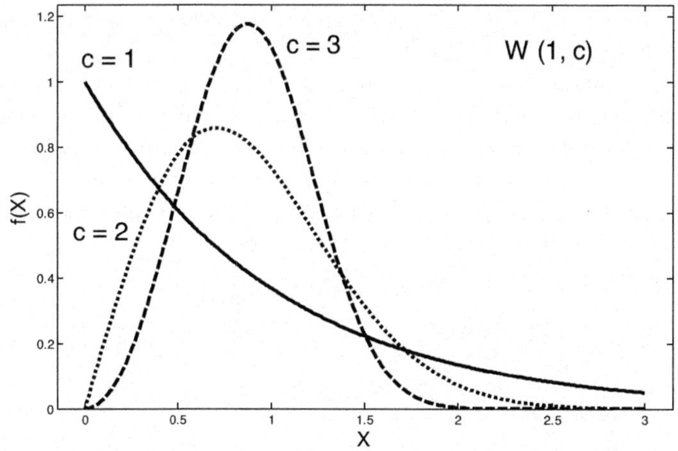

Fig. 1.18 Effect of shape parameter on the course of the Weibull distribution

the normalized probability density of the Weibull distribution $W(0,1,1)$ can be obtained, see Fig. 1.18. For calculations of statistical characteristics, the reduced variable U is advantageously used, which is a simple linear transformation of the original variable x.

The kth general moment has the form

$$M_k = E(U^k) = \Gamma_k \quad \text{where} \quad \Gamma_k = \Gamma\left(1 + \frac{k}{\beta}\right)$$

or

$$M_k = E(x^k) = \sum_{j=0}^{k} \binom{k}{j} A^j \alpha^{k-j}\, \Gamma_{k-j}$$

Here $\Gamma\left(1+\frac{j}{\beta}\right) = \int_0^\infty t^{j/\beta} e^{-t} \mathrm{d}t$ is gamma function. For its approximation the Stirling formula $M_k = E(x^k) = \sum_{j=0}^{k} \binom{k}{j} A^j \alpha^{k-j}\, \Gamma_{k-j}$ can be used

$$\Gamma(z) \approx \sqrt{2\pi}\, z^{z-1/2} e^{-z} \exp\left\{\frac{1}{12\,z} - \frac{1}{360\,z^3} + \frac{1}{1260\,z^5} - \cdots\right\}$$

The mean value of the Weibull distribution is then equal to $E(U) = \Gamma(1)$ or $E(x) = A + \alpha\Gamma(1 + 1/\beta) = A + \alpha\Gamma_1$.

Furthermore, the **mode** is equal $M_o = A + \alpha\{(\beta - 1)/\beta\}^{1/\beta}$.

and the **median** is $\tilde{x}_{0,5} = A + \alpha(\ln 2)^{1/\beta}$.

The mean value of $E(U)$ is only a function of the shape parameter β and depends on the shape of the gamma function (see Fig. 1.19).

The mean value of $E(U)$ increases over all limits for $\beta = 0$ and decreases to one for $\beta \to \infty$. The minimum value is $E(U) = 0.8856$ for $\beta = 2.16623$. A small value of the shape factor β leads to a large value not only of the mean value but also of the variance. Interestingly, there is no β for which the median is equal to both the mode and the mean. The mean value of $E(U)$ as a function of β can be expressed by an approximate relation based on the Taylor expansion

$$E(U) \approx 1 - \frac{0,57722}{\beta} + \frac{0,98905}{\beta^2} - \cdots$$

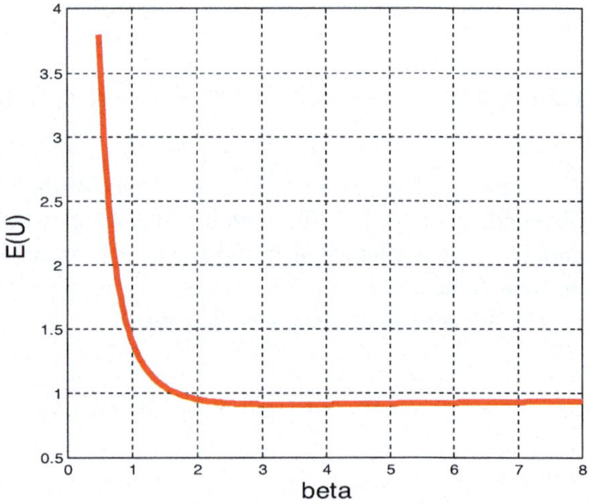

Fig. 1.19 The influence of the size of the shape parameter β on the mean of the reduced variable U

To determine the mean value of the x variable, is sufficient to multiply $E(U)$ by the scale parameter α and add the threshold value A. For the higher general moments of the Weibull distribution, is valid that $E\left(U^k\right) = \Gamma_k$ and thus

$$E(x^2) = A^2 + 2A\alpha\Gamma_1 + \alpha^2\Gamma_2\, E(x^3) = A^3 + 3A^2\alpha\Gamma_1 + 3A\alpha^2\Gamma_2 + \alpha^3\Gamma_3$$

The higher central moments of the Weibull distribution are

$$C_2(U) = \sigma_U^2 = \Gamma_2 - \Gamma_1^2,$$
$$C_3(U) = \Gamma_3 - 3\Gamma_2\Gamma_1 + 2\Gamma_1^3$$
$$C_3(U) = \Gamma_4 - 4\Gamma_3\Gamma_1 + 6\Gamma_2\Gamma_1^2 - 3\Gamma_1^4$$

The higher central moments of the Weibull distribution with respect to the variable x are equal

$$C_k(x) = \alpha^k C_k(U)$$

Thus, the **variance** with respect to the variable x is $\alpha^2\left(\Gamma_2 - \Gamma_1^2\right)$.
For the coefficient of variation then it is valid that

$$V(U) = \frac{\sqrt{\left(\Gamma_2 - \Gamma_1^2\right)}}{\Gamma_1} \text{ and } V(x) = \frac{\alpha\sqrt{\left(\Gamma_2 - \Gamma_1^2\right)}}{A + \alpha\Gamma_1}$$

The dependence of the variance on the shape parameter β can again be expressed by an approximate relation based on the Taylor expansion

$$\sigma_U^2 \approx \frac{\pi^2}{6\,\beta^2} \approx^\circ 1.64493/\beta^2 \text{or} \sigma^2 \approx \frac{\alpha^2\pi^2}{6\,\beta^2} \approx^\circ 1.64493\,\alpha^2/\beta^2$$

For skewness, a relationship can be derived based on the above moments

$$g_1 = \frac{\Gamma_3 - 3\Gamma_1\Gamma_2 + 2\Gamma_1^3}{\sqrt{\left(\Gamma_2 - \Gamma_1^2\right)^3}}$$

It can be found that for the shape parameter $\beta = 3.60$ is $g_1 = 0$. However, it is not exactly a symmetric distribution because the mean value is not equal to the median and mode.

The standard method for estimating the parameters of the Weibull distribution $W(A,\ \alpha,\ \beta)$ is the maximum likelihood method (see Sect. 1.7), which leads to a system of three non-linear equations (McLachlan & Peel, 2000)

$$-(\beta-1)\sum_{i=1}^{N}\frac{1}{x_i-A}+\frac{\beta}{\alpha}\sum_{i=1}^{N}\left(\frac{x_i-A}{\alpha}\right)^{\beta}=0$$

$$-\frac{N\beta}{\alpha}+\frac{\beta}{\alpha}\sum_{i=1}^{N}\left(\frac{x_i-A}{\alpha}\right)^{\beta}=0$$

$$\frac{N}{\beta}-N\ln\alpha+\sum_{i=1}^{N}\ln(x_i-A)-\sum_{i=1}^{N}\left(\frac{x_i-A}{\alpha}\right)^{\beta}\ln\left(\frac{x_i-A}{\alpha}\right)=0$$

This system can be simplified to the iterative solution of only two equations, namely

$$\frac{\beta-1}{\beta}\sum\nolimits_{i=1}^{N}\frac{1}{x_i-A}-N\frac{\sum_{i=1}^{N}(x_i-A)^{\beta-1}}{\sum_{i=1}^{N}(x_i-A)^{\beta}}=0$$

$$\frac{1}{\beta}+\frac{1}{N}\sum\nolimits_{i=1}^{N}\ln(x_i-A)-\frac{\sum_{i=1}^{N}(x_i-A)^{\beta}\ln(x_i-A)}{\sum_{i=1}^{N}(x_i-A)^{\beta}}=0$$

After finding the estimates, $\hat{A}$, $\hat{\beta}$, it is possible to put them into the equation

$$\hat{\alpha}=\left[\frac{1}{N}\sum_{i=1}^{N}\left(x_i-\hat{A}\right)^{\hat{\beta}}\right]^{1/\hat{\beta}}$$

There are a number of algorithms for the iterative solution of the above equations, but it is a complicated task, where in many cases no solution is found (Rinne, 2008). This is largely due to the fact that the global maximum on the likelihood function is for $A=x_{(1)}$.

It is quite simple to use parameter estimates of the 2-parameter Weibull distribution for a range of A_j values in the range $0 \le A_j \le x_{(1)}$. For each A_j, a shifted sample with elements $y_i = x_i - A_j$, which has a Weibull distribution dependent only on the parameters α_j and β_j, is determined. To estimate the parameter β_j, it is necessary to iteratively solve the equation

$$\frac{N}{\beta_j}+\sum_{i=1}^{N}\ln y_i-\frac{N\sum_{i=1}^{N}y_i^{\beta_j}\ln y_i}{\sum_{i=1}^{N}\ln y_i^{\beta_j}}=0$$

After finding the estimate $\hat{\beta}_j$, the estimate of the parameter α_j is determined by substituting into the equation

$$\hat{\alpha}_j=\left[\frac{1}{N}\left(\sum_{i=1}^{N}y_i^{\hat{\beta}_j}\right)\right]^{1/\hat{\beta}_j}$$

The optimal A_j is searched by maximizing the likelihood function, which has the form

$$\ln(A_j, x) = N\left\{\ln\left(\frac{\hat{\beta}_j}{\hat{\alpha}_j^{\hat{\beta}_j}}\right) - 1\right\} + (\hat{\beta}_j - 1)\sum_{i=1}^{N}\ln(x_i - A_j) \tag{1.7}$$

Advantageously, the graph of dependence $y = \ln(A_j)$ on x is created and its maximum is selected.

For $j = 1$, i.e. $A = 0$, is as the zero approximation of the shape parameter selected value

$$\beta_{0,1} = \frac{1}{N\ln x_{(N)} - \sum_{i=1}^{N}\ln x_i}$$

In the subsequent cycles ($j = 2, 3, \ldots$), the zero approximation is $\beta_{0,j} = \beta_{j-1}$.

It is always convenient to compare this solution with the boundary solution for $\beta = 1$, when

$$\hat{A} = x_{(1)}\,,\ \hat{\alpha} = \frac{1}{N}\sum_{i=1}^{N}(x_i - \hat{A})$$

with respect to reaching the maximum of the likelihood function (1.7).

For cases where $A = 0$ (or its value is known) the distribution $W(0, \alpha, \beta)$ results, for which the method of moments can be advantageously used. Different moment ratios are used, which depend only on the shape parameter β. An example is

$$R(\beta) = \frac{[E(x)]^2}{[E(x^2)]} = \frac{\Gamma^2(1 + 1/\beta)}{\Gamma(1 + 2/\beta)} \approx \frac{\left(\frac{1}{N}\sum_{i=1}^{N}x_i\right)^2}{\frac{1}{N}\sum_{i=1}^{N}x_i^2}$$

After finding the optimal $\hat{\beta}$, it is possible to insert it into the equation

$$\hat{\alpha} = \frac{\overline{x}}{\Gamma(1 + 1/\hat{\beta})}$$

Another possibility is to use the coefficient of variation V, where

$$V = \frac{\sigma}{E(x)} = \frac{\sqrt{\Gamma(1 + 2/\beta) - \Gamma^2(1 + 1/\beta)}}{\Gamma(1 + 1/\beta)} \approx \frac{s}{\overline{x}}$$

The method of moments can be easily extended to the 3-parametric Weibull distribution by adding, for example, the relation for the skewness g_1 (Rinne, 2008).

Relatively simple is to use of the so-called Weibull moments defined as

$$m_r = \sum_{i=0}^{N-1} (1 - i/N)^r \left(x_{(i+1)} - x_{(i)}\right) \text{kdy}\, x_{(0)} = 0$$

To estimate the parameters of the Weibull distribution, the following relations based on moments m_r can be used (Rinne, 2008).

$$A = \frac{m_1 m_4 - m_2^2}{m_1 + m_4 - 2m_2}$$

$$\alpha = \frac{m_1 - A}{\Gamma(1 + 1/\beta)}$$

$$\beta = \frac{\ln 2}{\ln(m_1 - m_2) - \ln(m_2 - m_4)}$$

A number of different estimators are based on order statistics. Relatively simple are the hybrid algorithms searching an independent estimate of the threshold A.

The estimate based on first-order statistic, i.e. $\hat{A}_o = x_{(1)}$ or $\hat{A}_o = 0.5\, x_{(1)}$, is usually chosen as a preliminary threshold value. Then the shape parameter can be estimated, based on the quantiles $\tilde{x}_{0.97}$ and $\tilde{x}_{0.17}$, according to the relationship

$$\hat{\beta}_0 = 2.989/\ln\left[\left(\tilde{x}_{0.97} - x_{(1)}\right)/\left(\tilde{x}_{0.17} - x_{(1)}\right)\right] \tag{1.8}$$

The mean value of the first-order statistic is $E(x_{(1)}) = A + \alpha N^{-1/\beta}\Gamma(1 + 1/\beta)$. The difference between the mean value of the first-order statistic and the mean value of the Weibull distribution is

$$E(x_{(1)}) - E(x) = \alpha\, \Gamma(1 + 1/\beta)\left(N^{-1/\beta} - 1\right)$$

This difference can be used together with $\overline{x}$ and s^2 for the estimation of the parameters A, α, β. By replacing the mean value with the arithmetic mean $\overline{x}$, the mean value of the first-order statistic with the value $x_{(1)}$ and using an estimate $\hat{\beta}_0$, the threshold value can then be estimated from the relationship (Wyckoff et al., 1980)

$$\hat{A}_1 = \overline{x} - \left(x_{(1)} - \overline{x}\right)/\left(N^{-1/\hat{\beta}_0} - 1\right)$$

Zenbil (1991) derived that the unbiased estimate of threshold A for known β is a function of the first two order statistics

$$\hat{A}_2 = x_{(1)} - (x_{(2)} - x_{(1)})/C(\beta) \quad \text{where}$$
$$C(\beta) = N\left[(N/(N + 1))^{1/\beta} - 1\right]$$

Again, if the shape factor β is not known, it is possible to use its estimate from Eq. (1.8), or another. Based on simulations, an estimate of the threshold value was found as a combination of the first two and the last order statistics in the form (Zanakis, 1979)

$$\hat{A}_3 = \left(x_{(1)}x_{(N)} - x_{(2)}^2\right)/\left(x_{(1)} + x_{(N)} - 2x_{(2)}\right)$$

Another simple threshold estimate is a combination of all ordinal statistics (Cooke, 1979).

$$\hat{A}_4 = 2x_{(1)} - (1-e)\sum_{i=1}^{N} x_{(i)}/e^i$$

In the book (Rinne, 2008), the choice of $\hat{A}_4$ is recommended. Linear regression based on Q-Q graphs is also used to estimate the parameters of the Weibull distribution. It is necessary to calculate ordered statistics (usually for $p_i = i/(N+1)$) and the regression model has the form

$$\ln\left[-\ln(1-p_i)\right] = \beta \ln\left(x_{(i)} - A\right) - \beta \ln(\alpha)$$

Knowing parameter, A, the estimates of α and β can be found by the method of weighted least squares with the weights defined in the book (McLachlan & Peel, 2000). If all three parameters are estimated, it is case of non-linear regression. Since the regression model is non-linear with respect to parameter A, only this parameter can be refined iteratively and the remaining parameters can be calculated by linear regression. Further modifications of regression procedures are presented, for example, in the book (McLachlan & Peel, 2000).

F. Normal distribution

The normal distribution is the most widely used continuous random variable distribution model in metrology and engineering statistics. When the same quantity is measured repeatedly under the same conditions (constant variance), random and uncontrollable influences cause deviations from the actual measured quantity, which usually have a normal distribution. A normally distributed random variable is also created by the composition (sum) of various random components, influences, and variables that are independent of each other, of which there are a larger number and each of them affects the resulting variable with approximately the same contribution. Note that the label "normal" does not mean that it is more normal than other distributions.

Normal (Gaussian) distribution $N(\mu, \sigma^2)$:

Probability density function $f(x) = \frac{1}{\sqrt{2\pi}} \exp\left(-\frac{(x-\mu)^2}{2\sigma^2}\right)$.

Examples of normal distribution $N(\mu, \sigma^2)$ for different parameters μ and σ^2 are shown in Fig. 1.20.

Distribution function $F(x) = \Phi\left(\frac{x-\mu}{\sigma}\right)$.

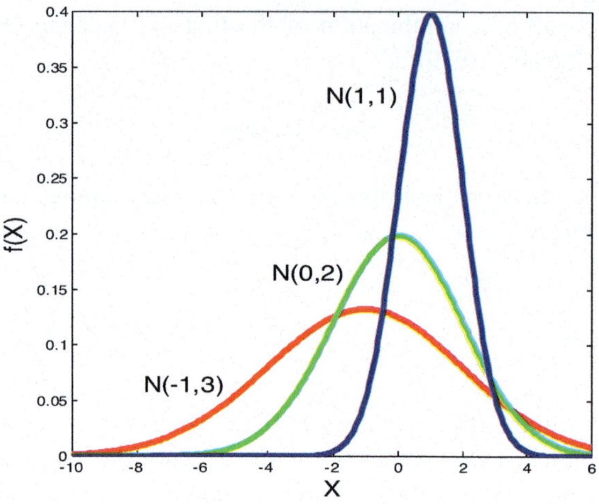

Fig. 1.20 Probability density functions of normal distribution $N(\mu, \sigma^2)$ for different parameters μ and σ^2

Function Φ(x) is given as $\Phi(x) = \frac{1}{\sqrt{2\pi}} \int\limits_{-\infty}^{x} \exp\left(\frac{-x^2}{2}\right) \mathrm{d}y$.

In textile metrology are **frequency and mass diagrams** corresponding to the probability densities of the normal distribution (for fiber lengths this is only a rough approximation not taking into account the finite range of lengths l) defined as

$$d_w(l) = \int\limits_{l}^{l_m} f_w(l)\mathrm{d}l = \frac{1}{2}\left[1 - \mathrm{erf}\left(\frac{(l - E_w(l))^2}{\sqrt{2D_w(l)}}\right)\right] \text{ and}$$

$$d_n(l) = \int\limits_{l}^{l_m} f_n(l)\mathrm{d}l = \frac{1}{2}\left[1 - \mathrm{erf}\left(\frac{(l - E_n(l))^2}{\sqrt{2D_n(l)}}\right)\right]$$

Here the function erf(x), i.e. the error function, is defined as $\mathrm{erf}(x) = \int_0^x e^{-t^2} \mathrm{d}t$

A mass diagram of cotton fiber lengths determined experimentally and approximated from the assumption of normality of the probability density function are very close (Azzouz et al., 2008).

Mean value: $E(x) = \mu$.

Variance: $D(x) = \sigma^2$.

Skewness: $g_1 = 0$.

Kurtosis: $g_2 = 3$.

The probability density of the standardized normal distribution is in Fig. 1.21.

It can be seen from Fig. 1.21 that, for example, only about 5% of all possible realizations of a random variable with a normal distribution lie outside the interval $\mu \pm 2\sigma$.

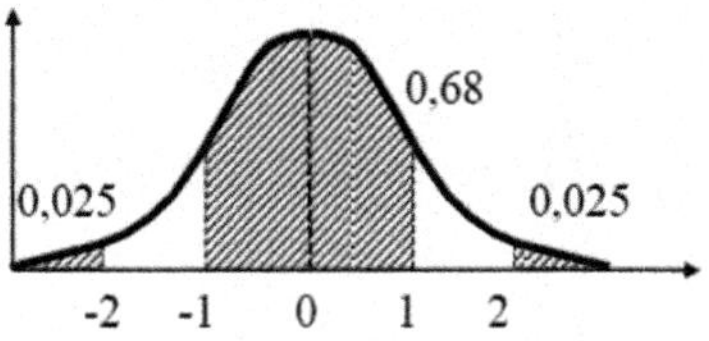

Standardized normal distribution $U \rightarrow N(0,1)$

P(-1.65 ≤ U ≤ 1.65) = 0.90	***f*(2) = 0.05**
P(-1.98 ≤ U ≤ 1.98) = 0.95	***f*(3) = 0.04**
P(-3 ≤ U ≤ 3) = 0.9973	***f*(4) = 0.00013**

Fig. 1.21 Probability density function of the standardized normal distribution

Parameter estimation: $\hat{\mu} = \overline{x}$, where $\overline{x} = \frac{1}{N}\sum_{i=1}^{N} x_i$ is arithmetic mean

$$\hat{\sigma}^2 = \frac{1}{N-1}\sum_{i=1}^{N}(x_i - \overline{x})^2.$$

Replacement of the binomial distribution by the normal distribution: It follows directly from the properties of the binomial distribution that $\mu = Np$ and $\sigma^2 = Np(1-p)$. It is also true that for $p = 0.5$ the binomial distribution is symmetric.

But the binomial distribution is skewed for $p \neq 0.5$ and the approximation by the normal distribution is then worse.

In some cases, the measurement results may actually be a selection from a mixture of two random variables with a normal distribution. This is usually the result of incorrect sample preparation, or incorrect identification of individual quantities (e.g. measurement of fiber geometry, which can be easily confused).

It is therefore important when a mixture with a proportion p of the first normally distributed component (parameters μ_1, σ_1) and a proportion $(1-p)$ of the second component normally distributed component (parameters μ_2, σ_2) can be distinguished from the resulting probability density function *fc*

$$f_c(x) = \frac{p}{\sqrt{2\pi\sigma_1^2}}e^{-(x-\mu_1)^2/2\sigma_1^2} + \frac{1-p}{\sqrt{2\pi\sigma_2^2}}e^{-(x-\mu_2)^2/2\sigma_2^2}$$

In the ideal case, when $p = 0.5$, $\mu_1 = \mu_2$, and $\sigma_1 = \sigma_2$, the *fc* is normal, i.e. symmetric and unimodal. However, if $p = 0.5$, $\sigma_1 = \sigma_2$ but $\Delta = |\mu_1 - \mu_2|/\sigma$ is nonzero the distribution looks quite different and for higher Δ (as $\Delta = 3$) is the probability density function bimodal (McLachlan & Peel, 2000).

If $\mu_1 = \mu_2$ it is true, for all p, that *fc* is unimodal.

The probability density function *fc* is unimodal for any p if (McLachlan & Peel, 2000)$(\mu_1 - \mu_2)^2 < \frac{27\sigma_1^2\sigma_2^2}{4[\sigma_1^2+\sigma_2^2]}$.

A sufficient condition for mixture bimodality for some p is that

$$(\mu_1 - \mu_2)^2 > \left(8\sigma_1^2\sigma_2^2\right)/\left(\sigma_1^2+\sigma_2^2\right)$$

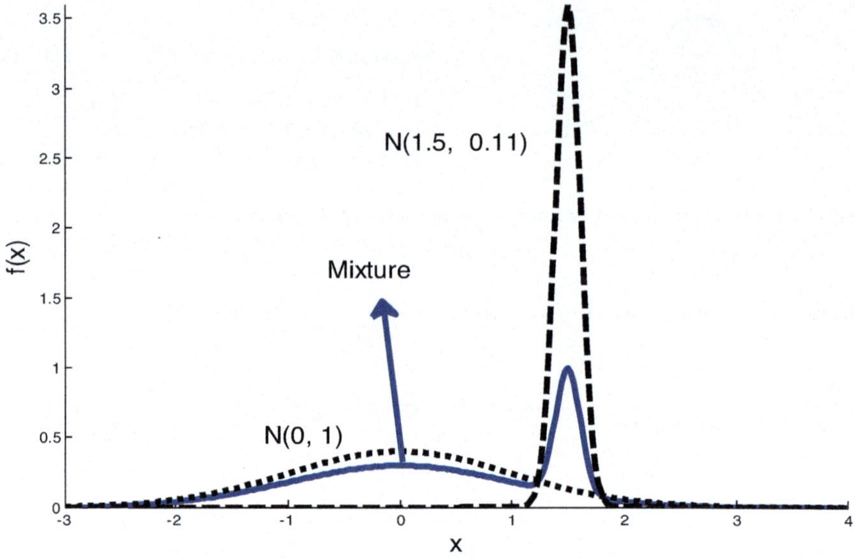

Fig. 1.22 Individual components and mixture of two normal pdf

However, there is always p for which the mixture distribution is unimodal. In some cases, the distribution of a mixture of two normally distributed random variables can be rather peculiar and bizarre [e.g. mixture 3/4 $N(0, 1) + 1/4\ N(1.5, 0.11)$] which is shown in Figs. 1.22 and 1.23.

1.7 Mathematical Statistics

Mathematical statistics deals with the analysis of samples obtained from basic sets (population). According to ČSN 010251, the term population includes the number of monitored cases, or objects that have the same properties from a given point of view. This amount can be limited, i.e. the population is **finite**, or **infinite**. In metrology and in technical practice, the basic population is often determined very vaguely, as a set of all possible measured values. It is fundamentally unavailable. However, in a deeper analysis, the population can be defined more precisely with regard to how the measurement results are required to be interpreted. The statistical analysis is performed on the basis of a representative sample from the basic population, i.e. a random sample. For a representative random sample is valid:

1. All elements of the sample (x_i) $i = 1, \ldots, N$, are understood as random variables that follow the same distribution law (homogeneity of the selection).

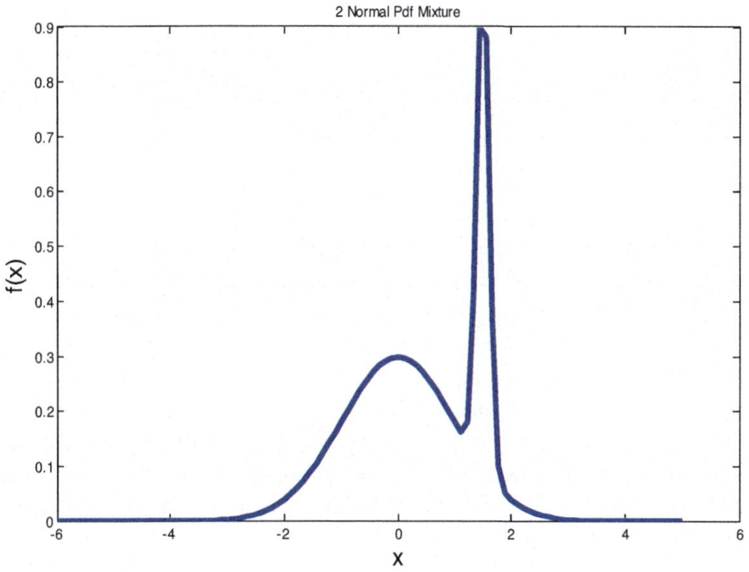

b)

Fig. 1.23 Mixture of two normal pdf, 3/4 $N(0, 1) + 1/4\ N(1.5, 0.11)$

2. The x_i values included in the sample are chosen independently of each other. Selecting any of them does not affect the value of any other selected element (independence of sample elements).
3. All elements of the population have an equal chance of being selected into the sample (this severely limits the definition of the base population).
4. The size of the sample is sufficient for a sufficiently accurate estimation of its parameters.

Univariate samples, the analysis of which is the focus of this chapter, are commonly created from populations for which the probability distribution is usually not fully known. Normality is commonly assumed, but must be verified based on the sampling distribution. The sample is characterized by standard information about the location and dispersion parameters along with their uncertainties. The complement of both pieces of information is information about the shape of the sampling distribution. The statistical characteristics of the location, dispersion, and shape of the sampling distribution are called sampling characteristics. The *statistical characteristics of the basic population* are then deduced from these selection characteristics. In statistical analysis, it is usually assumed that the distribution of the sample is the same as the distribution of the population.

For *continuous* random variables, the sampling distribution is characterized, for example, by the probability density function $f(x,\boldsymbol{a})$, and for *discrete* random variables, the probability density function is denoted as the probability function $p(x,\boldsymbol{a})$.

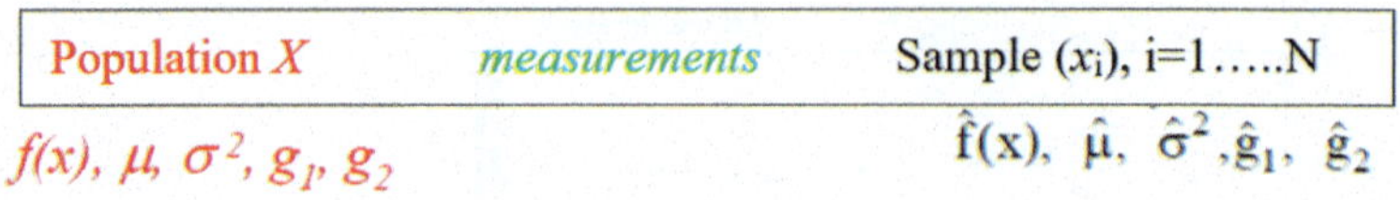

Fig. 1.24 The relationship between the population and the sample (the "∧" symbol indicates parameter estimates or probability densities from the data)

The probability density function depends on a vector of parameters $\boldsymbol{a}$, whose components are the location, dispersion, and shape parameters. The goal is to determine appropriate estimates $\hat{\mathbf{a}}$ of these parameters. Since the estimates $\hat{\mathbf{a}}$ are also random variables, it is necessary to determine their distribution or at least the characteristics of these distributions. Since the analysis primarily concerns information about the basic population, confidence intervals are constructed based on estimates, in which the components of the vector $\boldsymbol{a}$ lie with the given probability. The construction of confidence intervals is closely related to testing hypotheses about the "unknown" parameters $\boldsymbol{a}$ of the population. Schematically, the relationship between population and sample is shown in Fig. 1.24.

1.7.1 Point Estimates

A point estimate (the computed vector $\hat{\mathbf{a}}$) is an estimate of the parameters (of the non-random vector $\mathbf{a}$) associated with the distribution of the random variable of the sample data. Since it is a function of random variables (elements of sample), the estimate $\hat{\mathbf{a}}$ is also a random variable with a distribution characterized, for example, by the probability density function $f(\hat{\mathbf{a}})$. It is usually possible to use several different methods for point estimation of parameters $\mathbf{a}$, which are assessed by criteria characterizing their quality. The most well-known are the criteria of consistency (agreement), unbiasedness, and efficiency. To estimate the parameter $\mathbf{a}$, it is valid that:

1. An estimate $\hat{a}$ is consistent when, for a given sample size N, the probability that its distance from the true value a is arbitrarily small is equal to one.

   $$P(|\hat{a} - a| \leq \varepsilon) = 1$$

   This is usually true for sufficiently large sample size and for unbiased estimates. In many cases the estimates are only asymptotically consistent (for N approaching infinity).
2. An estimate $\hat{a}$ is unbiased when, for a given sample size N, if its mean value is equal to the population parameter a. Thus, the bias b of the estimate

   $$b = a - E(\hat{a})$$

is equal to zero ($b = 0$). In many cases the estimates are only asymptotically unbiased (for N approaching infinity).

3. An estimate $\hat{a}$ is efficient when its variance around the true value is minimal relative to all possible estimates of that parameter. Of the two-point estimates $\hat{a}_1$ and $\hat{a}_2$ of the parameter a, determined from the same random sample of size N, the one with the smaller variance is more efficient.

The variance $D(\hat{a})$ of an estimate is a characterization of its "accuracy". If the estimate $\hat{a}$ contains all the information about the sample, it is said to be sufficient. The best unbiased estimate is one that is unbiased, efficient, and sufficient at the same time. Figure 1.25 shows the probability densities of three estimates differing in bias and precision.

To express the overall quality of the estimate the **mean squared error** (MSE) is used

$$\mathrm{MSE}(\hat{a}) = E(\hat{a} - a) = [E(\hat{a}) - a]^2 + D(\hat{a}) = b^2 + D(\hat{a})$$

Thus, the mean square error of the estimate is generally the sum of the square of the bias of the estimate and its variance. In some cases, the estimate is dependent on an external parameter h "driving" its bias and variance. As h increases, the bias increases and the variance of the estimate decreases, usually more rapidly. One can then find the optimal estimate that minimizes the MSE at a certain bias. Thus, unbiasedness is not always optimal. It is often advantageous to define the above-mentioned errors as a proportion of the size of the parameter a, i.e. as a coefficient of variation also referred to in this context as a normalized error (Piersol, 2000). The normalized random error of the estimate e_n is defined as

$$e_n(\hat{a}) = \frac{\sqrt{D(\hat{a})}}{a} = \frac{\sqrt{E(\hat{a}^2) - (E(\hat{a}))^2}}{a}$$

The normalized error of the bias estimate e_b is then

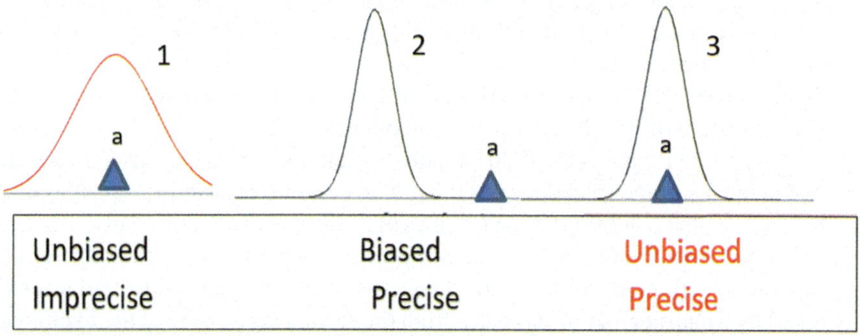

Fig. 1.25 The precisions and biases of the estimator $\hat{a}$ distribution determined by three different methods

$$e_b(\hat{a}) = b/a = E(\hat{a})/a - 1$$

and for the normalized mean square error of the estimate e_c is valid

$$e_c(\hat{a}) = \frac{E(\hat{a} - a)}{a} = \frac{\sqrt{b^2 + D(\hat{a})}}{a}$$

If the error e_n is small enough, it can be written that $\hat{a}^2 \approx a^2(1 \pm e_n(\hat{a}))$ or $\hat{a} \approx a\sqrt{(1 \pm e_n(\hat{a}))} \approx a(1 \pm e_n(\hat{a})/2)$ (Piersol, 2000). Then it is $e_n(\hat{a}^2) \approx 2e_n(\hat{a})$.

This means that for small random standardized errors of the estimates, the random standardized error of the squared estimate is twice the random standardized error of the estimate (Piersol, 2000). For the case where the bias of the estimate is negligible and the normalized mean squared error of the estimate e_c is small enough, i.e. $e_c < 0.1$, the probability density of the estimate $f(\hat{a})$ can be approximated by a normal distribution with mean $E(\hat{a}) \approx a$ and variance $\sigma^2(\hat{a}) \approx ae_c$

$$f(\hat{a}) \approx \frac{1}{ae_c\sqrt{2\pi}} \exp\left(\frac{-(\hat{a} - a)^2}{2(ae_c)^2}\right)$$

For the 95% confidence interval of the parameter a from the estimate $\hat{a}$, it is approximately true that (Piersol, 2000)

$$\frac{\hat{a}}{1 + 2e_c} \le a \le \frac{\hat{a}}{1 - 2e_c}$$

For small $e_c < 0.1$, this interval becomes even simpler, i.e. $\hat{a}(1 - 2e_c) \le a \le \hat{a}(1 + 2e_c)$. If $e_c > 0.1$, the estimation $\hat{a}$ distribution is usually non-symmetric [$\chi 2$ or more complicated (Piersol, 2000)]. A number of methods are available for constructing point estimates $\hat{\mathbf{a}}$ (a vector with m components) of the parameters of the probability density function $f(x, \boldsymbol{a})$ from a random sample of size N:

- Moment method: comparison of sample moments with theoretical ones. The result is a system of m equations with m unknowns. (For a normal distribution, the estimate of the mean is directly the arithmetic mean, and the estimate of the variance is the sample variance s^2).
- Quantile method: Sample and theoretical quantiles are compared. The result is again a system of m equations with m unknowns.
- Maximum likelihood method: The likelihood function is maximized (for independent elements of the selection, it is the product of their probability densities $f(x_i, \boldsymbol{a})$). Usually, the conditions of zero values of the first derivatives are solved, which again leads to a system of m equations with m unknowns. The advantage of the maximum likelihood method is the direct possibility of determining variances or covariance matrix $D(\hat{\mathbf{a}})$ of estimates from the second derivatives of the likelihood function.

For the method of moments and the quantile method, it is possible to use the computer-intensive Bootstrap method to determine the variances of the estimates (see Chap. 6).

Maximum likelihood method: The parameter estimate a is determined by maximizing the likelihood function. For the case of independent elements of the sample (independent measurements x_i), this function has the form

$$L(a) = \prod_{i=1}^{N} f(x_i)$$

Here $f(x_i)$ is the probability density function of the sample element x_i. The estimate of the parameter a is then determined by maximizing the likelihood function $\hat{a} = \max[L(a)]$, or preferably its logarithm $\hat{a} = \max[\ln L(a)]$. Since the logarithm is a monotonous transformation, the position of the extremum is preserved. The properties of the maximum likelihood method are described in more detail in Sect. 7. 4. Here, its use in estimating the parameters of the normal distribution is merely demonstrated. The logarithm of the likelihood function for a random independent sample from a normal distribution has the form

$$\ln(L) = (-N/2)\,[\ln(2\pi) + \ln(\sigma^2)] - \frac{\sum_i (x_i - \mu)^2}{2\sigma^2}$$

The derivative with respect to μ is $\frac{\partial \ln(L)}{\partial \mu} = 2\sum_i x_i - 2N\,\mu$.

The derivative with respect to σ^2 is $\frac{\partial \ln(L)}{\partial \sigma^2} = \frac{\sum_i (x_i-\mu)^2}{[\sigma^2\sqrt{2}]^2} - \frac{N}{2\sigma^2}$.

Estimates of the mean and variance can be directly determined from these equations $\hat{\mu} = \frac{1}{N}\sum_{i=1}^{N} x_i$ and $\hat{\sigma}^2 = \frac{1}{N}\sum_{i=1}^{N} (x_i - \mu)^2$.

It can be seen that the estimate of the variance is biased and uses an unknown mean μ, which can, however, be replaced in the calculation by an estimate $\hat{\mu}$. The variance of the estimate is generally defined by (Cramér, 1999).

The variance $D(\hat{\sigma}^2)$ of the estimate $\hat{\sigma}^2$ is generally defined by relation (Cramér, 1999) $D(\hat{\sigma}^2) = \frac{1}{N}(C_4 - C_2^2) - \frac{2}{N^2}(C_4 - 2\,C_2^2) + \frac{1}{N^3}(C_4 - 3\,C_2^2)$.

where C_r is the r-th central moment. For the case of a normal distribution is $C_4 = 3\,C_2^2$, and therefore

$$D(\hat{\sigma}^2) = \frac{2(N-1)}{N^2} C_2^2$$

The surface of the negative logarithm of likelihood function for the sample (10, 20, 30, 40, 50) is shown in Fig. 1.26.

It can be seen from Fig. 1.26 that the mean and variance estimates are correlated. The last term of the logarithm of the likelihood function for the normal distribution is the sum of the squares of the deviations from the mean value, which should be minimal. The estimation method minimizing S, or $\chi 2$

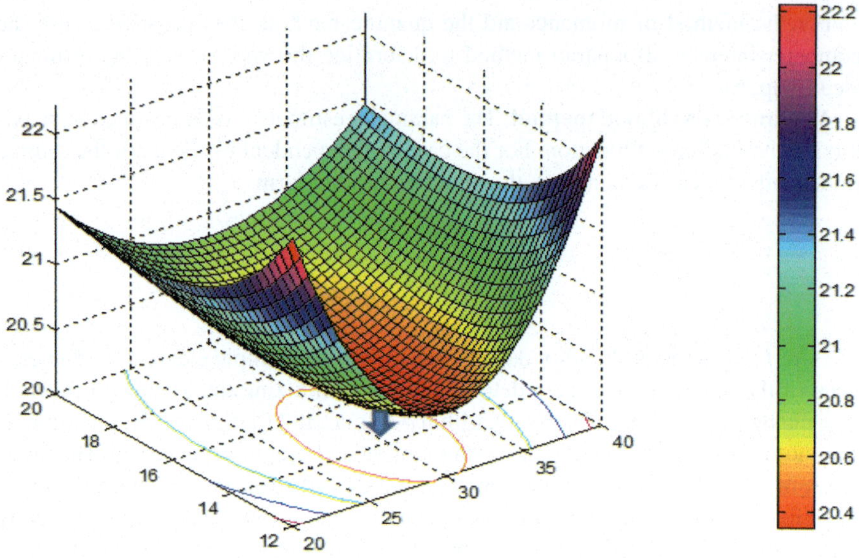

Fig. 1.26 Surface of likelihood function for sample (10, 20, 30, 40, 50)

$$S = \sum_i (x_i - \mu)^2 \text{ or } \chi^2 = \frac{\sum_i (x_i - \mu)^2}{\sigma_i^2}$$

is referred to as the **method of least squares** (LSM). For the normal distribution, this method is equivalent to the maximum likelihood method. Let each element of the sample have a different variance σ_i^2. The condition of LSM is then again zero of the first derivative, i.e.

$$\frac{\partial \chi^2}{\partial \mu} = 2\sum_i \frac{x_i}{\sigma_i^2} - 2\,\mu \sum_i \frac{1}{\sigma_i^2} = 0 \text{ then } \hat{\mu} = \sum_i \frac{x_i}{\sigma_i^2} / \sum_i \frac{1}{\sigma_i^2} = \sum_i x_i\, w_i$$

Here the weights w_i are normalized, i.e. their sum is equal to one. The estimate of the mean is therefore a weighted arithmetic average. For the classic unweighted average, the weights are $w_i = 1/N$. In this case, extremely large values are "preferred". The weights corresponding to the minimization of the squares of the relative deviations are $w_i = \frac{1}{x_i^2} / \sum_i \frac{1}{x_i^2}$. In this case, extremely small values are "preferred". The compromise criterion is the weight $w_i = \frac{1}{x_i} / \sum_i \frac{1}{x_i}$. This type of weight requires only positive data.

In Fig. 1.27 the differently weighted mean estimates for sample data generated by the formula 10 + randn(10), where randn(x) is a normally distributed random number generator (with mean 10 and variance 1) are shown. One value of O varying from 1 to 60 were added. The "standard" curve corresponds to the unweighted arithmetic mean, the "relative" curve corresponds to the minimization of relative deviations, and the "compromise" curve corresponds to compromise weights (the product of absolute and relative deviations).

It can be seen that at higher O simulating the outlier, the compromise weighted means are very little different (thus robust).

For a *normal distribution*, it is valid that

Parameter μ estimate $\overline{x}$ variance of estimate $D(\overline{x}) = \frac{\sigma^2}{N}$.
Parameter σ^2 estimate s^2 variance of estimate $D(s^2) = \frac{2\sigma^4}{N}$.
For linearly dependent data, it is $D(\overline{x}) = \frac{\sigma^2}{N} + \frac{N-1}{N}\sigma^2$.

For asymmetric distributions, $\overline{x}$ and s^2 are correlated with a correlation coefficient

$$\rho(\overline{x}, s^2) = \frac{g_1(x)}{\sqrt{N}}$$

For positively skewed distributions, these estimates are correlated also positively.

The maximum outlier value x_k for normally distributed data is bounded by inequality

$$(x_K - \overline{x})^2 \leq \frac{(N-1)^2}{N} s^2$$

There is also a bound of the date range

$$(x_{(N)} - x_{(1)})^2 \leq 2\,(N-1)\,s^2$$

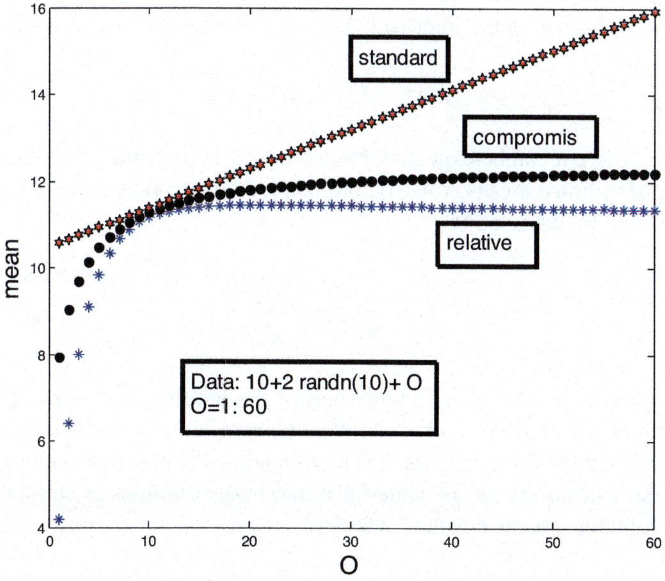

Fig. 1.27 Influence of O values on the size of the weighted averages

The arithmetic mean $\overline{x}$ is an effective estimate of the population mean μ (it has minimal variance) only for a normal distribution. For the case of the Laplace (two-tailed exponential) distribution, the effective is **sample median** $\tilde{x}_{0.5}$, for which the variance has the form

$$D(\tilde{x}_{0.5}) = \sigma^2/(2N)$$

For the Laplace distribution, the ratio $E_M = D(\overline{x})/D(\tilde{x}_{0.5}) = 2$, which means that the median is twice as efficient as the arithmetic mean. For the class of exponential distributions with parameter α, the EM ratio can generally be determined from the relation $E_M = \alpha^2\Gamma(3/\alpha)/\Gamma^3(1/\alpha)$ (Meloun & Militký, 2011). Thus, for a distribution with parameter $\alpha > 2$ (long tails), the median is more efficient than the arithmetic mean. For flat-peaked distributions, it is recommended to use the *quartile half-sum*, which has the form

$$P_F = (\tilde{x}_{0.75} - \tilde{x}_{0.25})/2$$

where $\tilde{x}_{0.75}$ is the upper quartile and $\tilde{x}_{0.25}$ is the lower quartile. For bounded distributions (arcsine and trapezoidal classes), the so-called *half-sum* is effective

$$\hat{x}_P = (x_{(N)} - x_{(1)})/2$$

where $x_{(N)}$ is the maximum and $x_{(1)}$ is the minimum sample element. The variance of the half-sum estimate is for a normal distribution equal to $D(\hat{x}_P) = (\pi^2\sigma^2)/(24\ln(N))$. For the uniform distribution, the variance of the half-sum estimate is equal to $D(\hat{x}_P) = (6\sigma^2)/[(N-1)(N-2)]$. For the arcsine distribution, the variance of the half-sum estimate is equal to $D(\hat{x}_P) = (5\pi^4\sigma^2)/N^4$. The arcsine distribution is defined in the interval (-*A*, *A*) and its probability density function is

$$f(x) = 1/(\pi\sqrt{A^2 - x^2})$$

In cases where it can be expected that there may be potential outliers in the data, the choice of trimmed means is also recommended. The α–percentage trimmed mean $\overline{x}(\alpha)$ is defined by the relation

$$\overline{x}(\alpha) = \frac{1}{n - 2M}\sum_{i=M+1}^{N-M} x_{(i)}$$

where $M = \text{int}(\alpha N/100)$ is the integer part of the number $\alpha N/100$ and $x_{(i)}$ are the order statistics (ascendingly sorted sample elements). For approximately symmetric distributions with possible outliers, it is recommended to choose the median $\tilde{x}_C$ from different estimates of the mean value for different distributions as an estimate of the center of symmetry (central value), i.e.

$$\tilde{x}_C = \text{med}\{\overline{x},\ \tilde{x}_{0.5},\ \hat{x}_P,\ P_F,\ \overline{x}(0.25)\}$$

Table 1.4 Properties of basic location and dispersion estimates

Parameter	Estimate	Variance of estimate	Effectivity	Distribution of estimate
μ	$\overline{x}$	σ^2/N	1	$N(\mu, \sigma^2)$
μ	$\tilde{x}_{0,5}$	$\sigma^2\pi/(2N)$	0,63	$N(\mu, \sigma^2)$
μ	$\tilde{x}_P$	$\sigma^2\pi^2/(24 \ln N)$	$24 \ln(N)/(\pi^2 N)$	$N(\mu, \sigma^2)$
σ^2	$\hat{\sigma}^2$	$2\sigma^4/N$	1*	$N(\sigma^2, D(\sigma^2))$
σ^2	s^2	$2\sigma^4/(N-1)$	1	$N(\sigma^2, D(\sigma^2))$
σ	$\hat{\sigma}$	$\sigma^2/(2N)$	1*	$N(\sigma, D(\sigma))$
σ	s	$\sigma^2/(2(N-1))$	1	$N(\sigma, D(\sigma))$
σ	R	$1.36\sigma^2/N$	0.368	$N(\sigma, D(\sigma))$
σ	d	$\sigma^2/(N(\pi-2))$	0.876	$N(\sigma, D(\sigma))$

where med{.} denotes the median of the elements in parentheses. To estimate the variance of the estimate $\tilde{x}_C$, it is possible to use interquartile lengths $k_{0.9} = (\tilde{x}_{0.95} - \tilde{x}_{0.05})/2$, when

$$D(\tilde{x}_C) = k_{0.9}^2/(2,72N)$$

The normality of the distribution from which the samples are drawn is commonly assumed. For different types of location and dispersion estimates, their properties, including distribution, are shown in Table 1.4.

1.7.2 Interval Estimates

From a statistical point of view, point estimates are less important because they do not say anything about the relationship between the estimate $\hat{a}$ and the true value of the parameter a. More information is provided by the *confidence interval*, which is the interval in which the true value of the given parameter will lie with a specified probability of $1 - \alpha$. The unknown parameter and distribution of the random variable $f(x,\mathbf{a})$ is then estimated not only by point estimation, but by two random values *LD* and *LH*. These form the end points of the confidence interval CI. The confidence interval will cover the unknown estimated parameter a with a previously chosen, sufficiently large confidence (expressing the degree of confidence) $c = 1 - \alpha$, which is called the confidence coefficient or statistical certainty. Three typical levels of c values are commonly used, namely 0.99, 0.95, and 0.9, which express the degree of confidence, usually as a percentage. Informally, these percentages express claims: virtually certain (99%), almost certain (95%), and partially certain (90%). Corresponding α 1, 5, and 10% are traditionally referred to as significance levels, similar to hypothesis testing. The confidence interval is defined by the relation

$$P(\text{LD} \le a \le \text{LH}) = 1 - \alpha$$

This equation means that the probability with which the actual value of the parameter a will lie in the random limits LD, LH is equal to $1 - \alpha$. However, it is not a probability interval in the true sense, but the reliability of coverage of a population parameter a and an interval with random limits LD, LH determined from samples of a given size. It is therefore the probability $P(A)$ of the statement $A\!: \mathrm{LD} \le a \le \mathrm{LH}$ (see Fig. 1.28).

The *probability interval* has, for example, the form

$$P(L_1 \le \hat{a} \le L_2) = 1 - \alpha$$

and expresses the probability with which a random variable (estimate) $\hat{a}$ lies within fixed limits L_1, L_2. The limits are chosen so that the probability of occurrence of the random variable a in this interval is $1 - \alpha$. It is therefore evident that the probability interval is not related to the actual parameter.

A simple interpretation of CI is that if 100 random samples are taken from a given distribution and the limits of $100c$ percent confidence interval LD_i LH_i are determined for each sample, the coverage of the true (unknown) value a will be appeared in the $100c$ samples.

For confidence intervals CI, it is valid that

(a) the larger is the sampling range N, the narrower is the confidence interval,
(b) the more accurate is the estimate and the smaller is the variance, the narrower is the confidence interval,
(c) the higher is the statistical certainty c, the wider is the confidence interval. Thus, the 95% CI is narrower than the 99% CI and wider than the 90% CI.

With regard to the requirement of the narrowest possible confidence interval, it is advisable to choose such estimates that are effective (for normally distributed data, see Table 1.3), when low dispersion can be achieved even for small sample sizes.

The standard confidence intervals of LD and LH are two-sided. In addition to these, a one-sided interval is also used in practice, either left-sided $[\mathrm{LD}, \infty)$ or right-sided $(-\infty, \mathrm{LH})$.

To construct a confidence interval, it is necessary to know either the distribution of the given parameter or its appropriate function. The procedure for constructing a confidence interval for the mean value μ of the normal distribution $N(\mu, \sigma^2)$ is

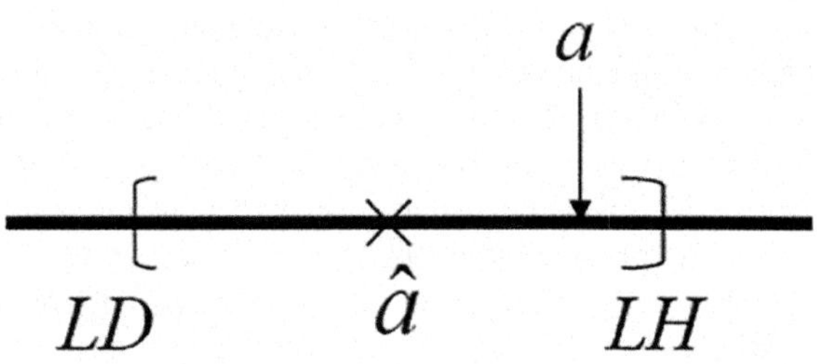

Fig. 1.28 The relationship between a confidence interval and the true value of parameter *a*

relatively simple. Since the best point estimate of the mean value μ is the sample arithmetic mean with distribution $N(\mu, \sigma^2/N)$, lies in the interval $\overline{x} \pm 2\sigma/\sqrt{N}$ of approximately 95% of the values of all possible averages of random variables from samples of the size N. The prerequisite is knowledge of the variance, which is unknown in practical tasks. Thus, the distribution of a suitable variable as function of estimates $\overline{x}$, s^2 is advantageously used.

If the data comes from a normal distribution $N(\mu, \sigma^2)$, the new variables can be defined (Pham, 2023).

$t = (\overline{x} - \mu)/s\sqrt{N}$, which has a Student's distribution with a parameter $df = N - 1$ (degrees of freedom),
$\chi^2 = (N - 1)\, s^2/\sigma^2$, which has a chi-square distribution χ^2 with a parameter $df = N - 1$ (degrees of freedom).

For the normal distribution, it is exactly valid that

$$P(-t_{1-\alpha/2} \le (\overline{x} - \mu)/s\sqrt{N} \le t_{1-\alpha/2}) = 1 - \alpha$$

Thus, after rearrangement, the $100(1 - \alpha)\%$ *confidence interval of the mean* is equal to

$$\overline{x} - t_{1-\alpha/2}\, s/\sqrt{N} \le \mu \le \overline{x} + t_{1-\alpha/2}\, s/\sqrt{N}.$$

The distribution of the sampling distribution $f(t)$ for a particular sample size N is shown schematically in Fig. 1.29.

For the general distribution characterized by the probability density $\varphi(x)$ it is valid that

$$P(t \le x) \approx \Phi(x) + \frac{g_1(x)}{\sqrt{n}} \frac{(2x^2 + 1)}{6} \varphi(x)$$

where $\Phi(x)$ is the distribution function of the standard normal distribution and $\varphi(x)$ is the corresponding probability density function. The confidence interval based on Student's distribution is shifted to higher values for skewed distributions.

When constructing a confidence interval for the parameter a from the estimates $\hat{a}, D(\hat{a})$ obtained by the maximum likelihood method, the asymptotic normality (for

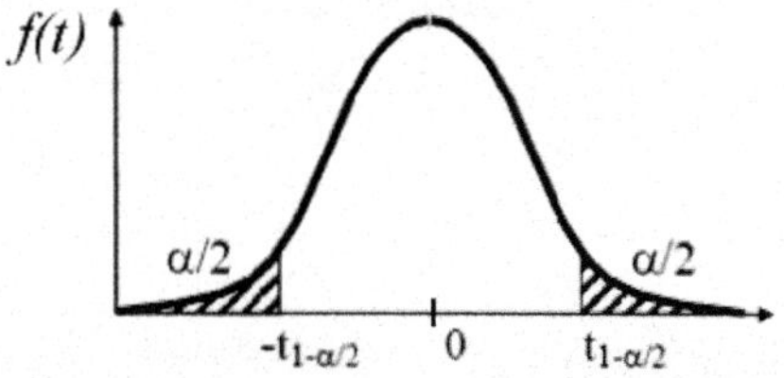

Fig. 1.29 Sampling distribution of parameter t for sample size N

$N \to \infty$) it can be used. The quantity $u = (\hat{a} - a)/\sqrt{D(\hat{a})}$ for having an N (0, 1) distribution is beneficial. The corresponding confidence interval then has the form

$$\hat{a} - u_{1-\alpha/2}\sqrt{D(\hat{a})} \leq a \leq \hat{a} + u_{1-\alpha/2}\sqrt{D(\hat{a})}$$

where $u_{1-\alpha/2}$ is the quantile of the standard normal distribution.

1.7.3 *Hypothesis Testing*

A statistical hypothesis is a statement (assumption) regarding the probability distribution of one or more random variables. This statement may refer to the parameters of the distribution of a random variable (e.g. the mean value of the length of the rod is 15 cm), or it may refer only to the distribution of the random variable (e.g. defects in goods have a Poisson distribution). A statistical hypothesis test is a rule that, based on the results obtained from a random sample, allows a decision to be made whether or not to reject the hypothesis being tested. When testing a statistical hypothesis, a distinction is made between the tested *null (basic)* hypothesis H_0 and the *alternative hypothesis* H_A. The test is to decide whether to reject the null hypothesis or not. The alternative hypothesis is the one we accept if we reject the null hypothesis. The entire testing procedure is actually the rejection of alternative hypotheses.

A certain test statistic is constructed to test the null hypothesis. If this statistic falls within the acceptance range, the null hypothesis is not rejected. However, if it falls into the critical region, the null hypothesis is rejected (see Fig. 1.30).

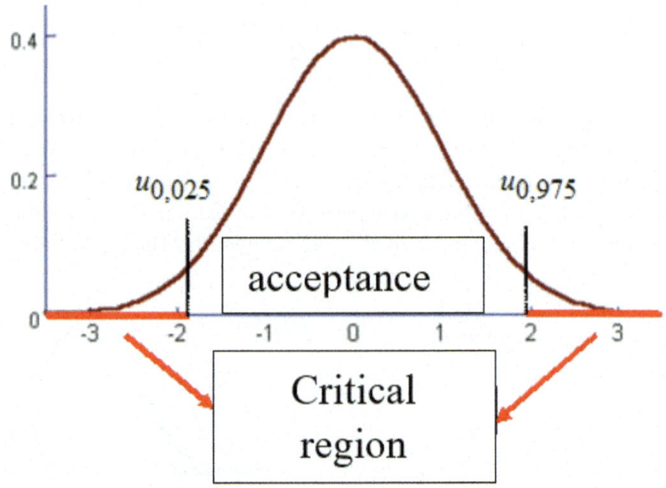

Fig. 1.30 Hypothesis test

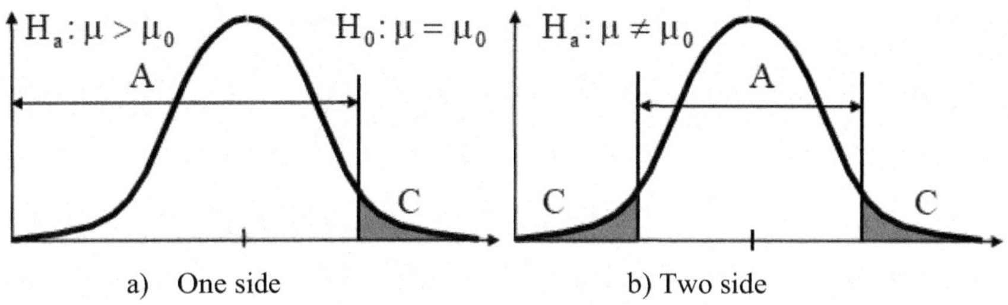

Fig. 1.31 Critical ranges for various alternative hypotheses

The probability of the test characteristic falling into the critical range is called the significance level α. It is often expressed as a percentage such as a 100 α% significance level, for example 5% or 1%. It is possible to define a critical field as two side or one side. Two-sided is defined when there is no reason why the test statistic should have either only a positive or only a negative sign. The significance level α is then divided into two parts of equal size $\alpha/2$ (see Fig. 1.31).

The procedure for testing a statistical hypothesis can be summarized in the following steps:

1. Formulation of the null hypothesis H_0 and the alternative hypothesis H_A according to the nature of the problem.
2. Choices of the level of significance α.
3. Choices of the test statistic $T(x_1, \ldots x_n)$, i.e. the function of values of a random sample with a known probability density function $f(T)$ for the case of validity and invalidity of the null hypothesis.
4. Determination of the critical range of the test characteristic based on its probability distribution and significance level.
5. Obtaining a random sample, calculating the test statistic and its quantiles forming the boundaries of the critical domain.
6. Deciding whether:
   (a) to reject the hypothesis H_0 and accept H_A if the calculated value of the test statistic falls into the critical range,
   (b) do not reject the hypothesis H_0, if the calculated value of the test characteristic does not fall into critical field.

When deciding whether to reject the null hypothesis, it should be considered that rejecting H_0 does not mean that the tested null hypothesis is not valid (however, it is unlikely with probability α). Rejection of the H_0 hypothesis only means that there is no confidence in its validity and in the next one the assumption will be made that H_0 is not valid and H_A can be accepted.

Failure to reject the H_0 hypothesis does not mean its automatic acceptance (however, it is very likely with a probability of $1 - \alpha$). It means that the test result did not show such a large discrepancy that would give sufficient reason to reject the hypothesis. Not rejecting H_0 means only that there is confidence in its validity, and

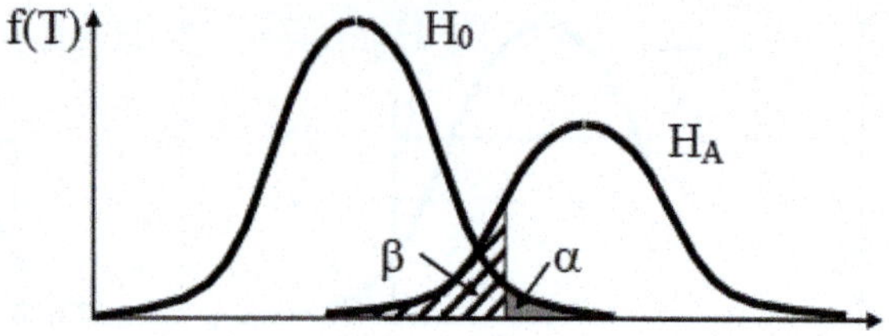

Fig. 1.32 Errors in hypothesis testing

in the following the assumption will be made that H_0 is valid or another hypothesis is valid, but it is not very different from H_0. When testing hypotheses, errors of two types arise (see Fig. 1.32):

Error of the first type $[\alpha]$: H_0 is valid but not accepted by the test (see Fig. 1.32 gray area).

Error of the second type $[\beta]$: H_0 is not valid, but was accepted by the test (see Fig. 1.32 hatched area).

In Table 1.5, possible variants of the probabilities α, β associated with the validity of the null hypothesis are presented.

The probability $1 - \beta$ is referred to as the power of the test.

Sometimes the so-called p-value is also used, where p is the smallest α at which, based on the given data, the null hypothesis H_0 is still rejected, i.e. the hypothesis H_0 is rejected when $p \leq \alpha$. Hypothesis tests are partly related to confidence intervals. In hypothesis tests, however, the distribution is generally centered on the value of the tested parameter (H_0), which is not the case for confidence intervals (see Fig. 1.33).

Table 1.5 Probabilities (α, β) associated with the validity of the null hypothesis

Statement	Acceptance range	Critical range
H_0 is valid	$1 - \alpha$	α
H_0 is not valid	β	$1 - \beta$

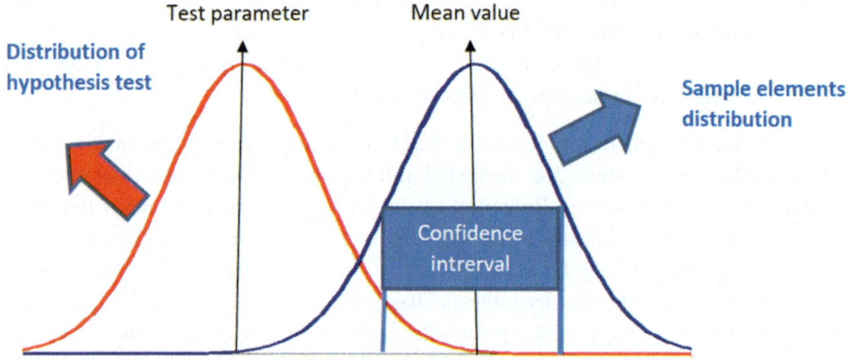

Fig. 1.33 Distribution of statistics for confidence interval construction and hypothesis testing

In *hypothesis testing*, for values of the distribution parameters (contained in the hypothesis H_0) that lie in a given confidence interval with the same level of significance, it usually turns out that H_0 is true. In order to use this relationship between confidence intervals and hypothesis tests (using the same statistics), the alternative hypothesis must be two-sided. Details on specific hypothesis tests of various types are given in books (Meloun & Militký, 2011; Pham, 2023).

1.8 Conclusion

This chapter covers only the basic facts needed to understand the specific procedures used in metrology and data treatment. Especially in metrology, mainly simplified procedures are still used without effective use of intensive computer-oriented methods and graphic-oriented techniques for verifying data behavior. These procedures and methods are discussed in particular in Chap. 6.

References

Azzouz, B., Ben Hassen, M., & Sakli, F. (2008). Adjustment of cotton fiber length by the statistical normal distribution: Application to binary blends. *Journal of Engineered Fibers and Fabrics, 3*(3), 155892500800300. https://doi.org/10.1177/155892500800300304

Cooke, P. (1979). Statistical inference for bounds of random variables. *Biometrika, 66*(2), 367–374. https://doi.org/10.1093/biomet/66.2.367

Cramér, H. (1999). *Mathematical methods of statistics (PMS-9).* Princeton University Press. http://www.jstor.org/stable/j.ctt1bpm9r4

Hosking, J. R. M. (1990). L-Moments: Analysis and estimation of distributions using linear combinations of order statistics. *Journal of the Royal Statistical Society Series b: Statistical Methodology, 52*(1), 105–124. https://doi.org/10.1111/j.2517-6161.1990.tb01775.x

Karvanen, J. (2006). Estimation of quantile mixtures via L-moments and trimmed L-moments. *Computational Statistics & Data Analysis, 51*(2), 947–959. https://doi.org/10.1016/j.csda.2005.09.014

McLachlan, G., & Peel, D. (2000). *Finite mixture models.* Wiley. https://doi.org/10.1002/0471721182

Meloun, M., & Militký, J. (2011). *Statistical data analysis.* In J. Militký & M. Meloun (Eds.). Woodhead Publishing Limited. https://doi.org/10.1533/9780857097200

Novitskii, P. V., & Zograf, I. A. (1985). Estimating the errors of means of measurement. *Énergoatomizdat.* (in Russian)

Pham, H. (2023). *Springer handbook of engineering statistics.* In H. Pham (Ed.). Springer, London. https://doi.org/10.1007/978-1-4471-7503-2

Piersol, J. S. B., & A. G. Piersol. (2000). Random data analysis and measurement procedures. *Measurement Science and Technology, 11*(12), 1825–1826. https://doi.org/10.1088/0957-0233/11/12/702

Rinne, H. (2008). *The weibull distribution.* Chapman and Hall/CRC. https://doi.org/10.1201/9781420087444

Rousseeuw, P. J., & Croux, C. (1993). Alternatives to the median absolute deviation. *Journal of the American Statistical Association, 88*(424), 1273–1283. https://doi.org/10.1080/01621459.1993.10476408

Vogel, R. M., & Fennessey, N. M. (1993). *L* moment diagrams should replace product moment diagrams. *Water Resources Research, 29*(6), 1745–1752. https://doi.org/10.1029/93WR00341

Wyckoff, J., Bain, L. J., & Engelhardt, M. (1980). Some complete and censored sampling results for the three-parameter Weibull distribution. *Journal of Statistical Computation and Simulation, 11*(2), 139–151. https://doi.org/10.1080/00949658008810399

Zanakis, S. H. (1979). A simulation study of some simple estimators for the three-parameter Weibull distribution. *Journal of Statistical Computation and Simulation, 9*(2), 101–116. https://doi.org/10.1080/00949657908810302

Zenbil, B. A. (1991). *Estimation techniques for a class of non regular distributions*. METU.

Chapter 2
Basics of Metrology

No instrument measures without errors.

Abstract Metrology is closely connected with the philosophy of science and the nature of measurement errors. This chapter contains basic information about the principles of general metrology, the statistical characteristics of measuring instruments, and the different components of measurement errors. The analysis of measurement results based on various typical metrological characteristics is explained using a rigorous statistical approach.

2.1 Introduction

Metrology is a science that deals with process of measurement and its evaluation in general. Measurement is understood as assigning numbers (digits) to attributes of objects, persons, or phenomena according to a certain classification rule (reliability and correctness). These rules must be verifiable. The numbers assigned to the objects/phenomena must reflect the relationships between them. In contemporary science, the principle of conservation of phenomena is postulated, which essentially means formulating theories that would not contradict intuitive ideas about phenomena. In the experimental sciences, it is a requirement that scientific theories do not conflict with observations. Data are not prioritized over all theoretical assumptions, but rather, there is a symbiosis between theory and data. Theory has a major role in determining what can be measured or otherwise observed primarily as data, which data are relevant, how the results of measurement are to be processed, how the data are interpreted, and how the interpreted data are legitimately used in the construction and confirmation of theories. In typical cases, some theoretical assumptions are sufficiently well established, but there is always a need for certain additional assumptions, which can, on the contrary, be supported by data. The theory of measurement and metrology includes a whole range of areas: general procedures and philosophies, measurement errors, types of measurement models, accuracy and correctness of measurements, calibration of measuring instruments, processing of results and their interpretation.

J. Militký et al., *Textile Material Research—Statistical Tools of Experimental Data Analysis*, https://doi.org/10.1007/978-981-97-6607-9_2

This chapter contains basic information about general metrology, measuring instruments, and measurement errors with a focus on the analysis of results and data processing (Crowder et al., 2020; Elishakoff, 1995; Meloun & Militký, 2012; Slaev et al., 2019; Thompson & Taylor, 2008; Wade, 2022).

2.2 Basic Terms from General Metrology

Metrology uses some special terms (Meloun & Militký, 2012):

Calibration—creation of a relationship between the desired quantity (poorly measurable) and the measured quantity (easily measurable). Calibration has two phases: model creation and model validation.

Adjustment—setting the device so that it measures correctly (using standards, tests, etc.). **Testing**—specific activity aimed at knowing the properties of materials and products. **Certification**—written official verification or certification (verification of conformity) by means of a certificate that the product or service is in conformity with the requirements of certain standards or technical conditions.

Quality Assurance—according to ISO, ASME, etc.

Auditing—an activity that verifies the ability to perform certain work.

Accreditation—official recognition that a testing or calibration laboratory, a certification body or an inspection body is qualified to perform certain tests or certain types of tests, calibration, certification, or inspection activities.

Accreditation Body—the governing body managing the accreditation system and granting accreditation in the relevant country, one state-one accreditation body (ČIA—Czech Institute for Accreditation).

Verification—confirmation by examination and provision of objective evidence that the specified requirements have been met (in the laboratory, verification of the suitability of the test technique).

Validation—confirmation of validity by examination (confirmation of agreement with the procedure of the test performed).

Measurements can be divided according to principle into three basic groups:

I. Comparative: The principle is to compare an unknown object with other objects of the same type using different comparators. An example is the measurement of weight on two pan scales by balancing with weights.
II. Direct: In direct measurements, results are obtained without the use of comparators or sensors. An example is the measurement of dimensions using calipers and the measurement of lengths using calipers. It is actually also a comparative measurement, but the comparison with the standard took place before the measurement.

III. Indirect: This is the basic principle of metrology. With the help of sensors, an easily measurable (electrical) quantity is obtained (which is related to the desired "measured" quantity of a physical nature). During calibration, this quantity is converted to a measured quantity.

The measurement is characterized by accuracy and correctness. Accuracy is related to the variability obtained from repeated measurements, and correctness is related to the distance from the true value (see Fig. 2.8). If samples obtained in different laboratories are analyzed by different methods (or operators), it is possible to proceed by dividing the total variance into clarified parts. It is an analogy with the variance analysis described, for example, in the book (Meloun & Militký, 2012).

The repeatability of the measurement s_o^2 expresses the variability caused by the device. It is measured as the variability of repeated measurements performed by one operator on the same instrument.

The reproducibility of the measurement s_r^2 expresses the variability caused by the operators. It is measured as the variability of repeated measurements performed by different operators on the same instrument.

Operators are labeled $j = 1 \ldots m$ and samples are labeled $i = 1 \ldots N$. The measurement result of the i-th sample by the j-th operator is x_{ij}. The overall average is

$$\tilde{x} = \left(\sum_{i=1}^{N} \sum_{j=1}^{m} x_{ij} \right) / (Nm) \tag{2.1}$$

and corresponding overall variance is

$$s^2 = \left(\sum_{i=1}^{N} \sum_{j=1}^{m} (x_{ij} - \tilde{x})^2 \right) / (Nm - 1) \tag{2.2}$$

The average of measurements obtained by the j-th operator is equal to

$$\overline{x}_j = \left(\sum_{i=1}^{N} x_{ij} \right) / N \tag{2.3}$$

The variance of the mean values of individual operators (*reproducibility of measurements*) is equal

$$s_r^2 = \left(\sum_{j=1}^{m} (\overline{x}_j - \tilde{x})^2 \right) / (m - 1) \tag{2.4}$$

The average repeatability of the measurements s_o^2 is then given by the relation

$$s_o^2 = s^2 - s_r^2 \tag{2.5}$$

In order to express the correctness of the measurement, it is usually necessary to use standards (etalons). Measurement standards also include the methods for their use and the units used. The actual specifications of the measurement procedure and measurement conditions together with the measurement units are standardized. It must be standardized especially in cases where there are several possible variants of measurement methods providing different results. The standard is (Foegelle, 2002; Meloun & Militký, 2012):

(a) nonobligatory regulation (for unregulated cases).
(b) regulation that is obligatory (for regulated cases, i.e. fire protection, health protection, ecology, safety).

Basic types of standards are:

(1) subject standards (characteristics of objects, raw materials, semi-finished products, products)—shape, parameters, dimensions, etc.
(2) regulatory standards (rules for technical activities).
(3) general standards (unification and definition of terms, designation of units, etc.).
(4) test standards—description of the test method, goal: reproducibility.
(5) nomenclature standards-interpretation of concepts-terms, definitions, names

There are standards of various companies, national standards (DIN—Deutsche Industrie Normen, GOST—Gosudarstvennyj standard, BS—British Standard, ASTM—American Standards Test Methods, ČSN—Czech State Standard) and international standards (ISO—International Standard Organization) (Foegelle, 2002; Meloun & Militký, 2012).

When processing the results of the measurement of physical quantities, it is generally necessary to proceed in accordance with the category to which the measured quantity belongs:

(a) **Extensive** physical quantities have values proportional to the size of the investigated system (mass m, volume V, enthalpy H, entropy S). For the overall system (c) which consists of two subsystems, or parts (0 and 1), is valid $m_c = m_1 + m_0$, $V_c = V_1 + V_0$ and $S_c = S_1 + S_0$.

Thus, the values of the extensive properties of the subsystems are added together.

(b) **Intensive** physical quantities have values independent of the size of the investigated system, but determine its "quality", or intensity (temperature T, pressure p, density ρ, stress σ).

The values of intensive properties of partial systems are "averaged" with respect to their proportions.

For example, the overall system density is $m_c = m_1 + m_0$ and $\rho_c V_c = \rho_0 V_0 + \rho_1 V_1$. The total density is then determined as a weighted arithmetic mean

$$\rho_c = \frac{\rho_1 V_1 + \rho_0 V_0}{V_0} = \rho_1 v_1 + \rho_0 v_0 \tag{2.6}$$

where $v_0 = \frac{V_0}{V_C}$ and $v_1 = \frac{V_1}{V_C}$ are the volume ratios (fractions) of subsystems.

In practice, it is usually easier to use the mass fractions $w_0 = \frac{m_0}{m_C}$ and $w_1 = \frac{m_1}{m_C}$ of the subsystems. In order to derive the total density as a function of the partial densities when the mass fractions are known, it is then necessary to proceed from the additivity of the volumes, i.e.

$$V_c = V_1 + V_0 \text{ and } \frac{m_c}{\rho_c} = \frac{m_0}{\rho_0} + \frac{m_1}{\rho_1} \text{ finally } \frac{1}{\rho_c} = \frac{w_0}{\rho_0} + \frac{w_1}{\rho_1}$$

$$\text{or } \rho_c = \frac{\rho_0\,\rho_1}{w_0\,\rho_1 + w_1\,\rho_0} \tag{2.7}$$

It is evident that in this case, it results in a weighted harmonic mean. For positive data, this is lower than or equal to the arithmetic mean valid for volume fractions. Of course, these relationships only apply to densities, and in the case of other intensive quantities, the type of weighted average must be found using an analogous procedure (for pressure or stress, the same weighted averages will come out as for density). Intensive quantities are almost always a proportion of extensive quantities

$$p = \frac{F}{S} \quad \rho = \frac{m}{V} \quad T = \frac{\Delta H}{\Delta S}$$

(c) **Specific** physical quantities F are extensive quantities E per unit of mass m (specific volume, specific heat). The values of the specific properties F of the subsystems are again "averaged" with regard to whether volume or mass fractions are used. For the mean values and variances of these properties, it is approximately valid

$$\overline{F} \approx \frac{\overline{E}}{\overline{m}} + \frac{\overline{E}}{\overline{m}^3} \quad s_F^2 = \frac{\overline{E}}{\overline{m}}(1 + v_m^2) \tag{2.8}$$

The variance of a specific quantity is therefore only related to the accuracy of the mass measurement expressed by the coefficient of variation v_m.

(d) **Molar** physical quantities are extensive quantities E related to the unit amount of substance [mol]. They are important for predicting properties from the composition of polymer chains.

In many cases, it is true that: ***Energy* = extensive multiplied by intensive**. Examples are:

Mechanical energy = elongation multiplied by tensile force
= volume multiplied by pressure
= area multiplied by surface tension
Thermal energy = entropy multiplied by temperature
Electrical energy = charge multiplied by potential.

In digital measurements, the primary data are obviously averaged in computer so that the distribution of the results has approximately the normal distribution even where the original data are strongly non-normally distributed. This is clearly visible from Fig. 2.1, where it is shown that even for a highly asymmetric data distribution, the averages of relatively small samples are approaching to be normally distributed.

In a complex metrological procedure, it is generally advisable to use the "experiment system", the diagram of which is shown in Fig. 2.2.

The individual phases of this system are:

PE (Preparation of Experiments): includes a number of activities that are necessary for the next stages of experimentation as well as for the interpretation of measurement results.

Main tasks:

(a) specification of the purpose of the experiment;
(b) determination of measured quantities;
(c) sampling strategy (placement of experimental points).

Usually, this stage is underestimated and only after a negative evaluation of the experiment must be specified.

RE (Realization of Experiments) i.e. metrology: related to the availability of measuring instruments and their components. For a number of modern devices

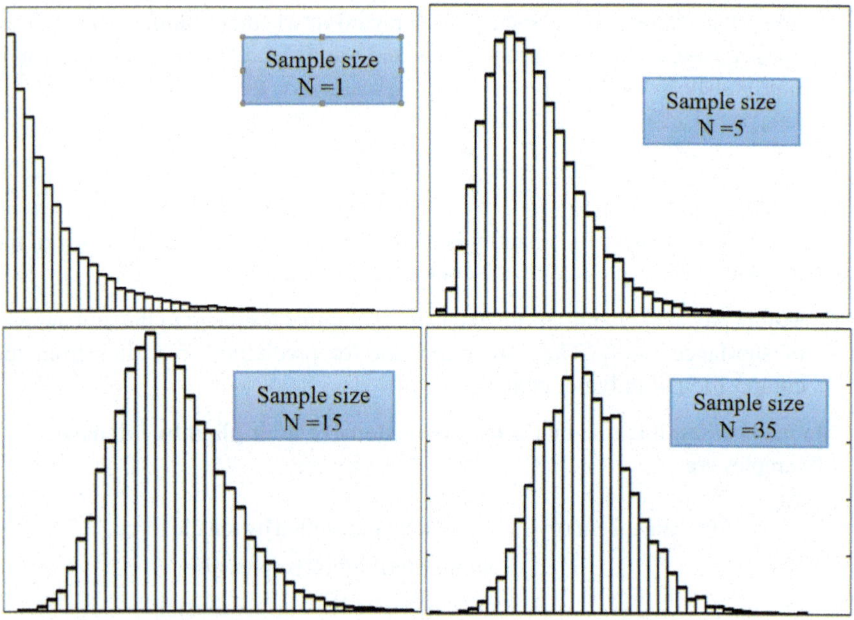

Fig. 2.1 Effect of the sample size N on the distribution of the arithmetic mean for data generated from Weibull distribution ($W(1, 1)$)

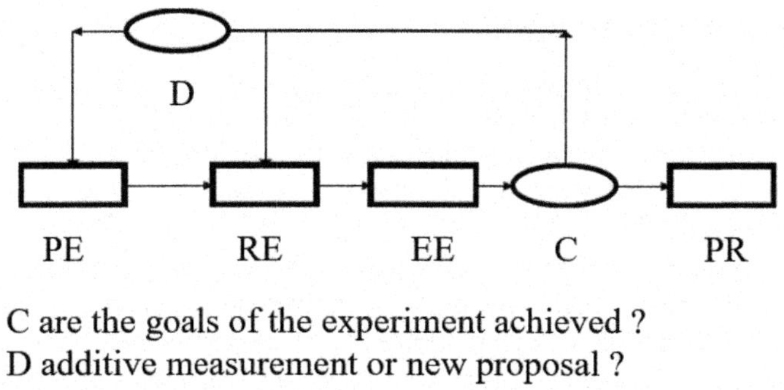

Fig. 2.2 Experiment system

connected to a computer, calibration, data collection, and their basic pre-processing (filtering, compression) are ensured. Main tasks:

(a) method selection;
(b) selection of measuring equipment and its calibration;
(c) organization of measurement and data collection;
(d) ensuring representative results.

EE (Evaluation of Experiments): is among the most discussed stages, which is related to the level of knowledge of the experimenter and the availability of suitable statistical tools or programs.

Main tasks:

(a) basic data processing;
(b) modeling;
(c) optimization.

PR (Presentation of Results): is important with regard to the interpretation and use of results in the professional community.

Main tasks:

(a) accuracy of results;
(b) comparison with known facts;
(c) interpretation and generalization.

Suitable algorithms and procedures are already available for individual tasks within these phases. However, it is always advisable to check whether a sufficiently detailed design of the experiment has been carried out before starting the measurement. Similarly, when presenting the results, it is necessary to consider both the realization of the experiment and the evaluation of the experiments. Decision points **C** and **D** allow to realize changes to ensure that the desired goals of the experiments are achieved.

2.3 Units of Measurement

At present, the units of the SI metric system are used practically all over the world. In this system, there are a total of seven basic and two additional units (angles) characterizing basic physical quantities (see Table 2.1) (Slaev et al., 2019; Thompson & Taylor, 2008; Wade, 2022).

In the SI system, these units are chosen so that no correction factors are needed for mutual conversions. There are also basic derived units and special units used in various fields (Thompson & Taylor, 2008). The characteristic of the textile is that it uses a number of special units. Basic-derived physical quantities and their units are listed in Table 2.2.

The kgf is a force induced by mass $m = 1$ kg in the Earth's gravitational field ($g = 9.81$ m s^{-2}). It is a fact that

$$F = \text{mg},\ 1N = 0.102\text{kgf},\ 1cN \approx 1\text{gf},\ 1\text{daN} = 1\text{kgf}$$

The unit *bar equal to* 105 Pa is also used to express pressure, especially in meteorology. Other derived measurement units that represent physical laws are related to the expression of measurement results, or by transforming primary data into an output quantity, they correspond to a combination of units of basic quantities. For example, the average rate of the material in textile process is the average of the lengths of processed material measured at constant time intervals divided by the length of

Table 2.1 Basic SI physical quantities and their units

Quantity	Unit	Mark	Symbolic
Length	Meter	m	[*L*]
Mass	Kilogram	kg	[*M*]
Time	Second	s	[*T*]
Intensity of electric current	Ampere	A	–
Temperature (thermodynamic)	Kelvin	K	–
Light intensity	Candela	cd	–
Amount (substance)	mol	mol	–

Table 2.2 Other derived SI physical quantities and their units

Quantity	Unit	Mark	Combination	Symbolic
Frequency	Hertz	Hz	s^{-1}	$[T^{-1}]$
Density	kg m^{-3}	kg m^{-3}	kg m^{-3}	$[M\ L^{-3}]$
Force	Newton	N	kg m s^{-2}	$[M\ L\ T^{-2}]$
Pressure	Pascal	Pa	N m^{-2}	$[M\ L^{-1}\ T^{-2}]$
Work (energy)	Joule	J	N m	$[M\ L^{2}\ T^{-2}]$
Power	Watt	W	J s^{-1}	$[M\ L^{2}\ T^{-3}]$

Table 2.3 Multiples and fractions of units

Multiple	Prefix	Fractions	Prefix
10^{18} exa	E	10^{-1} deci	d
10^{15} penta	P	10^{-2} centi	c
10^{12} tera	T	10^{-3} mil	m
10^{9} giga	G	10^{-6} mikro	u
10^{6} mega	M	10^{-9} nano	n
10^{3} kilo	k	10^{-12} piko	p
10^{2} hekto	h	10^{-15} femto	f
10^{1} deka	da	10^{-18} atto	a

Note that only multiples and fractions of seconds are not decimal. Multiples and fractions are used to express results within one order of magnitude, which is very beneficial from the point of view of statistical processing

these intervals. This directly follows from the units of velocity [m/s] = [m]/[s]. It is therefore evident that the laws of physics or the transformation of the primary data to the output quantity must be *dimensionally homogeneous.* Thus, the combination of input quantities must have the same combination of units as the output quantity. This also applies to different types of models using physical quantities. Dimensional homogeneity can also be achieved by switching to dimensionless groups of units of input and output quantities. When processing experimental data, the transition to dimensionless quantities is carried out by standardization, when

$$u = (x - E(x))/\sqrt{D(x)} \tag{2.9}$$

Here, $E(x)$ is the mean and $D(x)$ is the variance. Usually, instead of these population characteristics, their estimates are used (see Chap. 1) (Meloun & Militký, 2012). In some cases, the conversion to dimensionless quantities is performed by dividing some quantity by its value in the standard state. For example, the dimensionless temperature can be defined as $T_r = T / T_s$, where T_s is the temperature of the standard state (e.g. ambient temperature or 20 °C temperature). In some cases, it is better to use multiples and fractions of basic units. Their overview is given in Table 2.3.

2.4 Measuring Instruments and Metrology

Measurement is one of the basic methods of obtaining quantitative information about the state of monitored quantities. Due to the limited accuracy of the measuring devices and the variability of the measurement conditions, even on a homogeneous material, where the actual value of the measured quantity is constant, the same results are not obtained. Each measurement thus represents a random variable that contains not only information but also noise, i.e. measurement errors. The magnitude of measurement errors is a quality criterion of the measurement process or measuring instrument. Measurement theory deals with the analysis of measurement and instrument errors. Errors of measurement are a somewhat special case of a commonly uncategorized variable in a *cardinal scale.*

2.4.1 Basic Terms

In Fig. 2.3, a simplified diagram of *measuring steps* is shown. Most measuring devices do not measure the required physical quantity (in many cases, it is not even possible), but use a sensor to measure another, *easily measurable* analog quantity (usually electric voltage, resistance, current, etc.) and restore results in original units in measuring device (Wade, 2022).

This measured quantity is functionally related to the investigated physical quantity in accordance with the laws of physics. With digital measuring devices, the measured analog quantity is in the form of a signal that can be transmitted to the evaluation unit after amplification. In this unit, transformation into a digital quantity (data), data pre-processing (noise suppression by averaging and removing gross outliers) and conversion into a physical quantity (measurement result) using a calibration relationship (often a straight line) are realized. With analog measuring instruments, signal amplification and processing in the evaluation unit are often omitted. Calibration is carried out by marking the scales of the instruments already in the units of the physical quantity. In Fig. 2.4, a scheme of an analog and digital thermometer is shown.

The quality of the measuring device depends mainly on the quality of the sensor (measurement principle), the level of data pre-processing, and the accuracy of calibration. Measurement errors can therefore be caused by a number of factors.

According to the place of origin in the measurement chain, errors can be divided into four basic groups:

1. *Instrumental errors* are caused by the design of the measuring device, the sensor used, etc. and determine measuring device quality. For many devices, these errors are known and guaranteed by the device manufacturer.

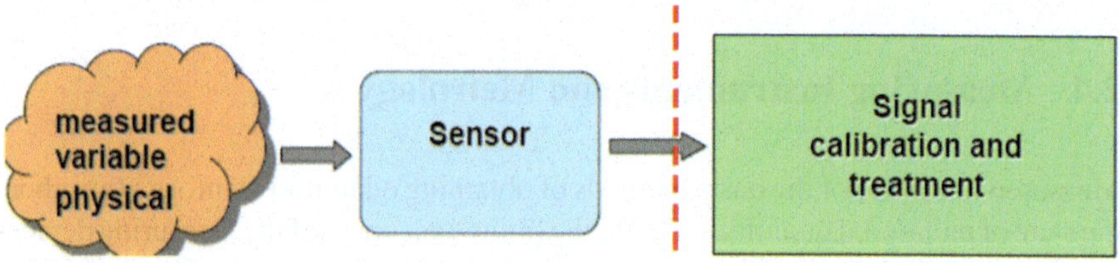

Fig. 2.3 Simplified diagram of the measuring steps

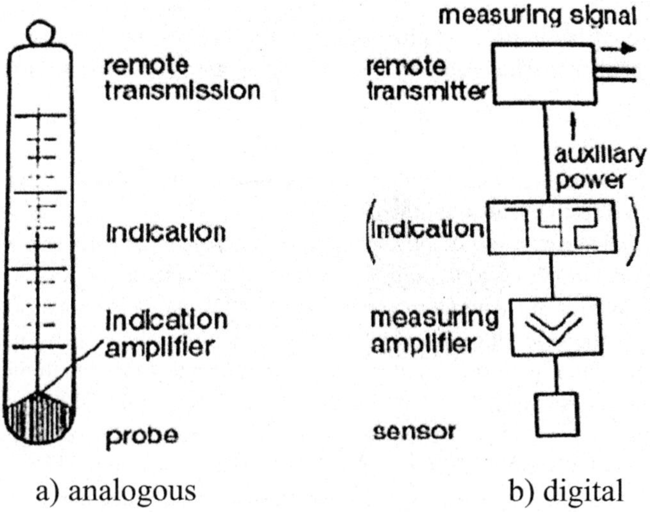

Fig. 2.4 Scheme of analog and digital thermometer

2. *Methodological errors* are related to the used methodology for determining measurement results, such as data registration, organization of measurements, elimination of external influences, etc.
3. *Theoretical errors* are related to the measurement procedure used. It mainly concerns measurement principles, physical measurement models, used parameters, physical constants, calibration models, etc.
4. *Data processing errors* are numerical calculation errors and are caused by the use of inappropriate statistical evaluation methods.

According to the causes of their occurrence, errors can be divided into three basic groups:

1. Random errors vary randomly in magnitude and sign during repeated measurements. These errors cannot be predicted and are described by a certain probability distribution. They are results of a number of causes that can be difficult to remove, or at least limit their influence.
2. Systematic errors affect the measurement result in a predictable way. They tend to be a function of time or parameters of the measurement process. They tend to have the same signs. Constant systematic errors reduce or increase the result of all measurements by a constant amount. They are often not visible by inspection of measurement results and can only be detected when these are compared with results from another (obviously more precise) device. There are also systematic errors with a time trend, caused by aging or wear of the measuring instrument. Systematic errors of the measuring device are divided into additive (zero value setting error) and multiplicative (sensitivity error). The type and size of device systematic errors are usually guaranteed by the manufacturer.

3. Gross errors are referred to as extraordinary deviating, or outliers are caused by an exceptional cause, sudden failure of the measuring apparatus, incorrect recording of the result, etc. They cause the given measurements to differ significantly from others.

In Fig. 2.5, the composition of overall error from systematic and random parts is shown (here probaility density function of measurements is shown and interrelation with so-called residuals is mentioned).

The absolute error Δ is defined as the difference between the measured quantity x and the true (unknown) value μ. For the additive measurement model $x_i = \mu + \varepsilon_i$, the statistical behavior of the random error is identical to the behavior of the measured variable x. If the true value μ is not known, a sample obtained by repeating the measurement under the same conditions is usually used, from which the arithmetic mean $\overline{x}$ and sample variance s^2 are calculated, i.e. $\{x_i, i = 1, \ldots N\} \Rightarrow \overline{x}, s^2$. Here, s^2 is a measure of measurement accuracy and $\overline{x} - \mu$ is a measure of measurement correctness (in this case μ is often replaced by mean value obtained by more precise device). Two basic options for expressing errors can be defined as:

$$\text{Absolute error} \quad \Delta_i = x_i - \mu \tag{2.10}$$

$$\text{Relative error} \quad \delta_i = \Delta_i / x_i \tag{2.11}$$

Depending on which of these errors is constant, measuring devices are also divided into categories (see Sect. 2.4.2).

Systematic errors are also commonly classified according to what information is available about the true value of the measured quantity (see Fig. 2.5) as:

(a) unknown (unrecognizable) error: the reference value is the actual value of the measured quantity (it is unknown and cannot be attributed to a measurement error).

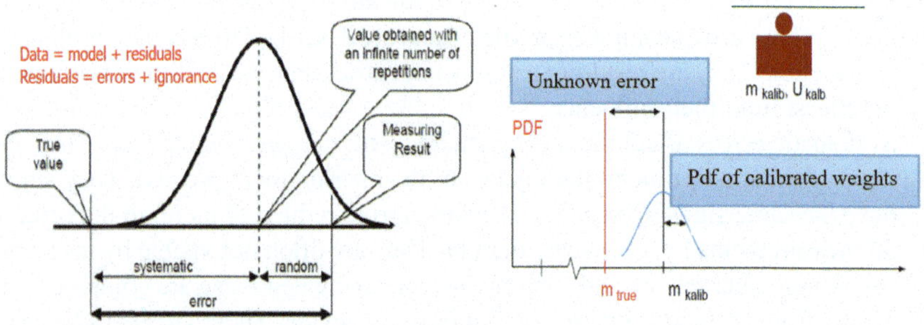

a) error components b) errors of weighting

Fig. 2.5 Different types of measurement errors

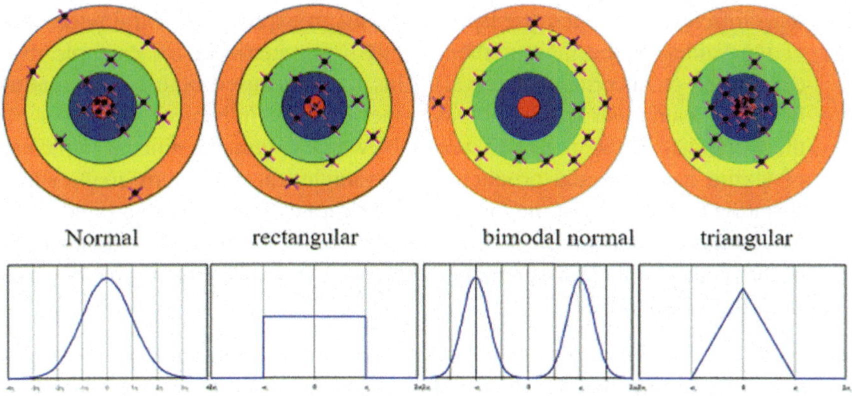

Fig. 2.6 Scatter of the measurement results for different error distributions (Foegelle, 2002)

(b) predictable (knowable) error: the reference value has a known size, it is, for example, a standard or a calibrated sample.

In Fig. 2.5a, the composition of overall error from systematic and random parts is presented. In Fig. 2.5b, the situation when measuring a calibrated weights on a scale is shown. It is obvious that even when weighing a calibrated weights, individual measurements will differ. If there is no systematic measurement error (see Fig. 2.5b), the average measurement value is equal to the mass of the calibrated weights. However, there may still be an unknown error caused by the value of the calibrated weights m_{kalib} being different from the correct value m_{true}. The value m_{true} can be estimated if, for example, a very accurate scale is available that can be used to re-calibrate the weights.

The specific distribution of errors (see probability density in Fig. 2.5b) determines their dispersion and concentration around the average value. This is shown for the case of zero systematic error in Fig. 2.6 (Foegelle, 2002).

It can be seen from Fig. 2.6 that for the bimodal U distribution, the measurement results will not be concentrated around the mean value. It is also noticeable that for a normal distribution (defined on the entire real axis), the measurement results will deviate more from the mean value than for a distribution defined only on the finite interval. This also shows the practical pitfalls in simulating errors.

2.4.2 Accuracy Characteristics of Measuring Devices

The quality of measuring instruments and measurement results is usually expressed in terms of corresponding inaccuracies (Thompson & Taylor, 2008). In many cases, it is necessary to distinguish:

(a) *inaccuracies of the measurement results* as an expression of the uncertainty of the values of the measurement results (characterized, for example, by a confidence or uncertainty interval). This can be caused both by fluctuations in the measurement conditions and other external influences, as well as by the variability of the measured materials, which is quite significant in the case of textile materials and products. For example, the minimum coefficient of variation (calculated) of the fineness of the yarn can range up to around 20%.
(b) *inaccuracies of the measuring device* (or process) as one of the characteristics of the quality of the measurement indicating a generally permissible deviation from the actual value.

It is obvious that the error of the measuring device is only one component affecting the error of the measurement results, and by choosing the accuracy of the measuring device, its influence can be greatly reduced. A typical function of the measuring device is shown in Fig. 2.7. The input here is the quantity x (*measured quantity*) and the output is the quantity y (*measurement result*). The mode of transformation $y = f(x)$ is known or is constructed using calibration (see Chap. 7).

When calibrating instruments, a series of output values (e.g. electrical voltage) is obtained for a series of values of the input quantity (e.g. temperature, humidity, etc.). By measuring, at different x, the corresponding set of y, creating points is obtained and functional transformation, i.e. function $y = f(x)$, is created. Schematic representation of variability of outputs is the band of uncertainty (Figs. 2.9, 2.11 and 2.12). The middle line in the uncertainty band is called the nominal characteristic y_{nom} and is usually indicated by the manufacturer. The nominal characteristic y_{nom} (or x_{nom}) differs from the real y_{real} (or x_{real}) by the error of the measuring device. To determine device errors, it is necessary to know the actual value of the measured quantity (e.g.

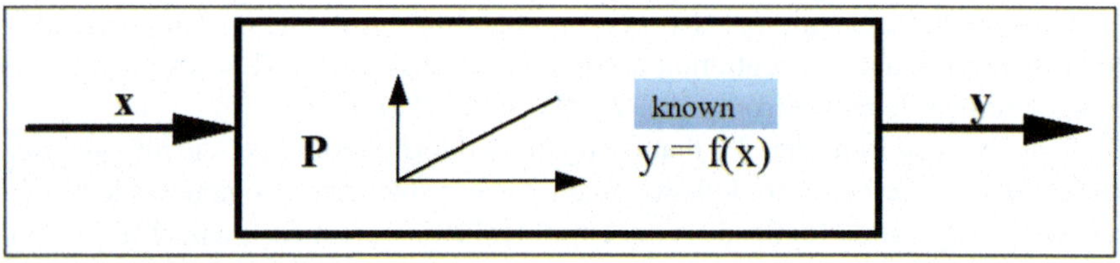

Fig. 2.7 Function of the measuring device

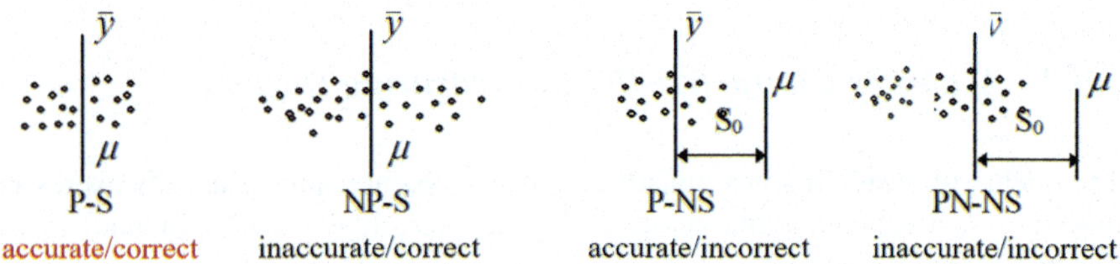

Fig. 2.8 Measurement correctness and accuracy

the so-called etalon) or to have another very accurate device available. In both cases, the value of μ or its estimate is available. The results of repeated measurements then make it possible to determine the levels of accuracy and correctness of the measurements.

The *accuracy of the instrument* is defined as the range of statistical uncertainty of the results. It is related to random errors and corresponds to the reproducibility of measurements. It is expressed as a variance around the mean value of a given tuple of measured results.

The *correctness of the instrument* indicates the average distance of the measurement results from the actual value and is related to systematic errors. It corresponds to the deviation of the measurement from the theoretical value.

For a simple measurement model $y_i = \mu + \varepsilon_i$, the mean and variance based on the measurements are determined, i.e. $\{y_i, i = 1, ...N\} \Rightarrow \overline{y}, s^2$. As a measure of measurement accuracy (P), the variance s^2 is chosen, which is identical to the variance of the measurement errors. The measure of measurement correctness (S) is the difference $\overline{y} - \mu$, which, however, can only be determined if μ is known (e.g. by using a standard). The differences between measurement accuracy and correctness are shown in Fig. 2.8.

A. Measuring instruments with constant absolute error

It is a case of a so-called constant additive error, i.e. an error of zero value (see Fig. 2.9). Here, the additive measurement error Δ_0 is constant (Zelený, 1976) (see Fig. 2.9).

The relative error is then a reciprocal function of the level of the measured variable x

$$\delta = \frac{\Delta_0}{x} \tag{2.12}$$

For practical calculations, it is advantageous to define the reduced relative error by the relation

$$\delta_R = \Delta_0/R \tag{2.13}$$

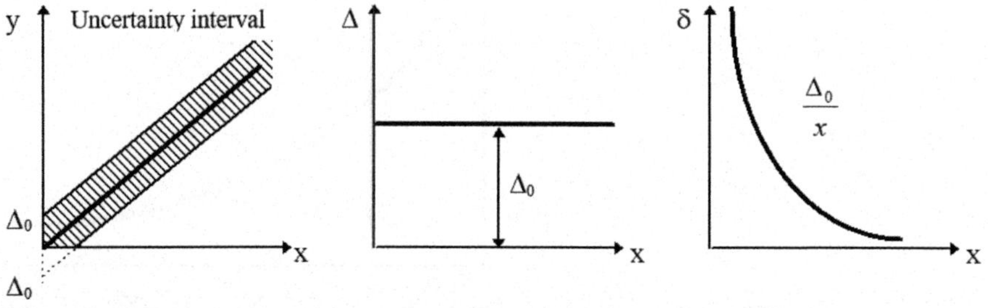

Fig. 2.9 Constant absolute measurement error

Here, the working range $R = x_u - x_L$, where x_L is the lower limit and x_U is the upper limit of the working interval. Considering the rapid growth of the relative error for low levels of the measured variable x, a sensitivity threshold is also defined, i.e. the value x_c for which the relative error is $\delta = 1$, or 100%

$$\text{thus } \frac{\Delta_0}{x_c} = \frac{\delta_R R}{x_c} = 1 \tag{2.14}$$

and then

$$x_c = \delta_R R \tag{2.15}$$

The significance of the sensitivity threshold can be seen from Fig. 2.10. It is also appropriate to determine the lower limit of the relative error $\delta = p$, [usually $p \sim 0.1$ (0.05)] for a low level of the measured variable x. The lower limit of the working interval is then equal

$$x_S = \Delta_0/p = x_c/p \tag{2.16}$$

Instruments operating with a constant absolute error are only of limited use for sufficiently high levels of the measured quantity x (Fig. 2.10).

B. **Measuring instruments with constant relative error**

This is a case of a constant *multiplicative error*, i.e. an error caused by the constant sensitivity of the device, where the sensitivity is understood as a relative deviation δ_S (Zelený, 1976) (see Fig. 2.11).

The absolute error is then a linearly increasing function of the level of the measured variable x

$$\Delta = P x \tag{2.17}$$

Fig. 2.10 Meaning of sensitivity threshold

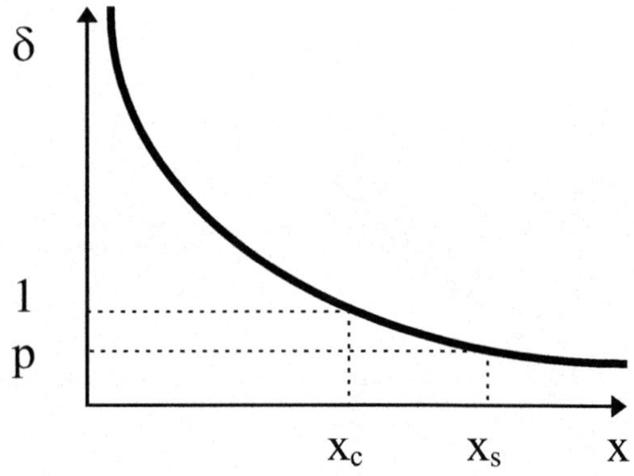

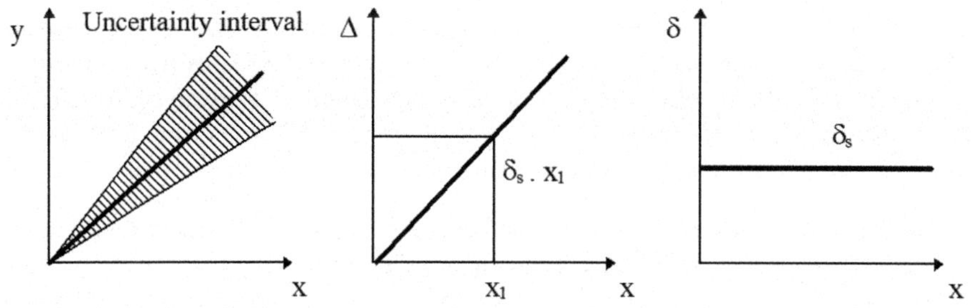

Fig. 2.11 Constant relative measurement error

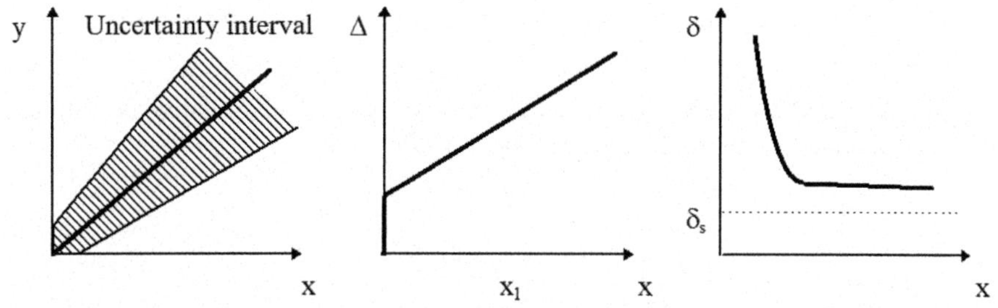

Fig. 2.12 Combined measurement errors

It is also possible to define the so-called limit accuracy Δ_m as the absolute error limit for the upper limit of the working interval

$$\Delta_m = \delta_S \, x_U \tag{2.18}$$

C. **Measuring instruments with combined error**

This is the case when the errors of the measuring device are the result of the so-called combined action of additive and multiplicative error (Zelený, 1976) (see Fig. 2.12). In this case, individual characteristics related to measurement errors are given by relations

$$\Delta = \Delta_0 + \delta_S x \tag{2.19}$$

$$\delta = \Delta_0 / x + \delta_S \tag{2.20}$$

For the reduced relative error, it is valid, that

$$\delta_R = \delta_0 + \delta_S x / R, \text{kde } \delta_0 = \Delta_0 / R \tag{2.21}$$

The classification criterion of the accuracy of the device in its entire measuring range is the *accuracy class*. The accuracy class is expressed by a positive dimensionless number from a specified number of series. This number is always greater than, or at most equal to, the largest absolute value of the reduced limit errors found under the given conditions in the entire measuring range of the device:

1. In the case of pure *multiplicative measurement errors*, there is constant relative sensitivity error (here directly the accuracy class of the device)

$$\delta_0 = 100\frac{\Delta_0}{x} \tag{2.22a}$$

This is usually valid only within a limited interval, indicated on the device by the manufacturer.

2. In the case of pure *additive measurement errors*, a reduced relative deviation is used (here equal to the accuracy class of the device)

$$\delta_0 = 100\frac{\Delta_0}{x_{\max} - x_{\min}} = 100\frac{\Delta_0}{R} \tag{2.22b}$$

where R is the measuring range. For instruments where measurement errors act additively, the relative deviation δ decreases hyperbolically as the quantity x increases.

3. For *combined measurement errors*, the uncertainty band is formed by the sum of the areas of the additive and multiplicative uncertainty bands. In contrast to cases of pure additive error, the growth of δ_R begins later, the larger is the ratio δ_S/δ_0. In these cases, two values are used to express the accuracy class δ_K: the reduced relative error δ_0 and the error at the upper limit of the measuring range δ_S.

The largest value of the reduced limit error detected in the entire measuring range is decisive for classifying the measuring device in one of the *accuracy classes*. The largest limit error must not exceed the $Rp/100$ value for the selected accuracy class p. The accuracy class is not indicated by a sign because the reduced limit error can be positive or negative with equal probability. The actual value of the measured quantity μ will then lie in the interval $\mu \pm pR/100$. According to the accuracy class, the devices are classified into the series $6, 4, 2.5, 1.5, 1.0, 0.5, 0.2, 0.1, 0.05, 0.02, 0.01$, 0.005, 0.002, and 0.001%, supplemented by the error type marker (multiplicative), (additive), or (combined) according to this scheme:

1. For pure *multiplicative errors*, the accuracy class is expressed by the sensitivity error δ_S and is presented by the number in circle.
2. For pure *additive errors*, the accuracy class is expressed by the reduced relative error δ_0, where R is the measuring range.
3. For instruments with a strongly uneven scale, the accuracy class is given as a ticked number and measuring range *R*.

4. In the case of simultaneous additive and multiplicative error, the accuracy class is given as a fraction δ_K/δ_0.

Based on the value of the accuracy class, the maximum deviation of the device and the device error can then be determined.

2.5 Measurement Errors

To express absolute measurement errors Δ, a probabilistic approach based on knowledge (or assumption of knowledge) of the probability law of their distribution expressed by the probability density function $f(\Delta)$ is most often used (Slaev et al. 2019; Thompson & Taylor, 2008; Wade, 2022). This allows for the inclusion of a systematic error component (as mean values) and a random error component (as the relative width of the distribution expressed by the variance or dispersion measures, respectively). A probabilistic approach based on interval analysis can also be used. For practical estimation of measurement errors Δ (including instrument errors), a whole range of characteristics calculated from sample error values $\Delta_i = x_i - \mu_s$ can be used, where μ_s is either the value of the standard or an estimate of the actual value of μ. The basic characteristics based on the knowledge of the error probability density $f(\Delta)$ are:

(a) *moments*
general $M_K(x) = \int x^K f(x)\mathrm{d}x$
central $c_K(x) = \int [x - M_1(x)]^K f(x)\mathrm{d}x$
(b) *special measures of dispersion* based on the choice of an appropriate approximating distribution. If only the error occurrence interval $-a \leq \Delta \leq a$ (or only the limit deviation a) is known, either a uniform or a triangular distribution is chosen (see Fig. 2.13).

In both cases, the mean value is $E(\Delta) = 0$. The standard deviation for a uniform distribution is equal to

$$\sigma_R = \frac{a}{\sqrt{3}} \approx 0,5774\,a \tag{2.23a}$$

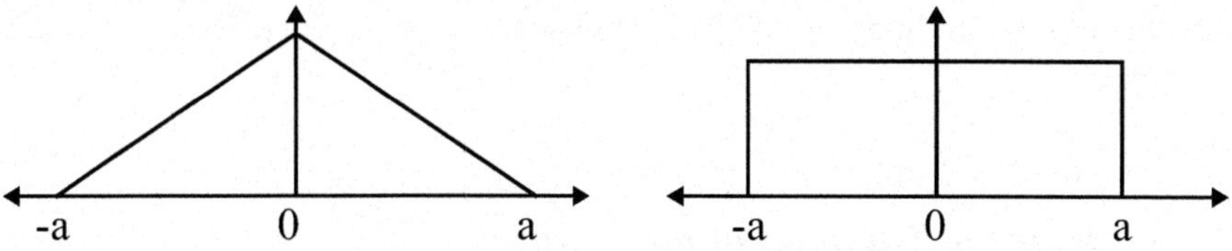

Fig. 2.13 Probability density functions for triangular and uniform distributions

and for the triangular distribution is valid

$$\sigma_T = \frac{a}{2\sqrt{6}} \approx 0.2041a \tag{2.23b}$$

Either σ_R or σ_T is then selected as a measure of dispersion, depending on which of these distributions better describes the given problem (see type B uncertainties).

(c) the *probability* $P(a \le \Delta \le b) = 1 - \alpha$ that the errors lie in the selected interval $[a, b]$. Let us say that the notation $P(s)$ denotes the probability with which the statement s is true.
(d) *quantiles*, i.e. error values $\tilde{\Delta}_\alpha$, for which it is true that $P(\Delta \le \tilde{\Delta}_\alpha) = \alpha$. This means that α% of all errors lie below the value of $\tilde{\Delta}_\alpha$.

Obviously, probabilities and quantiles are related to each other. For example, the assumption of a symmetric distribution allows the determination of limits $[a, b]$ as special quantiles for which $P(\Delta \le a) = \alpha/2$ and $P(\Delta \le b) = 1 - \alpha/2$.

In general, a certain interval $[a^-, b^+]$ of their possible values can be defined for these characteristics. If the probability density $f(\Delta)$ is known, this task can be solved using standard statistical methods (confidence intervals). Another option is to use the methodology for determining the uncertainty of measurement results. With a non-probabilistic approach, interval analysis can also be used.

In order to analyze the measurement results, it is generally necessary to know their probability distribution. However, this requires determination of a considerable number of measurements. Four basic types of error distributions are usually considered:

1. *Uniform distribution* refers to measurement results where errors arise from rounding or "trimming" (quantization).
2. The *normal distribution* refers to measurement where the resulting error is the sum of a number of partial errors and the measurements were made under conditions of constant variance.
3. The *Laplace* (two-sided exponential) *distribution* refers to measurement results where the variance of the measurement fluctuates around a certain mean value.
4. The *lognormal distribution* associated with the multiplicative measurement model refers to the situation where the resulting error is the product of partial errors. There are cases where the relative measurement error is constant.

If the error distribution is bounded by a finite interval (as in the case of a uniform distribution), the limiting error of measurement can be calculated as half of this interval.

2.5.1 Moment Estimates of the Errors

In general, knowing the errors Δ_i, $i = 1 \ldots N$, the variance σ_Δ^2 can be determined. Based on the idea that the measurement is expressed by a simple additive model

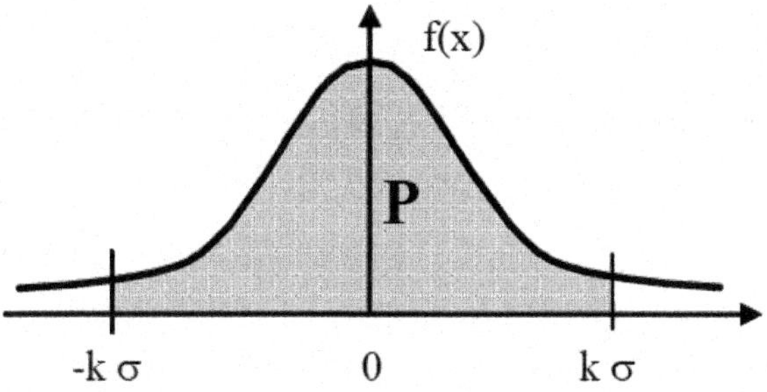

Fig. 2.14 Probability interval of error

$x_i = \mu + \Delta_i$, the variance of the measured quantity x can also be easily determined as σ^2. If it is true that $\overline{\Delta} = 0$ (the mean value of the errors $E(\Delta) = 0$, i.e. the systematic component of the error is zero), it is valid that $\sigma_\Delta = \hat{\sigma}$ where $\hat{\sigma} = \sqrt{\frac{1}{N-1}(x_i - \overline{x})^2}$.

Probability interval of error

Let the errors have a symmetric probability density function $f(\Delta)$ with mean $E(\Delta) = 0$. Let the corresponding distribution function $F(\Delta)$ be also known. The probability interval $[-k\sigma, k\sigma]$ where 100(1 − α) % of all errors lie (see Fig. 2.14) is then related to probability

$$\mathrm{P} = (-k\sigma \le \Delta \le k\sigma) = F(k\sigma) - F(-k\sigma) = 1 - 2F(-k\sigma) = 1 - \alpha \quad (2.24)$$

Here, − k is the α quantile and k is the 1 − α quantile of the error distribution. For the many of different distribution, abs(k) = 1.64 for $P = 0.9$. This choice of P thus eliminates the need to know the error distribution.

2.5.2 *Quantile Estimates of Errors*

In the case of the same situation as for moment estimates, i.e. $\sigma^2 = \sigma_\Delta^2$ is the simple characteristics of dispersion so-called quantile deviation

$$K_{1-q} = (\tilde{x}_{1-q/2} - \tilde{x}_{q/2})/2 \quad (2.25)$$

Here, $\tilde{x}_\alpha$ is the quantile of the error distribution (100 α % of all errors lie below this value). A value of $2K_{1-q}$ defines the width of the interval in which $P = 100$ (1 − q)% of the errors lie (Fig. 2.15).

For the chosen statistical certainty, $P = 1 - q$ is the *limit measurement error* equal to $\Delta_{\Delta P} = K_{1-q}$. This quantity corresponds to the boundary of the interval containing 100(1 − q) % of all errors Δ. Its size generally depends on the value of q and on the distribution of errors. It is assumed that the mean value of random measurement errors is zero, or the mean value of the measured quantity x is μ. For calculations, it is necessary to specify the statistical certainty P.

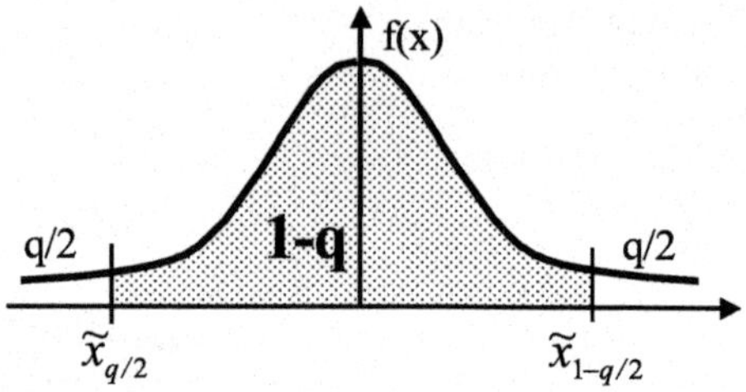

Fig. 2.15 Quantile interval of error

A special marking is used for selected P values:

The mean error ($P = 0.5$) is defined by the relation $\Delta_{\Delta 0.5} = (\tilde{x}_{0.75} - \tilde{x}_{0.25})/2$. For a normal distribution, it is true that $\Delta_{\Delta 0.5} = 0.68\,\sigma$. The mean error is used to express errors in very precise physical measurements.

The probable error ($P = 0.683$) is defined by the relation $\Delta_{\Delta 0.683} = (\tilde{x}_{0.8415} - \tilde{x}_{0.1585})/2$. For a normal distribution, it is true that $\Delta_{\Delta 0.683} = \sigma$.

The error for the unknown distribution ($P = 0.9$) is equal to $\Delta_{\Delta 0.9} = (\tilde{x}_{0.95} - \tilde{x}_{0.05})/2$. For a *number of distributions*, it is true that $\Delta_{\Delta 0.9} = 1.65\,\sigma$. In cases where the distribution of errors is not known, the choice of $P = 0.9$ is advantageous because it also allows adding partial quantile errors $\Delta_{\Delta 0.9i}$

$$\Delta_{\Delta 0.9} = \sqrt{\sum_{(i)} \Delta^2_{\Delta 0.9i}} \tag{2.26}$$

However, relation (2.26) is not correct for other values of P. In the case where the measurement errors have a normal distribution, it is possible to write

$$\Delta_{\Delta P} = u_{(1+P)/2}\,\sigma$$

where $u_{(1+P)/2}$ is the $100\,(1 + P)/2$-percent quantile of the standardized normal distribution. Also, for other distributions, the limit quantile error $\Delta_{\Delta P}$ can be expressed by the relation $\Delta_{\Delta P} = H\sigma$, where the value of H is a function of the statistical certainty P and the kurtosis g_2 of the error distribution, i.e. $H = \text{function}(P,\ g_2)$. For the generalized exponential and trapezoidal distributions, which include the normal ($g_2 = 3$), uniform ($g_2 = 1.8$), and Laplace ($g_2 = 6$) distributions, H can be in range $0,9 \leq P \leq 0,99$ estimated according to the approximate relation

$$H \approx 1.62\left[3.8(g_2 - 1.6)^{2/3}\right]^Z \text{ where } Z = \log\left[\log\left(\frac{1}{1-P}\right)\right] \tag{2.27}$$

In some cases, probabilistic information about the causes of variable uncertainty is missing. It is only possible to express a bound for each variable x_j in the form $[a_j^L \leq x_j \leq a_j^U]$. By using this interval, it is possible to define so-called interval variables. It is generally valid for interval variables

$$A = [a_1, a_2] \quad \text{for} \quad x \in \{a_1 \leq x \leq a_2\}$$

where $a_j^L a_j^L$ is the lower limit value and a_j^U is the upper limit value of the variable x_j. The use of interval variables is described, for example, in the work (Elishakoff, 1995).

2.6 Measurement of Random Variables

In a number of cases, the realization of a random variable characterized by the probability density $f(x)$ is measured on an instrument with an instrument error characterized by the probability density f (x, y), see Fig. 2.16.

The probability density function of the measurement result is then expressed as a convolution

$$f(y) = \int f^*(x, y) f(x)\,\mathrm{d}x \tag{2.28}$$

If it is chosen that $f(x) \approx N(\mu, \sigma^2)$ and $f^*(x, y) \approx N(x, \tau^2)$, it is after integration

$$f(y) \approx N(\mu, \sigma^2 + \tau^2) \tag{2.29}$$

There is, therefore, an inflation of the error variance by the variance of the measuring device. This means that only the sum of the variances $\sigma^2 + \tau^2$ is obtained from the data $(y_1, \dots y_N)$. In some cases, the resulting variance is a combination of variances from different sources, then it is generally true that

$$\sigma_V^2 = \sum_{(i)} \sigma_i^2 + \sum_{i=1}^{m} \sum_{j>i} \mathrm{cov}(i, j) \tag{2.30}$$

where σ_i^2 is the variance due to the i-th source and $\mathrm{cov}(i, j)$ is the covariance between the i-th and j-th source of errors. There are two limiting situations:

The *lower limit* corresponds to independent variances (geometric mean of absolute errors, i.e. standard deviations), when

$$\sigma_{VN} = \sqrt{\sum_{(i)} \sigma_i^2} \tag{2.31}$$

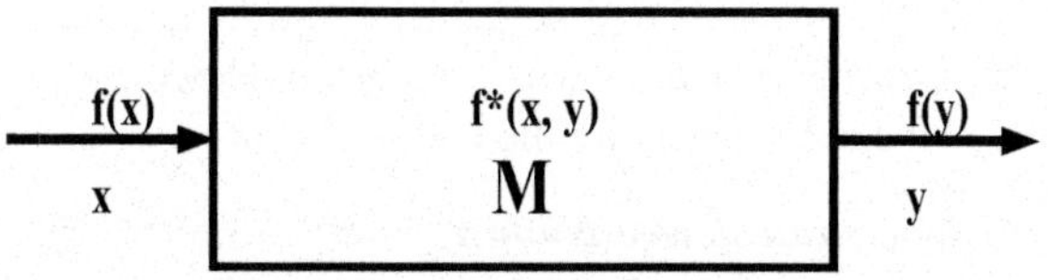

Fig. 2.16 Convolution of the error of the random variable and the instrument error

The upper limit corresponds to linearly dependent variances (arithmetic mean of absolute errors, i.e. standard deviations), when

$$\sigma_{VL} = \sum_{(i)} \sigma_i \tag{2.32}$$

For linearly dependent data, it is $\text{cov}(i, j) = \sigma_i \sigma_j$ and, it follows from the well-known triangle inequality that

$$\sqrt{\sum_{(i)} \sigma_i^2} < \sum_{(i)} \sigma_i \tag{2.33}$$

In accordance with the fact that a worse option is always accepted (see measurement uncertainties), it is appropriate to choose σ_{VL} as the overall standard deviation for cases where the error sources are correlated.

The result of the convolution for the measurement of random variables can be used for indicative selection of the measuring device. The total measurement error for the case that the measurement and measuring instrument variances come from independent sources is equal

$$\Delta_V = \sqrt{\sigma^2 + \tau^2} \tag{2.34}$$

where τ^2 is the variance of the measuring device and σ^2 is the variance expressing the variability of the measurements. The following device categories can be defined as:

1. *Precise device* $\Delta_V \approx \sigma$

Here, it is no longer possible to further increase the accuracy of the measurement by using a more accurate device. Accuracy can only be increased by an increasing number of repeated measurements.

2. *Optimal device* $\tau \approx \sigma/3 \quad \Delta_V = \sqrt{\sigma^2 + \sigma^2/3} = \sigma\sqrt{10/9} \approx \sigma$

Here, the total error Δ_V is only insignificantly higher compared to the case of an absolutely accurate device $\tau^2 = 0$.

3. *Device with comparable errors* $\boldsymbol{\tau} \approx \boldsymbol{\sigma} \quad \Delta_V = \sigma\sqrt{2} = 1.44\,\sigma$

When the measurement is here repeated multiple times and the arithmetic mean is used, the error Δ_V will be reduced N-fold, and thus also the random component of the instrument error τ^2 will be reduced. However, the systematic component cannot be reduced by repeating the measurement.

4. *Inaccurate device* $\boldsymbol{\Delta_V} \approx \boldsymbol{\tau}$

Repeated measurements will not bring any new information here, and significant measurement accuracy can only be achieved by using a more accurate device.

Considering the influence of instrument error, it is therefore advisable to choose instruments from category 2, which in the textile and similar industries will normally significantly reduce the price of the measuring instrument (you pay for the accuracy of the measuring instrument). In some cases, it is expedient to also define the *sensitivity of measuring devices*. Sensitivity is generally understood as the response of a change in the output y to a unit change in the input x (i.e. as the first derivative)

$$S = \lim_{\Delta x \to 0} \frac{\Delta y}{\Delta x} = \frac{\mathrm{dy}}{\mathrm{dx}} \approx \frac{m_y}{m_x} tg\alpha \tag{2.35}$$

This is shown schematically in Fig. 2.17.

Sensitivity can also be defined for the case of a combination of measuring devices, or partial elements (see Figs. 2.18, 2.19, and 2.20).

(a) Serial arrangement

$$y = f_n[f_{n-1}, \ldots f_2[f_1(x)]]$$
$$S = \frac{\mathrm{dy}}{\mathrm{dx}} = \frac{\mathrm{dy}_1}{\mathrm{dx}} \frac{\mathrm{dy}_2}{\mathrm{dy}_1}, \ldots \frac{\mathrm{dy}_n}{\mathrm{dy}_{n-1}} = \prod_{(i)} S_i \tag{2.36}$$

(b) Parallel arrangement

$$y = \sum_{i=1}^{n} y_i$$
$$S = \sum_{i=1}^{n} S_i \tag{2.37}$$

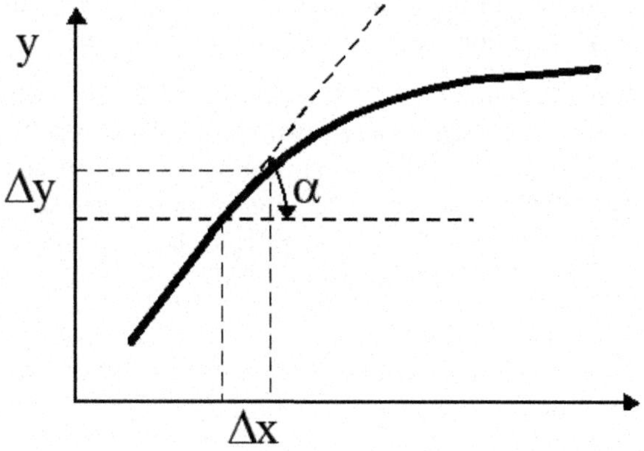

Fig. 2.17 Definition of sensitivity

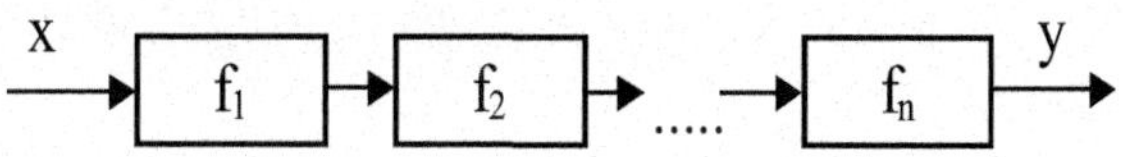

Fig. 2.18 Definition of instrument sensitivity

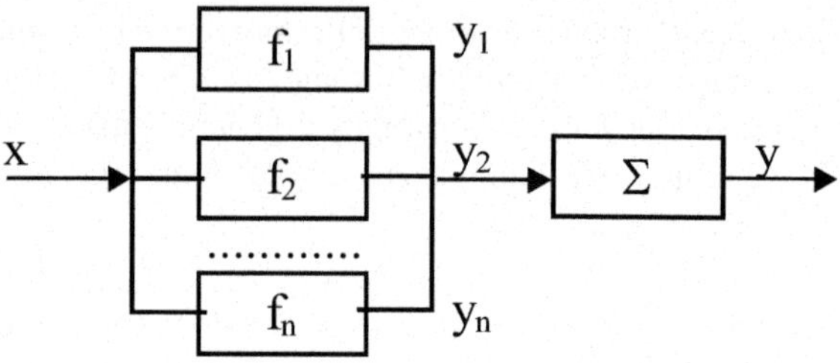

Fig. 2.19 Definition of instrument sensitivity

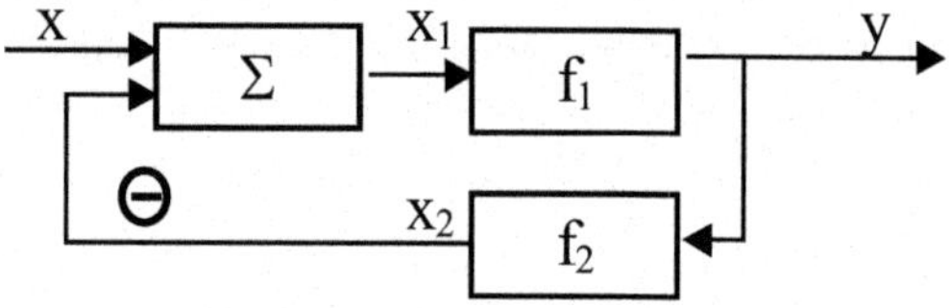

Fig. 2.20 Definition of instrument sensitivity

(c) Compensation loop

$$y = f_1[x - f_2(y)], x_1 = x - x_2$$
$$S = S_1\left[1 - \frac{dx_2}{dy}\frac{dy}{dx}\right] = S_1[1 - S/S_2] \quad (2.38)$$

An example of a compensation loop is weighting.

Information about errors or variability of measuring devices can also be used to evaluate differences between devices, or between laboratories.

Repeatability in the laboratory is defined as the value below which lies the absolute magnitude of the difference between two measurement results obtained under the same conditions with a specified probability (0.95).

Reproducibility in laboratories is the value below which lies the absolute magnitude of the difference between two measurement results performed on the same types of instruments by different operators in different laboratories obtained on the same sample under the same conditions with a specified probability (0.95).

Let the within-laboratory variance be s_w^2 and the between-laboratory variance be s_L^2, then the repeatability is equal to

$$R_P = \frac{2}{\sqrt{2}} s_W \quad (2.39)$$

and *reproducibility* is equal to

$$R_E = \frac{2}{\sqrt{2}}\sqrt{s_W^2 + s_L^2} \quad (2.40)$$

2.7 Comparison of Two Measuring Devices

Let there be two measuring instruments ***a*** and ***b*** for measuring the same quantity x with a normal distribution $N(\mu, \sigma^2)$. The conformity of two measuring instruments is often verified based on the sample $(z_1, \dots z_N)$ obtained by measuring instrument ***a*** and the sample $(y_1, \dots y_N)$ obtained by measuring instrument ***b***. Various modifications of the t-tests of the hypotheses Ho: $\mu_z = \mu_y$, or Ho: $\mu_z - \mu_y = 0$ depending on whether it is possible to assume that the measurement took place at the same place (by non-destructive tests) can be used. Before realization of t-tests, it is necessary to verify the equality of the variances σ_z^2 and σ_y^2 using the F-test and to verify the validity of the assumption that both samples come from a normal distribution. An overview of suitable methods is presented in the work (Bouleau, 2021; Meloun & Militký, 2012; Taylor, 1982; Zelený, 1976). In general, however, systematic instrument errors must also be considered (see Fig. 2.21). Let it hold that

The device **a** is measured with a systematic error (bias) B_a and the measurement errors ε_a have a normal distribution $N(\mu, \sigma_a^2)$.
The device ***b*** is measured with a systematic error (bias) B_b and the measurement errors ε_b have a normal distribution $N(\mu, \sigma_b^2)$.

Simple additive measurement models can usually be used:

Device a

$$y_i = B_a + x_i + \varepsilon_{ai} \tag{2.41}$$

Device b

$$z_i = B_b + x_i + \varepsilon_{bi} \tag{2.42}$$

Let measurements be uncorrelated, i.e.

$$\text{cov}(\varepsilon_{ai}, \varepsilon_{bi}) = \text{cov}(\varepsilon_{ai}, x_i) = \text{cov}(\varepsilon_{bi}, x_i) = 0 \tag{2.43}$$

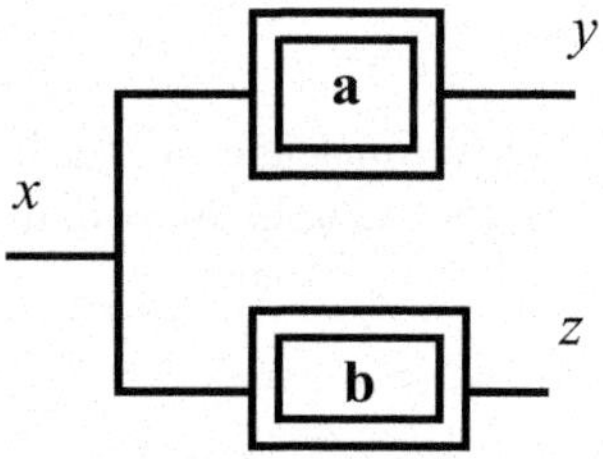

Fig. 2.21 Measurement on two devices

Under these assumptions, it is valid, that

$$E(y) = B_a + \mu \quad D(y) = \sigma^2 + \sigma_a^2 \tag{2.44}$$

$$E(z) = B_b + \mu \quad D(z) = \sigma^2 + \sigma_b^2 \tag{2.45}$$

$$\begin{aligned} \operatorname{cov}(z, y) &= E(z\,y) - E(z)\,E(y) \\ &= B_a\,B_b + \mu\,(B_a + B_b) + E(x^2) \\ &\quad - (B_a - \mu)\,(B_b - \mu) = E(x^2) - E^2(x) = \sigma^2 \end{aligned} \tag{2.46}$$

The covariance between the results of the devices is therefore equal to the variance of the measured quantity.

Consider measurements on the same samples (y_i, z_i) $i = 1 \ldots N$. The $\overline{y}, \overline{z}, s_z^2, s_y^2$ can be calculated by a standard way and covariance is calculated as

$$\hat{c}(z, y) = \frac{1}{N-1} \sum_{i=1}^{N} (y_i - \overline{y})\,(z_i - \overline{z}) \tag{2.47}$$

The estimate of the variability of the measured material σ^2 is

$$\hat{\sigma}^2 = \hat{c}(z, y) \tag{2.48}$$

Difference of systematic deviations (bias) is

$$\hat{B}_a - \hat{B}_b = \overline{y} - \overline{z} \tag{2.49}$$

Measurement error estimate for the instrument ***a*** is

$$\hat{\sigma}_a^2 = s_y^2 - \hat{c}(z, y) \tag{2.50}$$

Measurement error estimate for the instrument ***b*** is

$$\hat{\sigma}_b^2 = s_z^2 - \hat{c}(z, y) \tag{2.51}$$

To estimate the mean μ, it must be known at least one bias or one bias is negligible. Let, for example, $B_a = 0$, then

$$\hat{\mu} = \overline{y} \quad \widehat{B}_b = \overline{z} - \overline{y} \tag{2.52}$$

Various types of tests can be implemented using these estimates. To compare the accuracy of the devices, the hypothesis $H_0\colon \sigma_a^2 = \sigma_b^2$ is verified, i.e. both devices are equally accurate. It is advantageous to specify auxiliary quantities

$$u_i = y_i + z_i = B_a + B_b + 2\,x_i + \varepsilon_{ai} + \varepsilon_{bi} \tag{2.53}$$

a

$$v_i = y_i - z_i = B_a - B_b + \varepsilon_{ai} - \varepsilon_{bi} \tag{2.54}$$

It is easy to determine then that

$$D(u) = 4\,\sigma^2 + \sigma_a^2 + \sigma_b^2 \;\; \mathrm{D}(v) = \sigma_a^2 + \sigma_b^2 \mathrm{acov}(u, v) = \sigma_a^2 - \sigma_b^2 \tag{2.55}$$

For the correlation coefficient, it is true that

$$\rho(u, v) = \frac{\sigma_a^2 - \sigma_b^2}{\sqrt{4\,(\sigma^2 + \sigma_a^2 + \sigma_b^2)\,(\sigma_a^2 + \sigma_b^2)}} \tag{2.56}$$

A hypothesis test $H_0{:}\,\rho(u, v) = 0$ is identical to a hypothesis test $H_0{:}\,\sigma_a^2 = \sigma_b^2$. A test statistic can then be formulated under the assumption of normality

$$T = \frac{\hat{\rho}(u, v)\,\sqrt{N - 2}}{\sqrt{1 - \hat{\rho}^2(u, v)}} \tag{2.57}$$

The statistic T has a Student's distribution with $N - 2$ degrees of freedom, assuming the validity of the H_o hypothesis. Using the variable v, a hypothesis $H_0{:}\,v = 0$ can be tested, which corresponds to the hypothesis of equal instrument bias.

$$T_v = \frac{\hat{v}\,\sqrt{N}}{\hat{\sigma}_v} = \frac{(\hat{B}_a - \hat{B}_b)\,\sqrt{N}}{\sqrt{\hat{\sigma}_a^2 + \hat{\sigma}_b^2 - 2\,\hat{\sigma}^2}} \tag{2.58}$$

The statistic T_v has a Student's distribution with $N - 1$ degrees of freedom, assuming the validity of the H_0 hypothesis.

It is also possible to test the hypothesis whether one of the devices is accurate (σ_a^2 or $\sigma_b^2 = 0$). For the case of the hypothesis $H_0{:}\,\sigma_a^2 = 0$, the test statistic has the form

$$C_0 = -N\,\ln\left[\frac{\hat{\sigma}_a^2\hat{\sigma}_b^2 - (\hat{\sigma}^2)^2}{\hat{\sigma}_a^2\,(\hat{\sigma}_a^2 + \hat{\sigma}_b^2 - 2\,\hat{\sigma}^2}\right] \tag{2.59}$$

If the hypothesis H_0 is valid, the statistic C_0 has a χ^2 distribution with one degree of freedom (it is valid, that $\chi^2_{0.95} = 3.84$).

2.8 Generalized Mean

The power transformation is commonly used to improve the distribution of data and is described in detail in Chap. 6. Here, the task of estimating the mean value using the *generalized mean* is solved for situations where the measurement results y have non-constant variance and non-symmetric distribution. It is assumed that during data processing, the original measured quantity x was transformed into the result of measuring y according to the relationship

$$y = f(x) = x^P \tag{2.60}$$

If quantity x has a symmetric distribution with constant variance σ_x^2 (so it is approximately normal), the distribution of y is non-symmetric with non-constant variance equal to

$$\sigma_y^2 \approx \left(\frac{\partial x^P}{\partial x}\right)^2 = P^2\, x^{2\,(P-1)}\, \sigma_x^2 \tag{2.61}$$

For known P, an inverse transformation (symmetrizing transformation) $Z = f(x)^{1/P} = y^{1/P}$ can be used with advantage. Since the quantity Z has an approximately normal distribution, the arithmetic mean can be used as an estimate of the mean value

$$\overline{Z} = \frac{1}{N}\sum Z_i \tag{2.62}$$

For measurement results (y_i) $i = 1, N$, the arithmetic mean corresponds to general mean (power average) [5]

$$\overline{y}_P = \left[\frac{1}{N}\sum y_i^{1/P}\right]^P \tag{2.63}$$

For different P, different power averages are obtained, see Table 2.4.

Table 2.4 Power averages

$P \to -\infty$	$\overline{y}_{-\infty}$	Lowest sample element $y_{(1)}$
$P = -1$	$\overline{y}_{-1}$	Harmonic mean
$P = 0^*$	$\overline{y}_0$	Geometric mean
$P = 1$	$\overline{y}_1$	Arithmetic mean
$P = 2$	$\overline{y}_2$	Quadratic mean
$P \to \infty$	$\overline{y}_\infty$	Highest sample element $y_{(N)}$

*The case when $P = 0$ is interesting. Here, it can be shown that the corresponding transformation $x^P \approx^\circ \ln(x)$, i.e. a geometric mean result

The geometric mean is defined by the relation

$$\overline{y}_0 = \exp\left[\frac{1}{N}\sum \ln(y_i)\right] \tag{2.64}$$

These power averages are generally applicable to positive data and the inequality $\overline{y}_P \geq \overline{y}_{P-1}$ also holds for any P. Thus, power averages are closely related to the inverse transform, which can make the decision about using them.

2.9 Conclusion

It is clear that errors caused by measurement and measuring instruments can be expressed in a number of different characteristics. For their calculation from the measurement results, it is advantageous to use the proper standard software system (e.g. MATLAB), which enables the verification of assumptions and the use of alternative techniques. For routine calculations, it is sufficient to use a suitable spreadsheet editor of the EXCEL type.

References

Bouleau, N. (2021). *The mathematics of errors*. Springer International Publishing. https://doi.org/10.1007/978-3-030-88575-5

Crowder, S., Delker, C., Forrest, E., & Martin, N. (2020). *Introduction to statistics in metrology*. Springer International Publishing. https://doi.org/10.1007/978-3-030-53329-8

Elishakoff, I. (1995). Convex modeling-a generalization of interval analysis for nonprobabilistic treatment of uncertainty. In *Proceedings of interenational conference on APIC 95*.

Foegelle, M. D. (2002, March 26). *Introduction to measurement uncertainty*. IEEE 802.11-06/0333r0. https://mentor.ieee.org/802.11/dcn/06/11-06-0333-00-000t-introduction-to-measurement-uncertainty.ppt

Meloun, M., & Militký, J. (2012). *Interactive statistical data analysis*. Carolinum. (in Czech)

Slaev, V. A., Chunovkina, A. G., & Mironovsky, L. A. (2019). *Metrology and theory of measurement*. De Gruyter. https://doi.org/10.1515/9783110652505

Taylor, J. R. (1982). *An introduction to error analysis* (3rd ed.). University Science Books.

Thompson, A., & Taylor, B. N. (2008). *Guide to use the international system of units*. NIST Special Publication 811.

Wade, H. A. (2022). *The ASQ metrology handbook*. ASQ Excellence.

Zelený, F. (1976). *Basic properties of measurement devices*. SNTL. (in Czech)

Chapter 3
Measurement Scales

One good measurement is worth a thousand expert judgments.

Abstract Data treatment is usually realized for results of measurement in cardinal scale. In textile metrology, some results are still in nominal or ordinal scales due to the use of subjective evaluation of complex characteristics, such as the appearance, hand, or comfort of textile products or intermediates. The basic properties of random variables in basic measurement scales are presented. The overview of the corresponding statistical characteristics used in data treatment and the correct statistical processing of the results are discussed.

3.1 Introduction

Data processing is generally dependent on the type of measurement scale. Despite the strong trend to use only the cardinal scale containing the maximum amount of information, data in other scales is still found in practice. The basic scales are:

A. **Nominal**. The numbers represent the names of the categories (classifications). *Examples*: Gender, nationality, technology, and device type. *Properties*: element/class identification only. *Possible mathematical operations*: None (only element counts within classes).
B. **Ordinal**. Numbers indicate the order of observations sorted by significance. *Examples*: Hardness scale, the intensity of pain, feel, fastness to light, the appearance of fabric, feeling when wearing. *Features*: order of importance. *Possible mathematical operations*: Algebra of inequalities. *Transformation*: $Z = g(X)$ (where $g(X)$ is a monotonically increasing function). Invariant under monotonic transformation.
C. **Interval cardinal (numeric)**. Equivalent intervals between values with no relation to a natural zero—do not represent an absolute quantity. *Examples*: Calendar years, IQ, temperature in °C. Properties: order and equidistance. *Possible mathematical operations:* Addition, subtraction, multiplication, and division of intervals only.

J. Militký et al., *Textile Material Research—Statistical Tools of Experimental Data Analysis*, https://doi.org/10.1007/978-981-97-6607-9_3

Transformation: $Z = a + bX$, where $(b > 0)$, a is the origin of the scale, and b is the coefficient related to the units of measurement. Invariant under linear transformation.

D. **Proportional cardinal (numeric)**. Numbers are units with equivalent intervals relative to the natural zero. *Examples.* Geometric and mechanical quantities, age, time, mass, and concentration. *Properties*: order, equidistance, natural zero. *Possible mathematical operations:* all. Transformation: $Z = aX$, where $(a > 0)$. Invariant under multiplicative transformation.

It is valid that a higher type scale includes the preceding scales. In this chapter, only an overview of the properties of the measurement scales and the possibilities of corresponding statistical processing of the results are discussed.

3.2 Nominal Scale

In this scale, there are only two operations, i.e. determination of equality or non-equality. It is actually a splitting of a set of objects (characters) into disjoint classes (categories) between which there are no relations. Classes can be named arbitrarily (numbers are understood as names in this scale). The nominal scale is mainly used for sorting and classifying objects. The following requirements must be met when choosing classes:

1. unambiguity of classification—this means that each object belongs to only one class;
2. existence—this means that at least one object exists in each class.

Examples of nominal scales are type of technology, manufacturing defects, subjective color of the product, type of machine, type of fiber, garment components, etc.

A special case is a dichotomous nominal item, where there is only a yes/no indication $\prec 0.1 \succ$ (e.g. male/female).

3.2.1 Descriptive Information

The same procedures as for other categorized variables are used. A categorized variable is generally a set of items for which:

1. no two items can occur simultaneously
2. the set of items is complete (occurs at least once)
3. each of the items has a meaning (can occur)
4. each item is identifiable
5. during identification, it is clearly determined which item occurred.

The items $\{a_1, \ldots a_k\}$ are divided into K classes (categories) ($K \ll k$). The result of the investigation (experiment) is then K absolute frequencies (number of items in the class) $\{n_1, n_2, \ldots .n_K\}$. The basic characteristics are: Sample size $N = \sum n_i$ and relative frequencies $f_i = n_i/N$; it is true that $\sum_{i=1}^{K} f_i = 1$. These frequencies are actually estimates of the probability of occurrence of the nominal variable in the given category. For a graphical representation of the frequency distribution, it is possible to use, for example, a bar or segment graph (see Fig. 3.1).

Attention: The categories are not ordered in any way (for nominal variables), and therefore, the creation of cumulative frequencies is meaningless. Also, graphs are generally only suitable for illustration, as individual classes (categories) can be grouped arbitrarily. For a clear graphic representation, it is generally necessary to introduce some criterion for sorting categories (i.e. ordinalization). It is advantageous to use so-called Pareto charts, where the individual categories are sorted in descending order according to their frequencies n_i. It is assumed that a category with a higher frequency is "more significant". An example of a Pareto chart to show the frequencies of different types of machine failures is in Fig. 3.2. Since the arrangement is carried out according to an external criterion (frequency of failures), it is also possible to show the cumulative frequencies (see the piecewise linear line in Fig. 3.2).

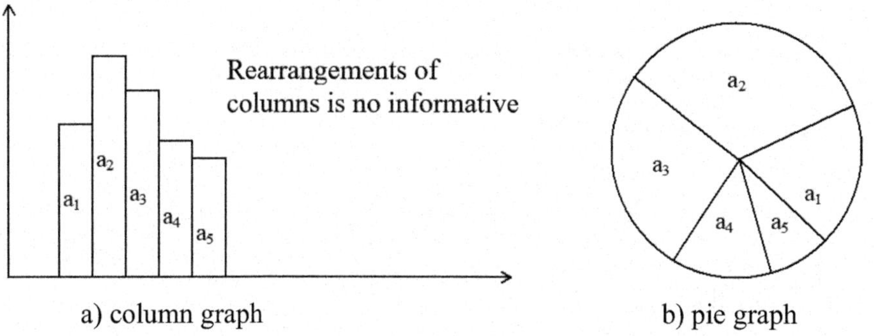

Fig. 3.1 Graphic representation of the frequency distribution of the nominal variable

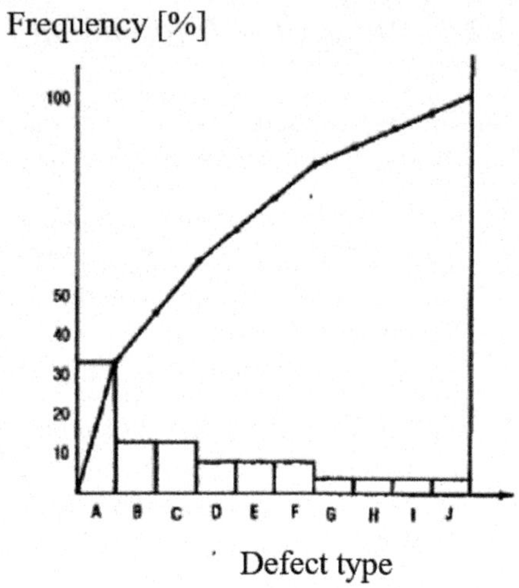

Fig. 3.2 Pareto graph for different defect

3.2.2 Characteristics of Distribution

Individual categories can only be treated separately.

Location: Modal category M_o and respective frequencies. For the modal category is $n_{M_o} = \max_{1 \le j \le K} n_j$, and the corresponding relative frequency is

$$f_{M_o} = \max_{1 \le j \le K} f_j \tag{3.1}$$

Variability:

Variance ratio

$$v = 1 - \frac{n_{M_o}}{N} = 1 - f_{M_o} \quad \left\langle 0, \frac{K-1}{K} \right\rangle \tag{3.2}$$

Nominal variation

$$\text{nomvar} = 1 - \sum_{j=1}^{K} f_j^2 = \frac{N^2 - \sum n_j^2}{N^2} = \sum_{j=1}^{K} f_j \left(1 - f_j\right) \quad \left\langle 0, \frac{K-1}{K} \right\rangle \tag{3.3}$$

Normalized nomvar

$$n\,\text{nomvar} = \frac{K\,\text{nomvar}}{K-1} \quad \langle 0, 1 \rangle \tag{3.4}$$

Properties of variability characteristics:

1. They are null if all items are in the same category;
2. They are maximal for a uniform distribution $f_j = 1/K, j = 1, K$ (maximal dispersion);
3. The higher the heterogeneity, the higher the value of the variability characteristic.

3.2.3 *Comparison of Frequencies and Confidence Intervals*

The basic assumption is the validity of a binomial distribution $Bi(N, p_i)$ for each category with absolute frequency (number of items) n_i. The maximum likelihood estimate of the probability of p_i is the relative frequency $f_i = n_i/N$. The corresponding variance is

$$D(f_i) = \frac{f_i\,(1 - f_i)}{N} \tag{3.5}$$

Confidence interval for the probability (theoretical relative frequency) p_i is defined as

$$P(p_D \le p_i \le p_H) = 1 - \alpha \tag{3.6}$$

1. **Small samples with** $N \le 30$, and $n_i < 5$. The level of significance α and statistical certainty $\omega = (1 - \alpha/2)$ are chosen, usually as $\alpha = 0.05$ and $\omega = 0.975$. For the lower bound of the confidence interval, it is true that

$$p_D = \frac{n_i}{(N - n_i + 1)\,F_\omega(q_1, q_2) + n_i}, \text{ kde } q_1 = 2(N - n_i + 1) a\, q_2 = 2n_j \tag{3.7}$$

Here, $F\omega(q_1, q_2)$ is the quantile of the F-distribution with q_1 and q_2 degrees of freedom. For the upper bound of the confidence interval, it is true that

$$p_H = \frac{(n_i + 1)\,F_\omega(r_1, r_2)}{N - n_i + (n_i + 1)\,F_\omega(r_1, r_2)}, \text{ kde } r_1 = 2(n_i + 1) a\, r_2 = 2(N - n_i) \tag{3.8}$$

2. **Samples with** $30 \le N \le 50$ and $n_i \ge 5$. It follows directly from the asymptotic approximation of the binomial distribution by the normal distribution that

$$[p_D, p_H] = \left(f_i - z_{1-\alpha/2}\, s_i, f_i + z_{1-\alpha/2}\, s_i\right) \tag{3.9}$$

Here $z_{1-\alpha/2}$ is the quantile of the standardized normal distribution, for $\alpha = 0.05$ is $z_{1-\alpha/2} = 1.96 \approx 2$, and the standard deviation is

$$s_i = \sqrt{\frac{f_i\,(1 - f_i)}{N}} \tag{3.10}$$

This confidence interval is used in the vast majority of works, although it is not always suitable, and its quality depends on the specific values of p and N. It is curious that the real statistical certainty for the confidence interval defined by Eq. (3.9) oscillates in the range of approx. 0.80–0.96, depending on the combination of N and p (Brown et al., 2001). A number of alternative procedures for calculating confidence intervals are presented in the work (Brown et al., 2001). Agresti and Coull (Agresti & Coull, 1998) recommend a modified confidence interval

$$[p_D, p_H] = \left[\tilde{f}_i - z_{1-\alpha/2}\sqrt{\frac{\tilde{f}_i(1-\tilde{f}_i)}{N+4}}, \tilde{f}_i + z_{1-\alpha/2}\sqrt{\frac{\tilde{f}_i(1-\tilde{f}_i)}{N+4}}\right]$$

$$\text{where } \tilde{f}_i = \frac{n_i+2}{N+4} \tag{3.11}$$

For smaller samples, this interval is significantly more efficient than the interval based on asymptotic normality (Agresti & Coull, 1998).

Among the best is the confidence interval proposed by Wilson (1927), for which it is true that

$$f_i\left(\frac{N+z}{N+z^2}\right) + 0.5\left(\frac{z^2}{N+z^2}\right) \pm \sqrt{\frac{f_i(1-f_i)}{N}\frac{N^2 z^2}{\left(N+z^2\right)^2} + 0,25\frac{z^4}{\left(N+z^2\right)^2}} \tag{3.12}$$

where $z = z_{1-\alpha/2}$ is the quantile of the standard normal distribution. Alternatively, the confidence interval limits p_D, p_H can be determined by solving a quadratic equation $(n_i - Np)^2 = z_{1-\alpha/2} Np(1-p)$ with respect to p.

3. **Samples with** $N \geq 50$ and $n_i \leq 5$

$$[p_D, p_H] = \left[\frac{1}{2N}\chi^2_{1-\alpha/2}(2n_i), \frac{1}{2N}\cdot\chi^2_{1-\alpha/2}(2n_i+2)\right] \tag{3.13}$$

where $\chi^2_\alpha(k)$ is the α-quantile of the chi-square distribution with k degrees of freedom.

4. **Samples with** $N \geq 50$ and $N - n_j < 5$

$$[p_D, p_H] = \left[1 - \frac{1}{2N}\chi^2_{1-\alpha/2}(2N - 2n_i + 2), 1 - \frac{1}{2N}\chi^2_{1-\alpha/2}(2N - 2n_i)\right] \tag{3.14}$$

5. **Samples with** $N \geq 50$ and $n_i = 0$, resp. $n_i = N$

For $n_i = 0$ is

$$[p_D, p_H] = \left[0, -\frac{\ln\alpha}{N}\right] \tag{3.15}$$

For $n_i = N$ is

$$[p_D, p_H] = \left[1 + \frac{\ln \alpha}{N}, 1\right] \tag{3.16}$$

Example 3.1 *Car color preference*

Car color preference was assessed based on a questionnaire completed by $N = 100$ young men. The results of the questionnaire are: Color white $n_1 = 20 f_1 = 0.2$, Color red $n_2 = 35 f_2 = 0.35$, Color black $n_3 = 45 f_3 = 0.45$.

Question: Is there a significant difference between the red and black color of the car?

Determination of 95% confidence intervals for p_2 red and p_3 black: The relation (3.9) is used when choosing $\left(z_{1-\alpha/2} \approx 2\right)$. The standard deviations are

$$s_2 = \sqrt{\frac{0.35(1-0.35)}{100}} = 0.047\, s_3 = \sqrt{\frac{0.45(1-0.45)}{100}} = 0.0497$$

For confidence intervals, then, it is true that

$$0.254 \le p_2 \le 0.445 \quad 0.351 \le p_3 \le 0.549$$

Because the confidence intervals overlap, there is no significant difference between preference red or black car colors.

3.3 Ordinal Scale

In this scale, in addition to determining equality or non-equality ($\neq, =$) there are also finding of the relations larger/smaller ($>, <$). It is again a matter of dividing a set of objects (items) into disjoint categories (classes), but between which there are already relations of the type smaller (less significant) and larger (more significant). By default, the convention is that classes are ordered from least important (worst) to most important (best). The classes can either be appropriately named (worst, below average, average, best) or numbered (1, 2, 3, 4, 5). The numbers here only determine the significance of the class and are only special symbols (there are no indicate mathematical operations or distance expressions). So, it cannot be said that the difference between class 3 and class 1 is two. In textile metrology, ordinal scales are mainly used for subjective evaluation (hand, appearance, feelings when worn, color fastness, light fastness, etc.). In general, base information is obtained by scoring or grading.

3.3.1 Descriptive Information

The same requirements as for the nominal scale must be met when choosing classes. The N items are again divided into K classes (categories) $\{a_1, ...a_K\}$ $(K \ll N)$. In addition, a clear arrangement of classes according to an external rule (e.g. the degree of conformity) from the worst (non-conforming) to the best class (fully conforming) must apply in order to make sense of the summation of relative frequencies and the calculation of the location characteristic. The result of the experiment (scoring or grading) is then K absolute frequencies (number of items in the class) $\{n_1, n_2, ...n_K\}$.

The basic characteristics are:

Sample size $N = \sum n_i$ and relative frequencies $f_i = n_i/N$. It must be valid that $\sum_{i=1}^{K} f_i = 1$.

These frequencies are again estimates of the probability of occurrence of an ordinal variable in a given category. Given the arrangement of the classes from worst to best, it makes sense to also define the cumulative frequencies:

$$F_1 = f_1, F_2 = f_1 + f_2, F_j = \sum_{i=1}^{j} f_j, F_k = 1\cdot$$

In Table 3.1 examples of different labeling of categories used in quality assessment are shown.

Commonly used are 5-point scales (e.g. hand, appearance, color fastness), 8-point scales (e.g. light fastness), or 11-point scales (hand, appearance). In general, a scale with more categories requires more data evaluators (raters) and good pre-training of them. A bar graph of relative frequencies, which can be interpreted as a histogram can be used to graphically represent frequency distributions.

Table 3.1 Typical categories of quality

Not acceptable	1	A
Poor	2	B
Average	3	C
Above average	4	D
excellent	5	E

3.3.2 Characteristics of Distribution

Individual categories can be evaluated separately or in relation to others.

Location:

1. The median Me category (category containing 50% of the data cumulated from all previous categories) is defined by the inequalities.

$$F_{\mathrm{Me}-1} < 0.5 \text{ and } F_{\mathrm{Me}} \geq 0.5$$

2. Ordinal median

$$\tilde{x}_{0.5} = \mathrm{Me} + 0.5 - \underbrace{\frac{F_{\mathrm{Me}} - 0.5}{f_{\mathrm{Me}}}}_{c} \tag{3.17}$$

Here, c is the fraction of median category data assigned to the upper half. Properties of the ordinal median:

1. $1 \leq \tilde{x}_{0.5} \leq K$
2. For the case $f_1 = 1$ is $\tilde{x}_{0.5} = 1$
3. For the case $f_k = 1$ is $\tilde{x}_{0.5} = K$
4. For the case $\sum_{i=1}^{\mathrm{Me}-1} f_i = \sum_{i=\mathrm{Me}+1}^{K} f_i$ is $\tilde{x}_{0.5} = Me$
5. The value $\tilde{x}_{0.5}$ shows the shift of the 50% dividing point; the higher it is, the more the data is concentrated in higher categories.

Variability:

Discrete ordinal variance, dorvar

$$\mathrm{dorvar} = 2\sum_{i=1}^{K} F_i\,(1 - F_i) = 2\left[\sum_{i=1}^{K} F_i - \sum_{i=1}^{K} F_i^2\right] \tag{3.18}$$

Properties:

1. $0 \leq \mathrm{dorvar} \leq (K-1)/2$
2. $0 \leq \mathrm{dorvar} \leq (K-1)/2$ for case that $f_i = 1$
3. $\mathrm{dorvar} = (K-1)/2$ for case that $f_1 = f_K = 0.5$
4. The more scattered the data, the bigger its dorvar.

3.3.3 Construction of a Confidence Interval for the Median Med

The procedure is suitable for $N \geq 30$. The construction is carried out in these steps:

(a) Cumulative frequencies $\left(F_D^*, F_H^*\right)$ are calculated from eqn.

$$\left[F_D^*, F_H^*\right] = 0.5 \pm \frac{0.5\, z_{1-\alpha/2}}{\sqrt{N}} \tag{3.19}$$

where $z_{1-\alpha/2} \approx 2$ is the quantile of the standard normal distribution for $\alpha = 0.05$

(b) Category D where is F_D^* and category H where is F_H^* are determined
(c) Corrections d and h are calculated from eqns.

$$d = \frac{F_D^* - F_{D-1}}{f_D} \text{ and } h = \frac{F_H^* - F_{H-1}}{f_{DH}}$$

(d) The limits S_M and H_M of the 95% confidence interval are calculated according to relations $S_M = D - 0.5 + d$ and $H_M = H - 0.5 + h$

Example 3.2 *Subjective evaluation of hand*

100 girls were asked to make a subjective assessment of the hand of the clothing fabric. Aim of data treatment was creation of 95% confidence interval for ordinal median. The results of hand evaluation and the basic characteristics of the subjective hand are shown in Table. 3.2.

Based on above relations it was calculated

$$\text{Median category: Me} = 4 \quad \text{Ordinal median } \tilde{x}_{0.5} = 4 - 0.5 + \frac{0.5 - 0.32}{0.48} = 3.8$$

For determination of the lower bound S_M of the confidence interval, the following variables were calculated

$$F_D^* = 0.5 - \frac{0.52}{\sqrt{100}} = 0.4 \quad D = H = 4 \quad d = \frac{0.4 - 0.32}{0.48} = 0.17$$

and finally, $S_M = 4 - 0.5 + 0.17 = 3.67$.

Table 3.2 Basic characteristics of subjective hand

Label	Class	n_i	f_i	F_i
Not acceptable	1	2	0.02	0.02
Poor	2	5	0.05	0.07
Average	3	25	0.25	0.32
Above average	4	48	0.48	0.80
Excellent	5	20	0.2	1

For determination of the upper bound H_M of the confidence interval, the following variables were calculated

$$F_M^* = 0.5 + \frac{0.52}{\sqrt{100}} = 0.6 \quad D = H = 4 \quad h = \frac{0.6 - 0.32}{0.48} = 0.583$$

and finally, $H_M = 4 - 0.5 + 0.583 = 4.083$.

Thus, the confidence interval of the ordinal median Med has the form

$$3.67 \leq \text{Med} \leq 4.083\cdot$$

Properties measured in ordinal scales are gradually converted to a higher type, i.e. cardinal scales.

3.3.4 *Preference Data*

A special type of ordinal scale are preference data, which are divided into only three categories (see Table 3.3).

The basic characteristic of this type of data is asymmetry.

Natural asymmetry $A* = f_+ - f_-$ has these properties.

$A* = 1$ it is case of $f_+ = 1$.
$A* = -1$ it is case of $f_- = 1$.
$A* = 0$ it is case of $f_+ = f_-$
$A* > 0$ it is case of dominance $+$
$A* < 0$ it is case of dominance $-$.

Asymmetry with respect to tails $A = \frac{f_+ - f_-}{f_+ + f_-}$ has these properties.

$A = 1$ it is case of $f_- = 0$.
$A = -1$ it is case of $f_+ = 0$.
$A = 0$ it is case of $f_- = f_+$
$A > 0$ it is case of dominance $+$
$A < 0$ it is case of dominance $-$.

The confidence interval for the asymmetry with respect to the tails $A(N > 30; n-, n+ > 5)$ has the form

$$P(A_D < A < A_H) = 1 - \alpha.$$

Table 3.3 Categories of items and preference data

Deterioration	Neutral	Improvement
$-$	0	$+$
$n-, f-$	n_0, f_0	$n+, f+$

Table 3.4 Responses of respondents

	+	0	–
n	17	21	10
f	0.354	0.438	0.208

For the lower bound of A_D, it is valid, that

$$A_D = \text{tgh}(a_D) \quad a_D = a - z_{1-\alpha/2}s_a \quad a = \frac{1}{2}\ln\left(\frac{f_+}{f_-}\right).$$

For the upper bound of A_H, it is valid, that

$$A_H = \text{tgh}(a_H) \quad a_H = a + z_{1-\alpha/2} \quad s_a = s_a = \frac{1}{2}\sqrt{\frac{1}{n_+} + \frac{1}{n_-}}$$

Example 3.3 *The effect of changing the pattern of the jacket on its wearing sensation*

A total of 48 respondents commented on whether the change in the pattern of the jacket improved the wearing sensation (+), the wearing sensation remained the same (0), or there was a deterioration of the wearing sensation (−). The results are shown in Table 3.4.

The asymmetry with respect to the tails is equal to $A = \frac{0.354-0.208}{0.354+0.208} = 0.2598$ and for the parameter a it is valid, that $a = \frac{1}{2}\ \ln\left(\frac{0.354}{0.208}\right) = 0.265$

Standard deviation is equal to $s_a = \frac{1}{2}\cdot\sqrt{\frac{1}{17} + \frac{1}{10}} = 0.199$ and limits of 95% confidence interval are

$$\begin{aligned} a_D &= 0.265 - 20.199 = -0.133 \quad A_D = \text{tgh}(-0.133) = -0.132 \\ a_H &= 0.265 + 20.199 = 0.663 \quad A_H = \text{tgh}(0.663) = 0.58\cdot \end{aligned}$$

The confidence interval of asymmetry with respect to tails is $-0.132 < A < 0.58$. It is therefore necessary to state that there was no significant improvement of jacket wearing sensation.

3.3.5 Transformation of Ordinal Scale

Using a suitable transformation, ordinal classes can be converted to a cardinal scale and then processed as for categorized data in a cardinal scale. A simple option is to use monotone transformations when expressing classes as integers 1, 2, … K. Yates (1948) recommended the transformation of the i-th class according to the relation

$$S_i = (2i - K - 1)/2 \quad \text{for}\, i = 1, 2, \ldots K \tag{3.20}$$

The variable S_i is considered as continuous in the interval $[-(K-1)/2, (K-1)/2]$. The S_i values are then the values representing the categories and the corresponding relative frequencies are f_i. Standard procedures for processing categorized cardinal data can then be used for statistical analysis (Meloun & Militký, 2012). So, the average is calculated from the relationship

$$\overline{x} = \sum_{i=1}^{K} S_i f_i. \tag{3.21}$$

The standard deviation is equal to

$$s_S = \sqrt{\sum_{i=1}^{K} f_i\, S_i^2 - \overline{x}^2}. \tag{3.22}$$

For $100(1-\alpha)\%$ interval of the mean value is approximately valid

$$\overline{x} \pm z_{1-\alpha-2} \frac{s_S}{\sqrt{N}} \tag{3.23}$$

Here again, the quantile of the standard normal distribution for $\alpha = 0.05$ equals approximately $z_{1-\alpha/2} \approx 2$. Also popular is the Ridit transformation defined by the relation (Donaldson, 1998)

$$R_i = F_{i-1} + 0,5 f_i \tag{3.24}$$

($F0 = 0$ is selected). Ridits are continuous random variables on the interval [0, 1]. Ridits are suitable for situations where a control group is compared with a modified group (e.g. the hand of raw and finished textiles) (Donaldson, 1998). For larger N, ridits are approximately normally distributed.

3.4 Cardinal Scale

Cardinal scale is the strongest type of scale, where values are numbers allowing mathematical operations and distance definition. In this scale there are results obtained by *measurements* on measuring devices. Also, most properties of textiles and characteristics of materials quality are expressed in cardinal scales. The data processing procedure is related to the sample size (number of measurements). For small samples, all elements of the sample are used (uncategorized data), and for large samples, already in the measurement phase, the results are sorted into classes (categorized data), which are represented by the so-called class symbol (usually the middle value of the class). In textile metrology, automatic categorization is common when measuring the properties of fibers and yarns using the so-called HVI (high volume instrument). In these cases, thousands of primary data are available and it is inconvenient to handle the entire sample, so the measuring device already uses categorization. Categorization is also automatic, e.g. when measuring fiber lengths.

3.4.1 *Treatment of Uncategorized Data*

When processing the univariate data $\{x_i\}$ $i = 1, \ldots N$ is commonly assumed to be a random sample composed of independent elements, homogeneous and normally distributed. To verify these assumptions, it is possible to use a whole range of exploratory data analysis methods (see Meloun & Militký, 2012 and Sect. 6.5).

The simplest *quantile graph* has quantile estimates, i.e. order statistics, created by sorting the elements of the sample by size in ascending order $x_{(1)} \leq x_{(2)} \leq \cdots \leq x_{(2)} \leq \cdots x_{(N)}$, plotted on the *y*-axis. It is clear that $x_{(i)}$ is the *i*th smallest element of the sample. The corresponding rank probabilities P_i are plotted on the *x*-axis, which are approximately equal to $P_i = i/(N+1)$. From the quantile graph, one can easily identify peculiarities of the data such as symmetry, skewness, and the presence of outliers (Meloun & Militký, 2012). To assess the agreement with the normal distribution, the quantiles of the normal distribution are also plotted in this graph as well. It is advantageous to use two theoretical normal quantile functions. One is based on estimates of the sample mean $E(x) \approx \overline{x}$ and sample standard deviation $\sqrt{D(x)} \approx s$, and the other uses robust estimates of location, i.e. the median $E(x) \approx \tilde{x}_{0.5}$, and the interquartile range $\sqrt{D(x)} \approx (\tilde{x}_{0.75} - \tilde{x}_{0.25})/1.349$. From the proximity of the quantile function of normal distributions to the sample quantiles, one can then judge how suitable the standard parameter estimates are and whether it is not better to use robust estimates. For illustration, Fig. 3.4 shows a quantile graph for a sample containing the amount of rainfall during one month (see Meloun & Militký, 2012).

It is clearly visible, that the distribution of the data is not symmetrical mainly due to one strongly skewed element. Classical estimates of location and dispersion (purple line) lead to a function that systematically falls outside most of the sample quantiles. A robust quantile function (red line) that describes the central part of the data well is much more advantageous. In this case, computer-intensive methods (e.g. Bootstrap—see Sect. 6.8) are suitable for determining confidence limits.

To verify normality, it is also advantageous to use the *Q-Q graph* for the normal distribution (*rankit graph*), where the order statistics $x_{(i)}$ are plotted on the *y*-axis and the quantiles of the standard normal distribution z_{Pi} are plotted on the *x*-axis, for rank probabilities $P_i = i/(N + 1)$. A linear trend here indicates the approximate normality of the sample. For the sample used in Fig. 3.3, the *Q-Q* plot is shown in Fig. 3.3.

A convex increasing dependence typical of a distribution skewed toward higher values and the presence of an outlier is evident. Computer-intensive methods (e.g.

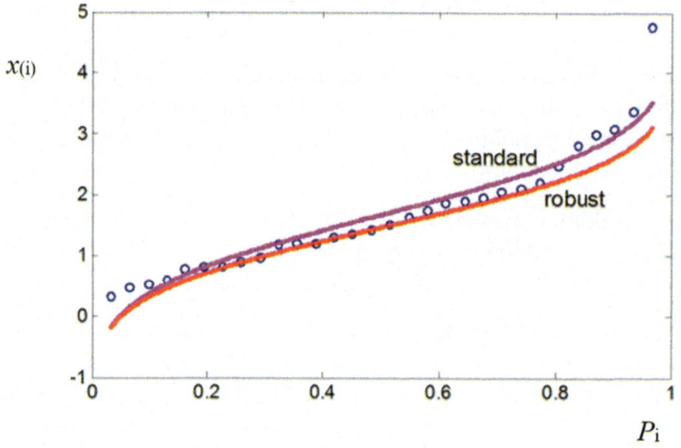

Fig. 3.3 Quantile plot for rainfall during one month

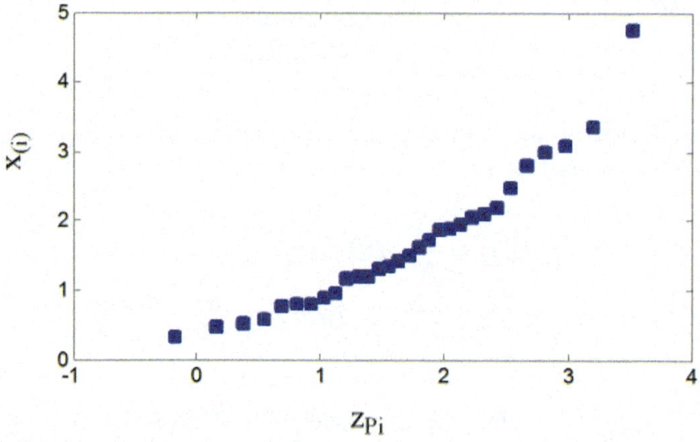

Fig. 3.4 Rankit graph for rainfall during one month

Bootstrap—see Chap. 6, Sect. 6.7) are also suitable for determining confidence limits in this case. Details on the construction and interpretation of *Q-Q* graphs are presented in Sect. 6.4.

To identify outliers, it is generally possible to use exploratory data analysis tools such as the *barrier method* (Meloun & Militký, 2012). The *moment method* is very simple, where it is valid that the outliers are those x_j for which is

$$\left|\overline{x}^{+} - x_j\right| \geq K_c\, s^{+} \tag{3.25}$$

where

$$K_c = 1.55 + \left[0.8\sqrt{g_2^{+} - 1}\,\log(N/10)\right] \tag{3.26}$$

In these relationships, $\overline{x}^{+}$ and s^{+} are the estimated mean and the standard deviation omitting the suspicious points, and g_2^{+} is the moment estimate of kurtosis computed without the suspicious points. For $N = 20$, and different distributions, the constant K_c is dependent on the kurtosis g_2.

- Normal distribution $g_2 = 3 \quad K_c = 1.89$
- Rectangular distribution $g_2 = 1.8 \quad K_c = 1.77$
- Laplace distribution $g_2 = 6 \quad K_c = 2.09$.

This approach is useful for sample sizes $N \geq 20$.

For estimates of the basic location and dispersion parameters and data in an uncategorized cardinal scale, the well-known relations apply (Meloun & Militký, 2012):

Location	Dispersion
Mean (arithmetic)$\overline{x}$	Variance s^2
Median $\tilde{x}_{0.5}$	Standard deviation s
	Variation coefficient $v = 100\, s/\overline{x}$

The accuracies of the estimates are characterized by using variances $D(.)$ (Meloun & Militký, 2012):

$$D(\overline{x}) = \frac{\sigma^2}{N} \quad D(\tilde{x}_{0.5}) = \frac{\pi\sigma^2}{2N}$$

$$D\left(s^2\right) = \frac{2\sigma^4}{N-1} \quad D(s) = \frac{\sigma^2}{2(N-1)}$$

$D(v) = \delta^2\left[\frac{N+\delta^2(2N+1)}{2N(N+1)}\right]$, where $\delta = \frac{\sigma}{\mu}$ is population median.

Classical data analysis assumes the normal distribution $x_i \sim N\left(\mu, \sigma^2\right)$ validity. Standard procedure is realized in two steps:

(a) determination of: $\overline{x}$, s^2 and 100 $(1-\alpha)$% confidence interval of the mean value (Meloun & Militký, 2012).

$$\overline{x} - t_{1-\alpha/2}(N-1)\frac{s}{\sqrt{N}} \leq \mu \leq \overline{x} + t_{1-\alpha/2}(N-1)\frac{s}{\sqrt{N}} \tag{3.27}$$

(b) determination of the coefficient of variation $v[\%]$ and the confidence interval for the population coefficient of variation δ, asymptotically it is (Meloun & Militký, 2012).

$$\frac{v}{100} - z_{1-\alpha/2}\sqrt{D(v)} \leq \delta \leq \frac{v}{100} + z_{1-\alpha/2}\sqrt{D(v)} \tag{3.28}$$

Robust data analysis is easily implemented using characteristics based on median and quantiles (Meloun & Militký, 2012). The estimate of the median $\tilde{x}_{0.5}$, the robust estimate of the standard deviation s_R, and the confidence interval for the population median Med are calculated.

$$s_R = \frac{x_{(N-k+1)} - x_{(k)}}{2\,z_{1-\alpha/2}} \quad \text{kde} \quad k = \frac{N+1}{2} - z_{1-\alpha/2}\sqrt{N/4}$$

Usually $\alpha = 0.05$ $(z_{1-\alpha/2} = 1.96)$ is chosen. The confidence interval of the population median is then asymptotically (Meloun & Militký, 2012)

$$\tilde{x}_{0.5} - t_{1-\alpha/2}(N-1)\,s_R \leq \text{Med} \leq \tilde{x}_{0.5} + t_{1-\alpha/2}(N-1)\,s_R \tag{3.29}$$

Also simple is the Hodges–Lehmann (HL) estimator of location, which is defined by the relation

$$\text{HL} = \text{med}\left\{\frac{x_i + x_j}{2},\ 1 \leq i < j \leq N\right\} \tag{3.30}$$

where med(.) denotes the median of all the elements in the parentheses. Interestingly, for $N = 4$, the HL estimate is identical to the arithmetic mean and is therefore not robust. As a robust estimator of variance, the median distance MAD from the median is often used, for which it is valid

$$\text{MAD} = 1.4826\,\text{med}\{|x_i - \tilde{x}_{0.5}|,\ i = 1 \ldots N\} \tag{3.31}$$

It should be considered that the robust behavior valid for medium and large samples may not be appeared for extremely small samples (typically for N smaller than 8) (Rousseeuw & Verboven, 2002).

Extremely small samples: ($N \le 20$)

Warning, with very small samples there is always a high uncertainty of the results and a large influence of outliers. Several types of estimators are available for small samples, which are generally less sensitive to the type of data distribution and robust to outliers. It is important that the corresponding mean square error of the estimate is sufficiently small even for sampling distributions different from normal (when the arithmetic mean has the smallest mean square error). As relatively effective for the wide class of distributions including the normal distribution, the Laplace distribution, the contaminated normal distribution, and selected non-symmetric distributions is the total median (TM), for which it is true that

$$\mathrm{TM} = \sum_{i=1}^{N}\sum_{j=1}^{N} \alpha_{ij}\left(\frac{x_{(i)} + x_{(j)}}{2}\right) \tag{3.32}$$

where the coefficients α_{ij} are a function only of the sample size N and do not depend on the sample distribution. It is valid that for $i = 1$ and N (Rousseeuw & Verboven, 2002).

$$\alpha_{11} = \alpha_{NN} = \frac{1}{N^N} \sum_{k=0}^{[(N-1)/2]} \binom{N}{k} (N-1)^k \tag{3.33}$$

For $i \neq j,\ 1 \le i < j \le N$ and N odd it is $\alpha_{ij} = 0$
For $i \neq j,\ 1 \le i < j \le N$ and N even it is

$$\alpha_{ij} = \frac{N!}{N^N((N/2)!)^2}\left(i^{N/2} - (i-1)^{N/2}\right)\left((N-j+1)^{N/2} - (N-j)^{N/2}\right) \tag{3.34}$$

and it is valid that $\alpha_{ij} = \alpha_{N-j+1,N-i+1}$.

For the case when $i = 1, j = N > 1$ is

$$\alpha_{1N} = \frac{N!}{N^N((N/2)!)^2} \tag{3.35}$$

The TM estimate can be written as a weighted arithmetic mean of order statistics with special weights a_i (Carolino & Barao, 2013; Dowd et al., 2015; Figueiredo & Gomes, 2004).

$$\mathrm{TM} = \sum_{i=1}^{N} a_i\, x_{(i)} \tag{3.36}$$

It can be seen that TM belongs to the class of estimates formed as a weighted average of order statistics (see Chap. 6). For a_i it is true that it does not depend on the type of data distribution and further that $a_i = a_{N-i+1}$, for $i = 1, \ldots, N$, and also $0 < a_1 < a_2 < \cdots < a_{[N/2]}$. Generally, it is valid.

$$a_i = \frac{1}{2}\left(\sum_{j=i}^{N} \alpha_{ij} + \sum_{j=1}^{i} \alpha_{ji}\right) \tag{3.37}$$

For larger N (up to 20), corresponding coefficients can be found in (Figueiredo & Gomes, 2004) For any N, it is also possible to use the general relations mentioned above to calculate these coefficients.

To calculate the variance of the TM estimate and possibly the limits of the confidence interval, it is possible to use the Bootstrap procedure, which is implemented in the following steps:

1. $B = 500 - 100$ Bootstrap samples $V_j = (x_{j1}, \ldots x_{jN})$ are generated from the original sample $Z = (x_1, \ldots x_N)$. The j-th sample is composed from the elements of the original sample randomly selected with return (i.e. some element may occur repeatedly and some may not occur at all).
2. For each sample V_j, the total median TM_j is determined.
3. The 5 and 95% quantiles and the arithmetic mean are determined from the empirical TM distribution (values TM_j).

The third central normalized moment g_1 can be used to characterize the skewness of the data distribution, and the expression (Figueiredo & Gomes, 2004) can be used to characterize the tails of the distribution

$$\tau = 0.5\left(\frac{F^{-1}(0.99) - F^{-1}(0.5)}{F^{-1}(0.75) - F^{-1}(0.5)} + \frac{F^{-1}(0.5) - F^{-1}(0.01)}{F^{-1}(0.5) - F^{-1}(0.25)}\right) / \left(\frac{u_{0.99} - u_{0.5}}{u_{0.75} - u_{0.5}}\right)$$

where $F^{-1}(p)$ is the $100p\%$ quantile of the distribution having the distribution function $F(x)$ and u_p is the quantile of the standard normal distribution.

With the help of skewness g_1, the most efficient estimate of the population mean value can then be found by selecting from the estimates: mean $\overline{x}$, median $x_{(0.5)}$, and total median TM. For skewed distributions it is true that (Carolino & Barao, 2013):

- $\overline{x}$ is best if $g_1 < 0.9$ regardless of sample size N or for $N > 16$
- TM is best if it is $0.9 \leq g_1 \leq 3.69$ and $N \leq 16$
- $x_{(0.5)}$ is best if $g_1 \geq 3.69$ and $N = 3$ or 4
- TM is best if $g_1 \geq 3.69$ and $N = 5$ to 16.

For symmetric distributions with different length of ends characterized by τ, it is true that (Carolino & Barao, 2013):

- $\overline{x}$ is best if $\tau < 1.01$ regardless of sample size N
- TM is best if it is $\tau \geq 1.8$ and $N \geq 3$
- $x_{(0.5)}$ is best if $1.01 \leq \tau \leq 1.8$ and $N \geq 5$.

It is therefore evident that knowing the type of distribution or at least its shape, it is possible to choose which of these estimates is the best. If the data distribution is completely specified, the maximum likelihood method can be used to estimate its parameters. Interestingly, globally the total median TM emerges as the best compromise estimator for arbitrary skewed and symmetric data distributions and very small samples, since the corresponding root mean square error is always at least close to the value for the efficient estimator (Dowd et al., 2015). In addition, TM is robust enough. In (Rousseeuw & Verboven, 2002), a number of non-parametric and robust methods for estimating location and dispersion parameters for extremely small samples are compared. From the results of the simulation studies, the authors formulated the following conclusions for the case where the location and dispersion parameter need to be estimated:

1. No reliable estimates can be found for $N = 2$.
2. For $N = 3$, it is recommended to use the median, but no robust estimator is available for a dispersion. It is recommended to use the average distance from median ADM, in the form

$$\mathrm{ADM} = \frac{1}{N}\sum_{i=1}^{N}|x_i - \tilde{x}_{0.5}|$$

or the median distance from the median MAD.

3. For $N > 3$, it is recommended to choose M—estimates of location T_r and dispersion S_M. Estimate T_r is a solution to the system of equations

$$\sum_{i=1}^{N}\Psi\left(\frac{x_i - T_r}{S_r}\right) = 0 \tag{3.38}$$

where is $S_r = \mathrm{MAD}$ and function $\Psi(.)$ has form $\Psi = \frac{e^x - 1}{e^x + 1}$.

For the iterative solution of the system of Eq. (3.38) by the Newton–Raphson method the zero iteration $T_r^{(0)} = \tilde{x}_{0.5}$ is selected and the next iterations have the form

$$T_r^{(K+1)} = T_r^{(K)} + S_r\frac{\sum_{i=1}^{N}\Psi\left(x_i - T_r^{(K)}\right)/S_r}{0.4132}.$$

In many cases, it is sufficient to determine only $T_r^{(1)}$, i.e. one-step approximation. An analogous procedure can be used for a robust dispersion estimate. A robust estimator of dispersion S_M is a solution to a system of equations

$$\sum_{i=1}^{N}\rho\left(\frac{x_i - T_r}{S_M}\right) = 0$$

where it is recommended to select $\rho(x) = \Psi^2(x/0.3739)$. In an iterative solution, it starts from $\left(S_M^{(0)} = \text{MAD}\right)$ and for the next iterations it is valid that

$$S_M^{(k+1)} = S_M^{(k)} \sqrt{2 \sum_{i=1}^{N} \rho\left(\frac{x_i - T_r}{S_M^{(k)}}\right)}.$$

Statistical analysis of extremely small samples is always performed with respect to the number of data N. From the various options described in the professional literature, a simple procedure verified by simulation experiments are here selected. The new approach to small samples inference based on so-called M statistics is described in the book (Demidenko, 2023).

Case $N = 2$

1. The arithmetic mean $\overline{x} = 0.5(x_1 + x_2)$ is determined. This corresponds to the total median TM and also the median $\tilde{x}_{0,5}$.
2. The 95% confidence interval of the mean is approximately equal.

$$\frac{x_1 + x_2}{2} - 12.7\frac{|x_1 - x_2|}{2} \leq \mu \leq \frac{x_1 + x_2}{2} + 12.7\frac{|x_1 - x_2|}{2} \tag{3.39}$$

In general, instead of coefficient 12.7, a coefficient depending on the type of distribution can be used (Meloun & Militký, 2012). If external information about the expected measurement result is known, it can be used as another value and calculated, for example, by analogy with the standard deviation using ADM.

Case $N = 3$

1. An estimate of the arithmetic mean $\overline{x}*$ is determined as the arithmetic mean of the two nearest values.
2. The 95% confidence interval of the mean is approximately equal

$$\overline{x} * -4.3 \frac{s}{\sqrt{3}} \leq \mu \leq \overline{x} * +4.3 \frac{s}{\sqrt{3}} \tag{3.40}$$

Instead of the standard deviation s, the average median distance ADM can be used with advantage.

Case $4 < N \leq 20$

Horn's procedure described in detail in the book (Meloun & Militký, 2012) is suitable for these extremely small samples. It is based on the order statistics $x_{(i)}$, which are elements of the sample in the ascending order, so that $x_{(i)} \leq x_{(i+1)}$.

E.g. for the sample (1, 3, 2, 6, 5} the corresponding order statistics are (1, 2, 3, 5, 6). Using Horn's procedure, the pivot depths are calculated first and then the pivots are evaluated. Depth of pivots is:

$$H = \frac{\text{int}[(N+1)/2]}{2} \quad H = \frac{\text{int}[((N+1)/2)+1]}{2} \tag{3.41}$$

An integer is selected from these two values. The lower pivot is equal to $x_D = x_{(H)}$ and the upper pivot is equal to $x_U = x_{(N+1-H)}$. Using the pivots, it is possible to easily determine the location and dispersion parameter estimates and from them the confidence interval of the mean:

1. Location

$$P_L = \frac{x_D + x_U}{2} \tag{3.42}$$

2. Dispersion

$$R_L = x_U - x_D \tag{3.43}$$

3. The 95% confidence interval of the mean (its estimate is P_L) is approximately equal to

$$P_L - R_L K_{0.975}(N) \le \mu \le P_L + R_L K_{0.975}(N) \tag{3.44}$$

The quantiles $K_{0.975}(N)$ for different N are shown in Table 3.5 (Meloun & Militký, 2012).

Example 3.4 *Strength of cotton fiber*

The five cotton fiber strength values were measured: (0.531; 0.677; 0.171; 0.065; and 0.848). The corresponding rank statistics are (0.065; 0.171; 0.531; 0.677; and 0.848). The purpose is to estimate the 95% confidence interval of the mean.

It is valid, that $\frac{N+1}{2} = 3$ and $\frac{N+1}{2} + 1 = 4$ and pivot depth is $H = \frac{4}{2} = 2$.Pivots then are $x_D = x_{(2)} = 0.171$ and $x_U = x_{(6-2)} = 0.667$.

Table 3.5 Quantiles $K_{0.975}^{(N)}$

N	$K_{0.975}^{(N)}$
4	0.738
5	2.094
6	1.035
7	0.72
8	0.564
9	0.915
10	0.668
15	0.466
20	0.397

Location parameter is then $P_L = \frac{0.171+0.667}{2} = 0.424$ and dispersion parameter is $R_L = 0.667 - 0.171 = 0.506$. The constant for the 95% confidence interval is equal to $K_{0.975}(5) = 2.094$. Thus, the lower bound of the 95% confidence interval of the mean is equal to $L = 0.424 - 2.094 \times 0.506 = -1.332$ and the upper bound of the 95% confidence interval of the mean is $U = 0.424 + 2.094 \times 0.506 = 2.18$. It is visible that this confidence interval includes zero and negative values, which is not physically possible but is a consequence of the small number of data.

3.4.2 *Treatment of Categorized Data*

Categorized data is formed by sorting numeric data into intervals that are the classes of the new items. We assign numerical values x_j (interval center) to individual classes. It is therefore a cardinal frequency categorization (pseudo-categorization). An example of categorization of lengths (selection size $N = 250$ fibers) is shown in Table 3.6.

During the construction of categorized data, the initial data are sorted and merged into K classes characterized by the class center x_i^* defined as

$$x_i^* = D_i + (H_i - D_i)/2 = (D_i + H_i)/2$$

and by the number of original sample elements n_i falling into this class bounded by the values of the lower limit D_i and the upper limit H_i, i.e. $\langle D_i, H_i \rangle$ (see Fig. 3.5).

The length of the class is equal $\Delta x_i = H_i - D_i$ and the total number of original sample elements is equal $N = \sum_{i=1}^{K} n_i$.

Table 3.6 Categorization of fibers lengths

Class	x_j	n_j	f_j	F_j
21–23	22	50	0.2	0.2
23–25	24	80	0.32	0.52
25–27	26	72	0.288	0.808
27–29	28	48	0.192	1

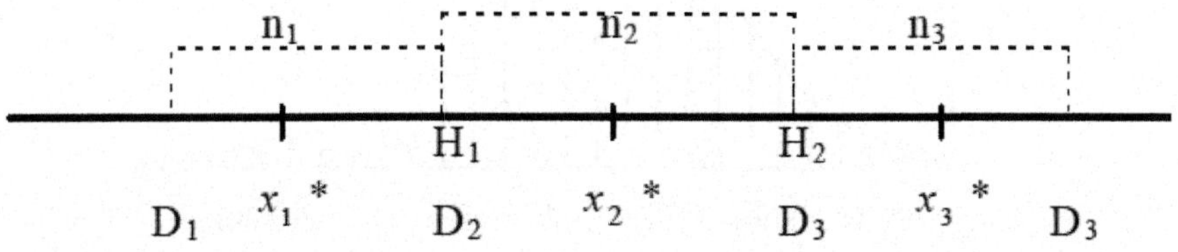

Fig. 3.5 Categorization to classes

When categorizing into equidistant (equally sized) classes, it is sufficient to choose the lower bound of the first class D_1, the number of classes K and the length of the classes Δx. Simple rules ensuring that all sample elements are included in the classes and limiting the difference between the bounds of the first and Kth class have form:

$$0 \leq K\Delta x - \left[x_N - x_{(1)}\right] \leq \Delta x \quad D_1 + K\Delta x > x_{(N)} \quad 0 < x_{(1)} - D_1 \leq \Delta x$$

There are several relations for choosing the appropriate number of classes K (Meloun & Militký, 2012):

$$\begin{array}{ll} N > 100 & K = \mathrm{int}\left[10\log(N)\right] \\ 40 \leq N \leq 100 & K = \mathrm{int}\left[2\sqrt{N}\right] \\ N < 100 & K = \mathrm{int}[1 + 1.44\ln(N)] \\ \text{universal} & K = \mathrm{int}\left[2.64(N-1)^{0.4}\right] \end{array}$$

where int[x] indicates the integer part of the number x. The disadvantage of dividing into classes with a constant length is that for most distributions, where the mode is significantly high, the probability of occurrence of data in the tail zone classes is significantly lower than the probability of occurrence of data in classes near the center (see Fig. 3.6).

It is therefore more advantageous to use unequal length classes so that the probability of data occurrence is approximately the same in each class (equiprobability principle). This requirement can be met quite simply by dividing the x-axis in the quantile graph into K equally long sections. The corresponding values on the y-axis are then directly the values of the class boundaries that meet the condition of approximately equal probability (Meloun & Militký, 2012).

After categorization, a new sample containing pairs $\left(x_i*, n_j\right)$ for $j = 1 \ldots K$ is available, from which the basic statistical characteristics are calculated.

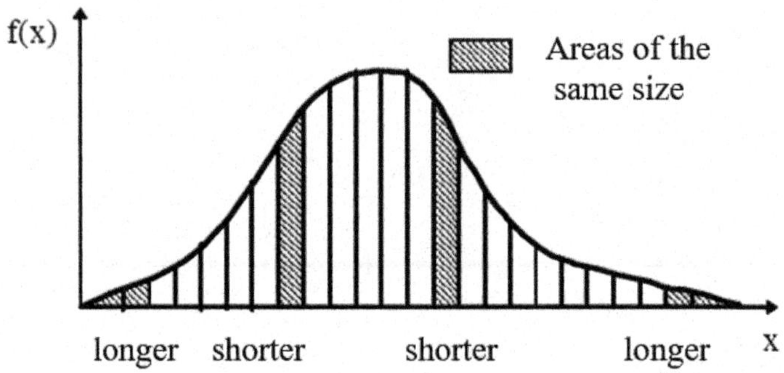

Fig. 3.6 Categorization ensuring approximately the same frequency in each class

Location characteristics:

Medián: $\tilde{x}_{0.5} = x_{\mathrm{Me}} - \Delta x_{\mathrm{Me}} \frac{F_{\mathrm{Me}} - 0.5}{f_{0.5}}$, kde Me je mediánová kategorie
Aritmetický průměr: $\bar{x} = \sum_{i=1}^{K} f_i x_i^* = \frac{1}{N} \sum_{i=1}^{K} n_i x_i^*$

Properties of the arithmetic mean:

1. If $\bar{x} = x_1^*$ it is $f_1 = 1$ and all the data are concentrated in the first class
2. If $\bar{x} = x_j^*$ it is $f_j = 1$ and all the data are concentrated in the j-th class
3. It is valid, that $\sum \left(x_i^* - A\right)^2$ is minimal, if $A = \bar{x}$.

Geometric mean: It is used only for positive data and can be calculated from the relation.

$$G = \prod_{j=1}^{K} \left(x_j^*\right)^{f_j} = \exp\left[\sum_{i=1}^{C} f_i \ln\left(x_i^*\right)\right] \tag{3.45}$$

Dispersion characteristics:

$$\text{Dorvar: Dorvar} = 2\,\Delta x \sum_{i=1}^{K} F_i(1 - F_i)$$

$$\text{Variance: } s^2 = \sum_{i=1}^{K} f_i \left(x_i^* - \bar{x}\right)^2 = \sum_{i=1}^{K} \left(f_i\, x_i^{*2} - \bar{x}^2\right).$$

Properties of the variance:

1. If $s^2 = 0$, it means that all data lie in one class, i.e. $f_i = 1$
2. The maximum value $s_{\max}^2$ is equal to $s_{\max}^2 = \left(x_K^* - x_1^*\right)/4$. This case occurs if it is $f_1 = f_K = 0.5$, i.e. half of the data lies in the first class and half of the data lies in the last class.
3. The larger s^2, the more the data move away from $\bar{x}$.

$$\text{Standard deviation: } s = \sqrt{s^2} = \sqrt{\sum_{i=1}^{K} \left(f_i\, x_i^{*2} - \bar{x}^2\right)}.$$

Characteristic of asymmetry: It is defined as the third normalized central moment (i.e. skewness). It is valid, that

$$A_s = \frac{\sum f_i \left(x_i^* - \bar{x}\right)^3}{s s^2} \tag{3.46}$$

$A_s = 0$ symmetry.
$A_s > 0$ skewed to the right (i.e. to higher values), see Fig. 3.7a.

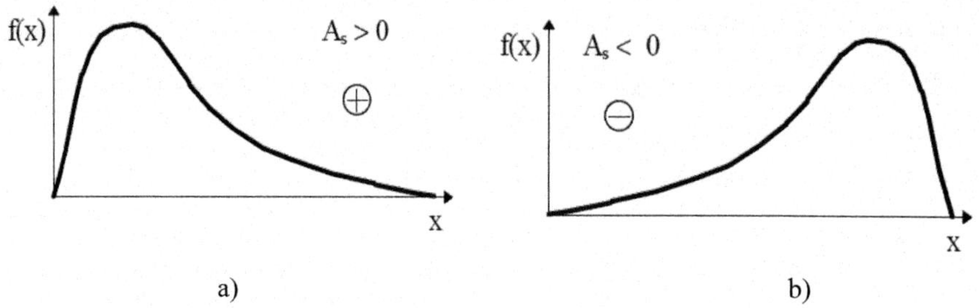

Fig. 3.7 Distributions with different asymmetry **a** skewed to the right, **b** skewed to the left

$A_s < 0$ skewed to the left (i.e. to lower values), see Fig. 3.7b.

Confidence interval constructions and hypothesis testing are performed in the same way as for uncategorized data.

3.5 Analysis of the Results of Indirect Measurements

Indirect measurements are actually calculated relationships that determine the resulting (required) quantity y as a non-linear function $y = g(\boldsymbol{x})$ of the results of direct measurements $\boldsymbol{x}$. From the vector of direct measurements $\boldsymbol{x} = (x_1, \ldots, x_m)$, the corresponding location and dispersion characteristics are usually determined, which are further used to calculate the location and dispersion characteristics of the required resulting quantity y. In Chap. 6 it is shown that each non-linear transformation changes the distribution of the data and causes non-constant variance. Data transformation methods that improve their distribution are discussed there. An inverse phenomenon occurs when analyzing indirect measurements. When the original data has a symmetric distribution, its functional transformation (based on physical models) has an asymmetric distribution shape, and instead of a constant variance, a random variable with a non-constant variance is created. It is therefore often interesting to also estimate the distribution of the measurement result.

3.5.1 *Monte Carlo Simulation*

A simple procedure is the Monte Carlo simulation method, which, for cases when y is a function of one directly measured quantity x, consists of the following steps:

1. Generation of random numbers u_i for $i = 1 \ldots m$ ($m = 500$–$10{,}000$) from the normalized theoretical distribution of direct measurements. A normal distribution is usually assumed, so the u_i are random numbers with an $N(0, 1)$ distribution.

2. Converting normalized generated random numbers to random numbers with a mean equal to the mean $\overline{x}$ and a variance equal to the sample variance s^2 using the relation $z_i = \overline{x} + u_i s$.
3. Calculation of the resulting simulated values y_i obtained by direct substitution, i.e. $y_i = g(z_i)$.
4. Construction of a suitable estimate of the probability density function $f(y)$, and possibly an estimate of the location and dispersion parameters.

This procedure, which is described in more detail in Sect. 3.5.1, depends critically on the choice of theoretical distribution and assumptions about the degree of dependence between direct measurements (expressed by the covariance). The generalization to cases where it is a vector of input quantities $\boldsymbol{x}$ is easy (Schwartz, 1975) where the problem is solved when both input parameters $\boldsymbol{x}$ and output parameters $\boldsymbol{y}$ are vectors and there are several partial transformations $y_j = g_j(\boldsymbol{x})$. Some other possible procedures for solving this task are described, for example, in work (Meloun & Militký, 2012).

This procedure is sufficiently universal and is also suitable for simulating the statistical behavior of complex systems. The general scheme for describing the behavior of complex systems breaks down into five phases.

A. <u>System model</u>

An appropriate model of the observed system requires knowledge of its structure. In contrast to classical modeling, attention should be paid here especially to those variables whose (random) variability affects the output quantity. The model does not have to be in the form of explicit expressions, but can be in the form of differential equations.

B. <u>Distribution of input quantities</u>

In some cases, it is possible to divide some (or all) of the inputs to be specified by some analytical distribution. Thus, in a number of cases where directly measured quantities are involved, the assumption of normality can be used and its parameters are estimated from a preliminary sample. If no sample is available, an estimate of the mean value can be determined as the expected most likely value. A rough estimate of the standard deviation has the form $\sigma_A \approx (U - L)/6$, where U is the value that the quantity (x_i) exceeds in a small number of cases (probability 0.15%) and L is the value below which the quantity does not fall with a high probability (probability 99.85%).

If it is not possible to find a suitable analytical distribution of the input variables, it is possible to use different systems of empirical densities (especially the Johnson and lambda systems are advantageous). Simple ways of choosing a distribution that do not require preliminary knowledge of the statistical behavior of the input variables are:

(a) If a symmetrical distribution of the data in the interval (A, B) can be expected and all values are approximately equally likely, the rectangular distribution is chosen.

(b) For symmetric distributions, where it can be expected that in the center area the probability of occurrence of individual realizations is higher, a parabolic distribution is chosen:

$f(x) = \frac{6}{(B-A)^2}(x - A)(B - x) \quad A \leq x \leq B.$

(c) In case when the data distribution is expected to be skewed toward lower values, the following triangular distribution can be used

$f(x) = \frac{2}{(B-A)}(x - A) \quad A \leq x \leq B.$

(d) In case when the data distribution is expected to be skewed toward higher values, the following triangular distribution can be used

$f(x) = \frac{2}{(B-A)}(B - x) \quad A \leq x \leq B.$

The use of these types of distribution only requires the specification of the limit values *A*, *B*, which the given quantity will not exceed under normal conditions. This procedure enables to specify the data probability distribution according to the amount of information about individual input variables.

C. Random number generation

Effective algorithms exist on virtually all computers for the generation of pseudo-random numbers with a uniform distribution $R[0, 1]$. For calculators, it is possible to use, for example, a recurrence relation

$$R_{i+1} = [\pi + R_i]^8 - \text{int}[\pi + R_i]^8$$

where int(x) is the integer part of the number x and $R_0 = 0$.

Given two independent random numbers R_i, R_{i+1} with a uniform distribution $R[0, 1]$, two independent random variables with a standardized normal distribution can be constructed using the Box-Müller transformation

$$Z_i = \sqrt{(-2 \ln R_i)} \sin(2\pi R_i + 1) Z_{i+1} = \sqrt{(-2 \ln R_i)} \cos(2\pi R_i + 1).$$

For a number of other distributions, the inverse transformation procedure can be used, which is applicable in cases where the quantile function Q(u) for a given distribution can be expressed in analytical form. For known Q(u), random numbers C_i from this distribution can be generated according to the relation $C_i = Q(R_i)$. There are a number of other options for random number generation, including the use of the lambda system of empirical probability densities.

D. Choice of number of simulations

The choice of the minimum number of simulations depends on which parameter is actually monitored. The rules are practically the same as for determining the sample size. However, it is usually recommended to choose thousands of simulations to ensure sufficient estimation efficiency.

E. Approximation of simulation results

With regard to universality, it is always advantageous to first find a suitable empirical probability density of the sample and only then to perform further statistical analysis. For this purpose, it is possible to use practically any system of empirical densities or (less advantageous) non-parametric estimates. The described procedure naturally also solves the problem of estimating mean values and other characteristics for functional transformations of random variables. Knowing the mean values $\overline{x}_i$, variances s_i^2, and covariances $C(x_i, x_j)$, it is possible to generate random vectors with a multivariate (normal) distribution and a specified correlation structure. In this way, the estimate of the distribution of $f(y)$ can be found accurately enough, and then the required characteristics can be obtained from it.

3.5.2 Methods Using Taylor Expansions

From a metrological point of view, it is usually sufficient to estimate the required statistical characteristics (transformed) from the basic statistical characteristics of the original data. Approximate procedures for solving this problem use replacement of function $g(\boldsymbol{x})$ by Taylor expansion into quadratic terms (for estimating the mean value) or just linearization (for estimating the variance). The procedure can be used in cases where the function $g(\boldsymbol{x})$ is differentiable and its replacement by a quadratic polynomial is sufficiently accurate.

In simplest case of one variable, the results of direct measurements form a sample (x_i) $i = 1, \ldots N$, from which $\overline{x}$ and ${s_x}^2$ are determined. The non-linear transformation to the required result is given by the well-known relation $y = g(x)$. For non-linear transformations is true, that

$$E(y) \neq g(E(x)) \tag{3.47}$$

thus, the mean value of a function is different from function of the mean value. The equality here only holds if $g(x) = ax + b$, i.e. a linear transformation. To find approximate estimates $\overline{y}$, ${s_y}^2$ of the characteristics of the required quantity y, the Taylor expansion of the function $g(x)$ in the neighborhood of $\overline{x}$ is used

$$g(x) \approx g(\overline{x}) + \frac{1}{1!}\frac{\mathrm{d}g(x)}{\mathrm{d}x}(x - \overline{x}) + \frac{1}{2!}\frac{\mathrm{d}^2 g(x)}{\mathrm{d}x^2}(x - \overline{x})^2 + \ldots + \frac{1}{n!}\frac{\mathrm{d}^n g(x)}{\mathrm{d}x^n}(x - \overline{x})^n \tag{3.48}$$

To estimate the mean value, Taylor expansion into quadratic terms is used, i.e.

$$\underbrace{E(g(x)) \equiv \overline{y} \approx g(\overline{x}) + \frac{\mathrm{d}g(x)}{\mathrm{d}x}\underbrace{E(x-\overline{x})}_{0} + \frac{1}{2}\frac{\mathrm{d}^2 g(x)}{\mathrm{d}x^2}\underbrace{E(x-\overline{x})^2}_{s_x^2}}_{\overline{y}\approx g(\overline{x})+\frac{1}{2}\frac{\mathrm{d}^2 g(x)}{\mathrm{d}x^2}s_x^2} \tag{3.49}$$

To estimate the variance, Taylor expansion into linear terms is used, i.e.

$$\underbrace{D(g(x)-g(\overline{x})) \equiv s_y^2 \approx D\underbrace{\left[\left[\frac{\mathrm{d}g(x)}{\mathrm{d}x}(x-\overline{x})\right]\right]}_{\left(\frac{\mathrm{d}g(x)}{\mathrm{d}x}\right)^2 \cdot s_x^2}}_{s_y^2\approx\left(\frac{\mathrm{d}f(x)}{\mathrm{d}x}\right)^2 s_x^2} \tag{3.50}$$

The following relationship can also be used to determine the variance estimate

$$D(g(x)) = E\left(\left[g(x) - E(g(x)\right]^2\right) = E\left(g(x)^2\right) - E^2(g(x)) \tag{3.51}$$

After approximation of $g(x)$ by Taylor expansion into quadratic terms Eq. (3.48) it is

$$D(g(x)) = E\left[\left(g(\overline{x}) + \frac{\mathrm{d}g(x)}{\mathrm{d}x}(x-\overline{x}) + \frac{1}{2}\frac{\mathrm{d}^2 g(x)}{\mathrm{d}x^2}(x-\overline{x})^2 - E(g(x))\right)^2\right] \tag{3.52}$$

Based on the Taylor expansion of the function $g(x)$ in the neighborhood of $\overline{x}$ and after using the mean value operator [see Eq. (3.49)], the following difference results

$$g(\overline{x}) - E(g(x)) \approx -\frac{1}{2}\frac{\mathrm{d}^2(g(x))}{\mathrm{d}x^2}s_x^2 \simeq -\frac{1}{2}\frac{\mathrm{d}^2(g(x))}{\mathrm{d}x^2}C_2(x) \tag{3.53}$$

where $C_2(x)$ is the second central moment of the quantity x (variance). After replacements and adjustments, it is

$$\begin{aligned} D(g(x)) \approx & \frac{1}{4}\left(\frac{\mathrm{d}^2 g(x)}{\mathrm{d}x^2}\right)^2\left(C_4(x) - C_2^2(x)\right) \\ & + \left(\frac{\mathrm{d}g(x)}{\mathrm{d}x}\right)^2 C_2(x) + \frac{1}{2}\frac{\mathrm{d}g(x)}{\mathrm{d}x}\frac{\mathrm{d}^2 g(x)}{\mathrm{d}x^2}C_3(x) \end{aligned} \tag{3.54}$$

Here $C_3(x)$ is the third central moment and $C_4(x)$ is the fourth central moment of x. For the case of symmetric distributions, the third central moment is zero, so the relation results

$$D(g(x)) \approx \frac{1}{4}\left(\frac{d^2g(x)}{dx^2}\right)^2 \left(C_4(x) - C_2^2(x)\right) + \left(\frac{dg(x)}{dx}\right)^2 C_2(x) \tag{3.55}$$

In order to use it, in addition to the mean $\overline{x}$ and variance $\hat{C}_2 = s_x^2$, it is necessary to determine also fourth central moment from the data

$$\hat{C}_4(x) = \frac{1}{N}\sum_{i=1}^{N}(x_i - \overline{x})^4 \tag{3.56}$$

In many cases, the transformation function is a power function, i.e. $y = ax^P$, where P is usually a small integer (− 1, 2, 3), or ½. In this case, the estimate of the mean is equal to

$$\overline{y} \approx a\overline{x}^P + \frac{P(P-1)}{2} s_x^2\,\overline{x}^{P-2} \tag{3.57}$$

Using the coefficient of variation $v = \frac{s_x}{\overline{x}}$, the estimate of the mean value in the transformation can be expressed as.

$$\overline{y} \approx a\overline{x}^P + \frac{P(P-1)}{2} v^2\,\overline{x}^2\,\overline{x}^{P-2} = a\overline{x}^P\left(1 + \frac{P(P-1)}{2} v^2\right) \tag{3.58}$$

For the variance in the transformation, it is valid, that

$$s_y^2 \approx P^2\,a^2\,\overline{x}^{2P-2} \approx P^2\,a^2\,\overline{x}^{2P}\,v^2 \tag{3.59}$$

So, it is clear that the "correction" for non-linearity is related to the coefficient of variation. E.g. for $P = 3$ (calculating the volume of a cube from measuring its edge) the correction factor is $1+3\,v^2$ and for $P = 2$ (calculating the cross-sectional area from measuring the radius) the correction factor is $1 + v^2$. For common measurements, the coefficient of variation ranges up to $v = 0.1$ (i.e. 10%). Then the correction factor for $P = 3$ is only 1.03 and for $P = 2$ is only 1.01. Thus, it can be seen that the refined estimates based on the Taylor expansion are not very different from direct substitution into the transformation equation, as long as the measurements are sufficiently accurate.

Example 3.5 *Cross section of circular fibers*

The radius of the circular fibers r_i ($i = 1, \ldots, N$) was measured and the purpose is to estimation of the mean value $\overline{s}$ of the cross-sectional area from the knowledge of mean radius $\overline{r}$, and corresponding standard deviation s_r. The transformation relation is $s = \pi\, r^2$ By direct substitution into the above relations, it is valid that

$$\overline{s} = \pi r^2 + \pi s_r^2 = \pi\left(\overline{r}^2 + s_r^2\right)a \quad s_S^2 \approx 4\pi^2\overline{r}^2 s_r^2.$$

It follows directly from the definition of variance $D(x) = E\left(x^2\right) - E(x)^2$ that $E\left(x^2\right) = D(x) + E(x)^2$, which corresponds to the approximate relation from the Taylor expansion and indicates that in this case it is not an approximation. Using the coefficient of variation $v = \frac{s_x}{\overline{x}}$, it is $\overline{s} = \pi\overline{r}^2\left(1 + v^2\right)$. From this equation, the influence of measurement accuracy on the need to use an approximate estimate of the mean value can be easily expressed.

Average accurate measurement: $v \approx 0.1$ and then $\overline{s} = 1.01\,\pi\overline{r}^2$, so there is no need to use the relation derived from the quadratic approximation using the Taylor expansion.

Very rough measurement: $v \approx 0.5$ and then $\overline{s} = 1.25\pi\overline{r}^2$, so there is beneficial to use the relation derived from the quadratic approximation using the Taylor expansion.

The case of a multidimensional transformation function

A frequent task is the case when there are more measured quantities in the transformation relation $g(x_1, \ldots, x_m)$, i.e. it is a function of m variables. Estimates of mean values and variances $\overline{x}_1, s_{x_1}^2, \ldots \overline{x}_m, s_m^2$ are known from measurements. The vector $\boldsymbol{x}$ is used instead of the scalar x. Thus, the vector of averages is defined as $\overline{x} = (\overline{x}_1, \overline{x}_2, \ldots, \overline{x}_m)$. For the Taylor expansion of the function $g(\boldsymbol{x})$ in the neighborhood $\overline{x}$ is valid

$$\begin{aligned} g(\mathbf{x}) \approx g(\overline{\mathbf{x}}) &+ \sum_{i=1}^{m} \frac{\partial g(\mathbf{x})}{\partial x_i}(x_i - \overline{x}_i) + \frac{1}{2}\sum_{i=1}^{m} \frac{\partial^2 g(\mathbf{x})}{\partial x_i^2}(x_i - \overline{x}_i) \\ &+ \sum_{i=1}^{m-1}\sum_{j>i}^{m} \frac{\partial^2 g(\mathbf{x})}{\partial x_i \partial x_j}(x_i - \overline{x}_i)\left(x_j - \overline{x}_j\right) + \end{aligned} \tag{3.60}$$

To estimate the mean value in the transformation, Taylor expansion into quadratic terms is again used, i.e.

$$\overline{y} \approx g(\overline{x}) + \frac{1}{2}\sum_{i=1}^{m} \frac{\partial^2 g(x)}{\partial x_i^2} s_{x_i}^2 + \sum_{i=1}^{m-1}\sum_{j>i}^{m} \frac{\partial^2 g(x)}{\partial x_i \partial x_j}\operatorname{cov}\left(x_i, x_j\right) \tag{3.61}$$

and Taylor expansion into linear and mixed terms is again used to estimate the corresponding variance

$$s_y^2 \approx \sum_{i=1}^{m}\left[\frac{\partial g(x)}{\partial x_i}\right]^2 s_{x_i}^2 + 2\sum_{i=1}^{m-1}\sum_{j>i}^{m} \frac{\partial g(x)}{\partial x_i}\frac{\partial g(x)}{\partial x_j}\operatorname{cov}\left(x_i, x_j\right) +$$

$$+\underbrace{\sum_{i=1}^{m-1}\sum_{j>i}^{m}\frac{\partial^2 g(x)}{\partial x_i \partial x_j} s_{x_i}^2 s_{x_j}^2}\tag{3.62}$$

usually is neglected.

Also, in this case, it is possible to achieve more accurate variance estimates using Eq. (3.51).

Example 3.6 *Calculation of fineness*

The purpose is to calculate the average fineness $\overline{T}$ from measurements of weights g_i and lengths L_i of fibers. Estimates $\overline{g}$, s_g^2, $\overline{L}$, s_L^2 are calculated from measurements and no correlation, i.e. $\mathrm{cov}(g, L) = 0$ is assumed. After substituting into Eq. (3.61) an expression for the mean fineness has the form

$$\overline{T} \approx \frac{\overline{g}}{\overline{L}} + \frac{\overline{g}}{\overline{L}^3} s_L^2 = \frac{\overline{g}}{\overline{L}}\left(1 + v_L^2\right)\tag{3.63}$$

The estimate of the mean value of the fineness is related only to the accuracy of the length measurement expressed by its coefficient of variation.

3.5.3 Two-Point Methods and Extensions

In many cases, the use of Taylor expansion, or simulation using the Monte Carlo method is limited by the complexity of the transformation relation $g(x)$, which may not be continuous everywhere, it may not be possible to analytically calculate its derivatives, etc. Then it is advantageous to choose the so-called two-point methods, when the moments of the function $y = g(x)$ are estimated from selected values of the input variables x and the corresponding "concentrated" probability densities. Rosenblueth two-point and three-point methods (Rosenblueth, 1975, 1981) use knowledge of the mean value $\overline{x}$, standard deviation s_x, and skewness coefficient g_1 of the input variable x to calculate selected moments of the output variable $y = g(x)$. The procedure for two-point method is applicable to any distribution of x and is exact for the first three general moments of y if $y = x$. It is generally only approximate. Two probability densities are chosen

$$P_{+}\delta(\overline{x} - x_{+}) \text{ and } P_{-}\delta(\overline{x} - x_{-})\tag{3.64}$$

where P_+ and P_- are the concentration coefficients, δ is the Dirac step function and x_+ x_- are the selected values of the quantity x. The following relation can then be used approximately for the nth general moment of the quantity y

$$E\left(y^n\right) \approx^\circ P_+ y_+^n + P_- y_-^n \tag{3.65}$$

where $y_+^n = (g(x_+))^n$ and $y_-^n = (g(x_-))^n$. For the parameters $P_\pm$ and positions $x_\pm$ the following relations must be valid

$$\begin{aligned} &P_+ + P_- = 1 \quad P_+x_+ + P_-x_- = \overline{x} \quad P_+(x_+ - \overline{x})^2 + P_-(x_- - \overline{x})^2 = s_x^2 \\ &P_+(x_+ - \overline{x})^3 + P_-(x_- - \overline{x})^3 = g_1\, s_x^3 \end{aligned} \tag{3.66}$$

The solution to this system of equations has the form

$$P_+ = \frac{1}{2}\left(1 \pm \sqrt{1 - \frac{1}{1 - (g_1/2)^2}}\right) \tag{3.67}$$

where the sign before the square root depends on the sign of the skewness (P_+ must be less than 1). For small skewness $g_1 \ll 1$, the relation $P_+ = 0.5 - g_1/4$ can then be used, and then

$$P_- = 1 - P_+ x_\pm = \overline{x} \mp s_x \sqrt{\frac{P_\mp}{P_\pm}} \tag{3.68}$$

If g_1 is not known and symmetric distributions can be assumed, they are $P_\pm = 0.5$ and $x_\pm = \overline{x} \pm s_x$. From Eq. (3.61) it is directly

$$\overline{y} \approx \frac{y_+ + y_-}{2} s_y \approx \left|\frac{y_+ - y_-}{2}\right| v_y = \left|\frac{y_+ - y_-}{y_+ + y_-}\right| \tag{3.69}$$

where ${}_v y$ is the coefficient of variation of the output variable y. If the first two derivatives of the function $g(x)$ exist, is $\overline{y}$ identical to the estimate obtained from the quadratic expansion of $g(x)$, i.e. Equation (3.61) and the estimates s_y, v_y are identical to the estimate obtained from the linearization of $g(x)$, i.e. Equation (3.62). If g_1 is approximately zero and the data are approximately normally distributed, the accuracy of the approximation (3.48) can be improved by choosing additional points. Since this is a symmetric distribution, the choice of positions must be symmetric about the mean. It is simple to add the position x_0 where the concentration is P_0 and then is $x_- = 2\overline{x} - x_+$. In this case, it is true that

$$\begin{aligned} &2P_+ + P_0 = 1\, P_+x_+ + P_-x_- = \overline{x}\, 2P_+(x_+ - \overline{x})^2 = s_x^2 \\ &2P_+(x_+ - \overline{x})^4 = 3s_x^4 \end{aligned} \tag{3.70}$$

Solving these equations leads to the values $P_+ = 1/6$, $P_0 = 2/3$, and $x_+ = \overline{x} + s_x/3$. In the case of multiple variables, the individual positions of the input variables are selected in the corners of the hypercube (Rosenblueth, 1975).

Further details can be found in the paper (Rosenblueth, 1975). Hong (Rousseeuw & Verboven, 2002) presents the solution of the two-point method for two input variables if the estimates of means, variances, skewness, and correlation coefficient are known. A comparison of two-point methods is made, for example, in work (He & Sällfors, 1994).

The extension of Rosenblueth methods (Rosenblueth, 1975, 1981) to cases of many correlated random variables is described in work of Harr (1989).

For three-point methods, in multidimensional cases, factorial experiment schemes 3^m are used. For each input variable x_j, the central level $\overline{x}_j$ and the upper or lower levels at locations $\overline{x} \pm s_j\sqrt{3/2}$ or $\overline{x} \pm s_j\sqrt{3}$ are selected. For specifications of the locations of points and corresponding probabilities the Taguchi's experimental design and its modification can also be used (Seo & Kwak, 2002).

These methods allow estimates of the mean, variance, skewness, and kurtosis of the function $y = g(x)$. If it is necessary to determine the distribution of the quantity y. Any system of four-parameter distributions (Pearson's, lambda, Johnson's, etc.) can be used. In work (Seo & Kwak, 2002), a procedure of the three-point method is proposed, which is suitable for cases where the distribution of input quantities is non-normal. It is based on the approximation of the nth moment of the function $g(\boldsymbol{x})$

$$E\left(\left[g(x)\right]^n\right) = \int_{-\infty}^{\infty} \left[g(x)\right]^n f(x)\mathrm{d}x \approx w_1\left[g(\overline{x} + \alpha_1 s_x)\right]^n + w_2\left[g(\overline{x} + \alpha_2 s_x)\right]^n + w_3\left[g(\overline{x} + \alpha_3 s_x)\right]^n$$

The parameters $(w_1, w_2, w_3, \alpha_1, \alpha_2, \alpha_3)$ are estimated from the central moments

$$C_k = \int_{-\infty}^{\infty} [x - \mu]^k f(x)\,\mathrm{d}x \approx w_1(\alpha_1 s_x)^k + w_2(\alpha_2 s_x)^k + w_3(\alpha_3 s_x)^k \tag{3.71}$$

for $k = 1, 2, 3, 4, 5$. Here C_k are the central moments estimated from the data of the input variable x. Detailed relations for the estimation of these parameters are presented in the paper (Che-Hao et al., 1995; Seo & Kwak, 2002).

3.6 Conclusion

It is obvious that the type of scale affects the entire subsequent analysis of the measurement results. The effort to use the cardinal scale runs into the problems of evaluating of textiles by consumers who select a suitable product based on subjective judgments. One way to account for this ordinal information is to measure various related quantities on a cardinal scale and find a multivariate calibration model to predict subjective judgment. This procedure was already proven in the assessment of hand and appearance.

References

Agresti, A., & Coull, B. A. (1998). Approximate is better than "exact" for interval estimation of binomial proportions. *The American Statistician, 52*(2), 119–126. https://doi.org/10.1080/00031305.1998.10480550

Brown, L. D., Cai, T. T., & DasGupta, A. (2001). Interval estimation for a binomial proportion. *Statistical Science, 16*(2), 101–117. http://www.jstor.org/stable/2676784

Carolino, E., & Barao, I. (2013). Robust methods in acceptance sampling. *REVSTAT—Statistical Journal, 11*(1), 67–82.

Che-Hao, C., Yeou-Koung, T., & Jinn-Chuang, Y. (1995). Evaluation of probability point estimate methods. *Applied Mathematical Modelling, 19*(2), 95–105. https://doi.org/10.1016/0307-904X(94)00018-2

Demidenko, E. (2023). *M-statistics: Optimal statistical inference for a small sample.* Wiley.

Donaldson, G. W. (1998). Ridit scores for analysis and interpretation of ordinal pain data. *European Journal of Pain, 2*(3), 221–227. https://doi.org/10.1016/S1090-3801(98)90018-0

Dowd, P. A., Pardo-Igúzquiza, E., & Egozcue, J. J. (2015). The total bootstrap median: A robust and efficient estimator of location and scale for small samples. *Journal of Applied Statistics, 42*(6), 1306–1321. https://doi.org/10.1080/02664763.2014.999650

Figueiredo, F., & Gomes, M. I. (2004). The total median in statistical quality control. *Applied Stochastic Models in Business and Industry, 20*(4), 339–353. https://doi.org/10.1002/asmb.545

Harr, M. E. (1989). Probabilistic estimates for multivariate analyses. *Applied Mathematical Modelling, 13*(5), 313–318. https://doi.org/10.1016/0307-904X(89)90075-9

He, J., & Sällfors, G. (1994). An optimal point estimate method for uncertainty studies. *Applied Mathematical Modelling, 18*(9), 494–499. https://doi.org/10.1016/0307-904X(94)90327-1

Meloun, M., & Militký, J. (2012). *Interactive statistical data analysis.* Carolinum. (in Czech)

Rosenblueth, E. (1975). Point estimates for probability moments. *Proceedings of the National Academy of Sciences, 72*(10), 3812–3814. https://doi.org/10.1073/pnas.72.10.3812

Rosenblueth, E. (1981). Two-point estimates in probabilities. *Applied Mathematical Modelling, 5*(5), 329–335. https://doi.org/10.1016/S0307-904X(81)80054-6

Rousseeuw, P. J., & Verboven, S. (2002). Robust estimation in very small samples. *Computational Statistics & Data Analysis, 40*(4), 741–758. https://doi.org/10.1016/S0167-9473(02)00078-6

Schwartz, L. M. (1975). Random error propagation by Monte Carlo simulation. *Analytical Chemistry, 47*(6), 963–964. https://doi.org/10.1021/ac60356a027

Seo, H. S., & Kwak, B. M. (2002). Efficient statistical tolerance analysis for general distributions using three-point information. *International Journal of Production Research, 40*(4), 931–944. https://doi.org/10.1080/00207540110095709

Wilson, E. B. (1927). Probable inference, the law of succession, and statistical inference. *Journal of the American Statistical Association, 22*(158), 209. https://doi.org/10.2307/2276774

Yates, F. (1948). The analysis of contingency tables with groupings based on quantitative characters. *Biometrika, 35*(1/2), 176. https://doi.org/10.2307/2332637

Chapter 4
Measurement Models

All models are wrong, but some are useful [G. E.P. Box].

Abstract Models of measurement are generally based on the type of action of the error components, assumptions about their variability, and the shape of their distribution. The basic types of measurement models, including their identification, characterization, and description, are broadly investigated. Additive and multiplicative measurement models are discussed in detail. This chapter also discusses suitable statistical analysis of the individual models of measurement.

4.1 Introduction

Data are usually the results of experiments or the results of information gathering. By default, it is assumed to be valid

$$\textbf{Data} = \textbf{model} + \textbf{residuals},$$

where residuals = errors + ignorance. Depending on the size of the range of experimental data, either an additive measurement model or a multiplicative measurement model can be chosen as a guideline. This chapter discusses the basic types of measurement models and their analysis methods. Different types of measurement models resulting from a combination of the type of action of the noise component, assumptions about the variability of the noise component, and the shape of its distribution are systematically described.

4.2 Basic Measurement Models

When analyzing experimental data, it is main goal an estimation of the parameters of a probabilistic model that defines the effect of disturbances, or relation to external variables influencing measurement variability. Here only simple probabilistic models

J. Militký et al., *Textile Material Research—Statistical Tools of Experimental Data Analysis*, https://doi.org/10.1007/978-981-97-6607-9_4

of so-called direct and indirect measurements are discussed. However, the described procedures can be used without problems directly for more complicated models, e.g. regression type. For typical experimental data, there are two limiting situations:

The range of the analyzed data is within one order of magnitude. This allows the use of standard statistical methods based on the assumption of constant variance or additive measurement model.

The range of the analyzed data is several orders of magnitude. A variance-stabilizing transformation or a multiplicative measurement model can then be used. This results often in a logarithmic transformation of the data. The disadvantage of the logarithmic transformation is the fact that at low concentrations the absolute measurement error is very small (close to 0) so that the relative error is constant, which contradicts reality.

The most frequent and *apparently* the simplest task is to determine the estimate of some defined quantity (e.g. the diameter of a rod from repeated measurements at the same place) from N repeated measurements forming a sample $(x_1, x_2, \ldots x_N)$. From a computational point of view, the equivalent task is the estimation of the location parameter (e.g. the diameter of the fibers from the measurement of the diameters of different fibers, where one measurement is made on each fiber) and the determination of its uncertainty (the confidence interval of the mean value). There are therefore samples whose elements are not repeated measurements but measurements on different specimens.

4.2.1 Additive Model of Direct Measurements

The basic additive model of direct measurements (e.g. determination of length, weight, etc.) has the form

$$x = \mu + \varepsilon \tag{4.1}$$

where μ is the true value of the measured quantity (rod diameter or fiber diameter) and ε is the random error of measurement. This model works quite well for the case of repeated measurements, but when it comes to different specimens, it often fails especially in cases where the sampling is biased (e.g. if the samples will be taken from fibers passing through different places, the probability of selection will depend on the length of the fibers, i.e. is larger for longer fibers). The parameter here is primarily the mean value μ (main quantity). The result of one measurement is x (for repeated measurements, the i-th repetition is denoted as x_i and the number of repetitions is N). The error is considered as a random variable with variance $D(\varepsilon) = \sigma^2$, which is also the second estimated parameter. The error variance σ^2 is identical to the measurement variance for this model. The assumptions of standard statistical analysis are:

- the mean value of measurement errors is zero, i.e. $E(\varepsilon) = 0$,
- the variance of measurement errors is constant, i.e. $D(\varepsilon) = \sigma^2$

- errors are mutually independent, i.e. $E(\varepsilon_i\,\varepsilon_j) = 0$
- errors have a normal distribution, i.e. $\varepsilon \approx N(0, \sigma^2)$.

The measurement variance is $D(x_i) = \sigma^2$. The estimate is the sample variance

$$s^2 = \frac{1}{N-1}\sum_{i=1}^{N}(x_i - \hat{\mu})^2$$

The estimate of the parameter μ is the sample mean

$$\hat{\mu} = \frac{1}{N}\sum x_i \text{ with variance } D(\hat{\mu}) = \frac{\sigma^2}{N}$$

This model is practically the only one used in "standard uncertainty" calculations. A discussion of the identification and procedures for violations of the first three assumptions is given in (Meloun & Militký, 2012). In the case when the mean value of the errors is not zero, but $E(\varepsilon) = k$, are results systematically biased estimates of the mean value, because $E(x) = \mu + k$. The assumption of constancy of variance is equivalent to the requirement of constant additive error of the measuring instrument $\Delta \approx \sigma$. Thus, for the additive model of measurements is relative error δ, equal to.

$$\delta = \Delta / x \tag{4.2}$$

The assumption of measurement independence is fulfilled in many cases. Difficulties can arise, for example, in kinetic experiments, where sampling on one system can lead to first-order autocorrelation

$$\varepsilon_i = \rho\,\varepsilon_{i-1} + u_i \tag{4.3}$$

Here u_i is a random variable with constant variance. It is true that $u_0 = 0$. The autocorrelation coefficient ρ is the correlation coefficient between the pairs x_j and $x_{j+1}, j = 1, \ldots, N-1$. It can be determined that

$$D(u) = D(x) = \frac{\sigma^2}{1-\rho^2} \tag{4.4}$$

It can be seen that in the case of significant autocorrelation there will be an increase in variance. The autocorrelation coefficient ρ is commonly estimated using the relation

$$\hat{\rho} = \frac{\sum_j^{N-1}(x_j - \overline{x})(x_{j+1} - \overline{x})}{[s^2(N-1)]} \tag{4.5}$$

where s^2 is the sample variance and $\overline{x}$ is the arithmetic mean. Roughly, if $\hat{\rho}$ lies in the interval

$$-2/\sqrt{N} \leq \hat{\rho} \leq 2/\sqrt{N} \tag{4.6}$$

can be considered ρ insignificant.

The assumption that errors have a normal distribution is necessary for the construction of confidence intervals (uncertainty of measurement results), or hypothesis testing. If sufficient data are available, the error ε distribution can be estimated from the distribution of x measurements, since the shape of the probability density function is identical for model (5.1). The normal distribution can be understood as one of the members of the class of elliptic symmetric distributions, which differ only in the length of the tails. In technical practice, where it is common to measure on different specimens, an asymmetric distribution of data skewed toward higher values is a frequent phenomenon. This distribution is common for data where there are order-of-magnitude differences in the specimens (e.g. environmental data).

A suitable transformation of $h(x)$ is often used to remove the asymmetry of the data distribution. However, if model (5.1) is valid, this leads to non-constant variance

$$D(h(x)) = \left[\frac{\mathrm{d}h(x)}{\mathrm{d}x}\right]^2 \sigma^2 \tag{4.7}$$

For example, for the commonly recommended logarithmic transformation $h(x) = \ln(x)$ it results

$$D(h(x)) = \left(\frac{\sigma}{x}\right)^2 = \delta^2 \tag{4.8}$$

This means that instead of a constant absolute error, there is a constant relative error (coefficient of variation) in this transformation, which contradicts the accepted measurement model. Correct analysis here requires the direct use of a skewed distribution and the construction of asymmetric confidence intervals.

4.2.2 *Multiplicative Model of Direct Measurements*

This model is based on the assumptions of constant relative error and non-negativity of measurements (these are physical quantities related to concentration or mass).

$$x = \mu \exp(\varepsilon) \tag{4.9}$$

Here ε has the same properties as in the additive model (Eq. 5.1). After a *correct* logarithmic transformation, this model leads to an additive model in logarithms

$$\ln(x) = \ln(\mu) + \varepsilon \tag{4.10}$$

The disadvantage of the multiplicative model is that for very low concentrations or small μ, the absolute measurement error is too small. If it is true that $\ln(x) \approx N(\nu, \tau^2)$, the random variable x has a *lognormal distribution* with parameters (Militký & Meloun, 1999).

$$\mu = \exp(\nu + \tau^2/2) \quad \sigma = \mu^2(\exp(\tau^2) - 1) \tag{4.11}$$

In addition to the mean value μ (an estimate is the arithmetic mean), the other location parameter as geometric mean is also used

$$\mu_G = \exp(E(\ln(x))) = \exp(\nu) \tag{4.12}$$

It can be easily proved that it is true

$$P(x \leq \mu_G) = P(x \leq \exp(\nu)) = P(\ln(x) \leq \nu) = 0.5 \tag{4.13}$$

This means that the geometric mean is here equal to the median

$$\text{med}(x) = \exp(\nu) = \mu_G \tag{4.14}$$

Data in a logarithmic transformation are here used for parameter estimates

$$\hat{\nu} = \frac{1}{N}\sum_{i=1}^{N} \ln x_i$$

$$\hat{\tau}^2 = \frac{1}{N-1}\sum_{i=1}^{N} (\ln x_i - \hat{\nu})^2$$

$$\hat{\mu} = \exp(\hat{\nu} + \hat{\tau}^2/2) \tag{4.15}$$

The variances of the estimates are in form

$$D(\hat{\mu}) = \frac{\mu^2}{N[\exp(\tau^2) - 1]} \quad D(\hat{\mu}_G) = E(\hat{\mu}_G)^2[\exp(\tau^2/N) - 1] \tag{4.16}$$

The mean value of the geometric mean is $E(\hat{\mu}_G) = \exp(\nu + \tau^2/2N)$. . Both averages are biased estimates. However, the geometric mean is preferable to the arithmetic mean (Bickel & Doksum, 1981). If an incorrect assumption about the error distribution is used, the parameters and consequently the entire statistical analysis will be distorted. For example, let the additive model (5.1) hold and an incorrect logarithmic transformation is applied to the data. Then it is

$$\ln(x) = \ln(\mu + \varepsilon) = \ln \mu + \ln(1 + \varepsilon/\mu) \tag{4.17}$$

By Taylor expansion, it is possible to approximate $\ln(1+x) = x - x^2/2 + x^3/3 + x^4/4$ and then

$$\ln(x) = \ln(\mu + \varepsilon) \approx \ln\mu + \varepsilon/\mu - 0,5\,(\varepsilon/\mu)^2 + (\varepsilon/\mu)^3/3 - (\varepsilon/\mu)^4/4...$$
$$\ln(x) \approx \ln(\mu + \varepsilon) \approx \ln\mu + \ln(1 + \varepsilon/\mu) \tag{4.18}$$

If a random variable S has distribution $N(\mu, \sigma^2)$, its central moments have form.

$$\mu_r = E[(S - \mu)^r]$$

It is valid, that

$$\mu_r = 0 \text{ for r odd or } \mu_r = \frac{r!}{(r/2)!}\frac{\sigma^r}{2^{r/2}} \text{ for } r \text{ even}$$

For small relative measurement errors $\delta = \sigma/\mu$, expressions for the mean value and variance of $\ln(x)$ can then be in the form

$$E(\ln x) = \ln\mu - 0.5\,\delta^2 - 0.75\,\delta^4$$
$$D(\ln x) = \delta^2 + 2,5\,\delta^4 + 4,66\,\delta^6 + 6\,\delta^8$$

It is therefore evident that using an incorrect assumption will affect both the mean and the variance. For μ greater than one, the *mean will be underestimated and the variance overestimated*.

If the resulting sample skewness is greater than 0.35, it is recommended to choose a three-parameter lognormal distribution with threshold value A. Knowing A, the $x - A$ can be used instead of x in all the above-mentioned relationships. Order statistics, i.e. ascendingly sorted sample values, can be used for quick estimation of parameter A.

$$A = \frac{x_{(1)}\,x_{(n)} - \tilde{x}_{0.5}^2}{x_{(1)} + x_{(n)} - 2\,\tilde{x}_{0.5}} \tag{4.19}$$

A more accurate estimate of the threshold A is given in the book (Deming et al., 1988).

4.2.3 *Power Class of Measurement Models*

This class of models is characterized by the fact that the transformation of both sides using the function $h(x)$ results in an additive model

$$h(x) = h(\mu) + \varepsilon \tag{4.20}$$

Preferably, the Box-Cox class of polynomial transformations is used as the function $h(x)$.

$$h(x) = \frac{(x-1)^{\lambda}}{\lambda} \quad \lambda \neq 0$$
$$h(x) = \ln(x) \quad \lambda = 0$$

where λ is the transformation parameter. For $\lambda = 1$ it is equal to an additive measurement model and for $\lambda = 0$ it is a multiplicative model. It can be shown that a suitable estimator of the μ parameter (mean value) is the *sample median*, which is invariant to a monotonic transformation.

Using the Taylor expansion, it is possible to express Eq. (4.21) in the form

$$x \approx \mu + \varepsilon/\mu^{1-\lambda} \tag{4.21}$$

In case when the $D(\varepsilon) = \sigma^2$ variance is small, it is in fact an additive model with non-constant errors for which a weighted arithmetic mean with weights proportional $\mu^{-(1-\lambda)/2}$ can be used as an estimator of μ.

The maximum likelihood method can be used to estimate the parameter λ. If the distribution of errors ε conforms to the assumptions of the additive measurement model (normality and independence), the log likelihood function has the form

$$\ln L(\lambda) = \sum (\lambda - 1)\ \ln(x_i) - \frac{1}{2\sigma^2} \sum [h(x_i) - h(\mu)]^2 \tag{4.22}$$

For fixed λ, a maximum likelihood estimate of the variance can be determined

$$\hat{\sigma}_c^2 = \frac{1}{N} \sum [h(x_i) - h(\mu)]^2 \tag{4.23}$$

where $h(\mu)$ is replaced by the arithmetic mean of the transformed data

$$h(\mu) \approx \frac{1}{N} \sum h(x_i) \tag{4.24}$$

After substituting into the likelihood function, the following relationship results

$$\ln L^*(\lambda) = \sum (\lambda - 1)\ \ln(x_i) - \frac{N\ \ln \hat{\sigma}_c^2}{2} \tag{4.25}$$

By maximizing of $\ln L^*(\lambda)$ according to λ (Smothers et al., 1999) it is possible to easily determine the maximum likelihood estimate $\hat{\lambda}$ of the transformation parameter λ. An inequality can be constructed based on the asymptotic $(1-\alpha)$ % confidence interval of this parameter is in the form

$$\ln L(\lambda) \geq \ln L(\hat{\lambda}) - 0,5\ \chi^2_{1-\alpha}(1) \tag{4.26}$$

All λ satisfying this inequality lie within the confidence interval and are therefore acceptable. In Eq. (5.27), the $\chi^2_{1-\alpha}(1)$ is quantile of chi-square distribution with 1 degree of freedom. A confidence interval based on Eq. (5.27) can be easily used to distinguish between additive and multiplicative measurement models. It is valid that:

- if the 95% confidence interval also contains one, the *additive model* is chosen.
- if the 95% confidence interval contains zero and not one, the *multiplicative model* is chosen.
- in other cases, it is possible to choose a general model in the form of Eq. (5.21) and use the procedure proposed in (Smothers et al., 1999) for further analysis.

Thus, power-law transformations can be advantageously used to distinguish between the two basic measurement models.

4.2.4 Combined Model of Direct Measurements

This model is a combination of additive and multiplicative measurement models on the form

$$x = \mu \exp(\varepsilon_1) + \varepsilon_2 \tag{4.27}$$

It is commonly assumed that both errors have a normal distribution $\varepsilon_1 \approx N(0, \sigma_1^2)$ and $\varepsilon_2 \approx N(0, \sigma_2^2)$. A simplifying assumption is that $\sigma_1^2 \approx \sigma_2^2$. This model was found for the case of monitoring pesticide levels ranging from 1 ppb to 100 units (Horowitz Trumpet) (Fig. 4.1).

The model (5.28) can be written in a logarithmic transformation

$$\ln x = \ln \mu + \varepsilon_1 + \ln\left[1 + \frac{\varepsilon_2}{\mu \exp(\varepsilon_1)}\right] \tag{4.28}$$

Taylor expansion leads to the relation

$$\ln x \approx \ln \mu + \varepsilon_1 + \left[\frac{\varepsilon_2}{\mu \exp(\varepsilon_1)}\right] \tag{4.29}$$

According to the ratios of error magnitudes, different special types of models are derived:

- for very small ε_1, the additive model in logarithms $\ln x \approx \ln \mu + [\varepsilon_2/\mu]$ with a non-constant variance $\delta = \sigma_2/\mu$ results,
- for very small ε_2 or for comparable magnitudes of ε_1 and ε_2 and large μ, the additive model in logarithms $\ln x \approx \ln \mu + \varepsilon_1$ with a constant variance σ_1^2 is used.

The variance of the measurement can be determined in the form

$$D(x) = \sqrt{\sigma_1^2 + \mu^2 \exp(\sigma_2^2)\,(\exp(\sigma_2^2) - 1)} \tag{4.30}$$

For *low values* of x, the assumption of normality can be used and approximate 95% confidence interval is $\overline{x} \pm 1.96\sqrt{D(x)}$.

For *high values* of x, the assumption of a lognormal distribution can be used. Details about determining the estimates of both variances can be found in (Meloun & Militký, 2012).

4.3 Typology of Measurement Models

There are a number of different models, which are divided according to the form of the error component:

Additive	A
Multiplicative	M
General	O.

According to assumptions about the variability of the error component ε, models can be divided into classes:

Constant variance	K
Non-constant variance	N
Autocorrelation	R.

Models of measurements are also dependent on the assumptions about the distribution of errors ε:

I. Symmetric distributions: They can generally be characterized by a class of probability density functions

$$p_\varepsilon(\varepsilon_i) = Q_N \exp\left(-\frac{|\varepsilon_i|^P}{\alpha}\right)$$

Here Q_N is the normalization factor, α is proportional to the variance and according to p the specific probability density is derived: $p = 1$ corresponds to Laplace distribution, $p = 2$ corresponds to Normal distribution, and $p \to \infty$ corresponds to rectangular distribution.

II. Asymmetric distributions: One of the ways to remedy this is to use a suitable symmetrization transformation $h(x)$ (so-called TBS models), where it is approximately$h(x_i) = h(\mu) + \varepsilon_i \quad E(\varepsilon_i) = 0,\ D(\varepsilon_i) = \sigma^2$

Then the following estimate can be used

$$\hat{\mu} = h^{-1}\left[\frac{1}{N}\sum h(x_i)\right]$$

For power transformation $h(x) = x^{\lambda}$ where $\lambda \neq 0$, or $h(x) = \ln x$ where $\lambda = 0$ is roughly valid

$$\hat{\mu} = \left[\frac{1}{N}\sum |x_i|^{\lambda}\right]^{1/\lambda}$$

Depending on the size of λ then different mean result:
Harmonic ($\lambda = -1$) $\leq$ geometric ($\lambda = 0$) $\leq$ arithmetic ($\lambda = 1$) $\leq$ quadratic ($\lambda = 1$).

Maximum likelihood method

Assuming independent ε_i, the likelihood function, or its logarithm is in the form

$$\ln L(\mu) = \sum \ln\ p(x_i)$$

The maximum likelihood estimate is then defined as.
$\hat{\mu} = \max(\ln\ L(\mu))$ calculated from the condition… $\frac{\partial \ln\ L(\mu)}{\partial \mu} = 0$.

General model of the errors

It includes both additive and multiplicative models, thus is $x_i = g(\mu, \varepsilon_i)\ i = 1....N$. When knowing $p_{\varepsilon}(\varepsilon)$ it is

$$p(x_i) = p_{\varepsilon}\left[g^{-1}\left(\mu, x_i, \varepsilon_i\right)\right]\left[\frac{\partial g^{-1}(.)}{\partial x_i}\right].$$

There are the following basic variants

A. **Additive models** $g^{-1}(.) = x_i - \mu \quad \frac{\partial g^{-1}(.)}{\partial x_i} = 1$.
B. **Multiplicative models** $g^{-1}(.) = \ln\ x_i - \ln\ \mu \quad \frac{\partial g^{-1}(.)}{\partial x_i} = \frac{1}{x_i}$.

For typical models of measurements with different form of the error (A, M) and different variability of the errors (K, N, R), see Sect. 5.3, it is possible to use the following treatment:

1. $AKp = 2$: The measurement model has the form $x_i = \mu + \sigma\varepsilon_i^*$, where $\varepsilon_i^* \approx N(0, 1)$.

   The measurement variance is $D(x_i) = \sigma^2$. The estimate of the parameter μ is $\hat{\mu} = \frac{1}{N}\sum x_i$ and its variance is $D(\hat{\mu}) = \frac{\sigma^2}{N}$.
   This model is practically the only one used in "standard uncertainty" calculations.

2. $ANp = 2$: The measurement model has the form $x_i = \mu + w_i\varepsilon_i^*$.

   The measurement variance is $D(x_i) = w_i^2$. The estimate of the parameter μ is

$$\hat{\mu} = \frac{\sum\left[\frac{x_i}{w_i^2}\right]}{\sum w_i^{-2}} \text{ and its variance is } D(\hat{\mu}) = \frac{1}{\sum w_i^{-2}}.$$

A special case is measurement with a constant relative deviation (relative accuracy) common to many measuring devices. Then there is a relative deviation equal to the $V = \frac{\sigma}{X}$ and variance is $\sigma^2 = V^2 X^2$. The weights are $w_i = V x_i$ and the measurement variance is $D(x_i) = V^2 x_i^2$. The estimate of the parameter μ is

$$\hat{\mu} = \frac{\left[\sum x_i^{-1}\right]}{\sum x_i^{-2}} \text{ and its variance is } D(\hat{\mu}) = \frac{V^2}{\sum x_i^{-2}}.$$

3. $ARp = 2$: The measurement model has the form $x_i = \mu + W_i$ where $W_i = \rho W_{i-1} + \sigma \varepsilon_i^*$.

The measurement variance is $D(x_i) = \sigma^2/(1 - \rho^2)$. The estimate of the parameter μ variance is

$$D(\hat{\mu}) = \frac{1}{N}\left[\frac{\sigma^2}{1 - \rho^2}\right]$$

4. $MKp = 2$: The measurement model has the form $x_i = \mu \exp(\sigma \varepsilon_i^*)$.

The measurement variance is $D(\ln\ x_i) = \sigma^2 \rightarrow D(x_i) \approx \sigma^2 x_i^2$. The estimate of the parameter μ is generally calculated as a weighted average (see $ANp = 2$). This model often better describes physical measurements when the results from measuring instruments are only positive.

For $p \neq 2$, i.e. *non-normal distribution* of the error component, the maximum likelihood method is more complicated. Simple expressions result for:

$$AKp = 1 \text{ where is } \hat{\mu} = \text{med}(x_i) \text{ and } AKp \rightarrow \infty \text{ where is } \hat{\mu} = \frac{1}{2}(x_{(1)} + x_{(n)}).$$

4.4 Conclusion

It is evident that statistical data processing has a number of specific features that must be considered. It is always advantageous to start by analyzing the appropriate type of measurement according to the nature of the data and the effect of measurement errors. After selecting the measurement model, it is possible to choose proper way of analysis, or verify model applicability to the data. In accordance with the concept of "statistical methods mining", it is often necessary to combine various approaches such as transformation, adaptive methods, and computer-intensive methods described in Chap. 6 to achieve reasonable results. The formal application of statistics, or modifications of the data to the needs of statistical analysis without a deeper analysis can lead to catastrophic results here.

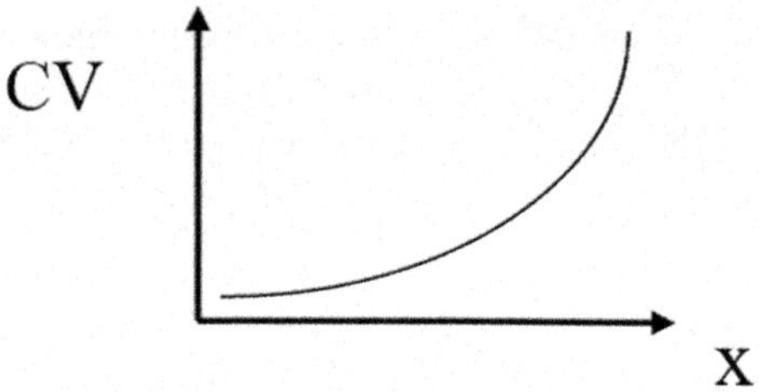

Fig. 4.1 Horowitz trumpet

References

Bickel, P. J., & Doksum, K. A. (1981). An analysis of transformations revisited. *Journal of the American Statistical Association, 76*(374), 296. https://doi.org/10.2307/2287831

Deming, S. N., Michotte, Y., Massart, D. L., Kaufman, L., & Vandeginst, B. G. M. (1988). *Chemometrics: A textbook*. Elsevier.

Meloun, M., & Militký, J. (2012). *Interactive statistical data analysis*. Carolinum. (in Czech)

Militký, J., & Meloun, M. (1999). Data treatment in trace analysis. *Conference Microelements, 99*. (in Czech)

Smothers, C. D., Sun, F., & Dayton, A. D. (1999). Comparison of arithmetic and geometric means as measures of a central tendency in cattle nematode populations. *Veterinary Parasitology, 81*(3), 211–224. https://doi.org/10.1016/S0304-4017(98)00206-4

Chapter 5
Measurement Uncertainties

The only relevant thing is uncertainty—the extent of our knowledge and ignorance. The actual fact of whether or not the events considered are in some sense determined, or known by other people, and so on, is of no consequence [Bruno deFinneti].

Abstract The measurement uncertainties in metrology are generally combining statistical characteristics of experimental results variability and mainly subjective not measurable sources of variability. The simplification of the metrological (ISO) approach and possible sources of inaccuracies due to the limiting validity of some a priori assumptions about the behavior of the data are discussed. The deep comparison of the relationships between the measurement uncertainties calculated in accordance with ISO standard and the uncertainties based on statistical data analysis are presented. A simulation-based method for calculation of uncertainties of indirect measurements is shown.

5.1 Introduction

It is known that measurement and interpretation of measurement results is the basis of both natural and technical sciences. No measurement is completely perfect, as it takes place on instruments with limited accuracy constructed according to approximate measurement principles, and a number of non-constant conditions occur during the measurement. In many cases, an integral part of the measuring chain is also a person as a source of subjectivity, or inaccuracies. In practice, therefore, measurements are burdened by a whole range of different noises, commonly referred to as errors, or systematic deviations (bias). These noises then cause the measured values to scatter and are a source of inaccuracy in the results. The method of combining individual errors is specified by the model of their action. The purpose of measurement is, in the simplest case, the determination of one (measured) quantity. The measurement results are then expressed using a suitable estimate of the true (unknown) value and the corresponding degree of uncertainty related to the error action model, or deflection. The main goal is the quantification of the measured variables and the

J. Militký et al., *Textile Material Research—Statistical Tools of Experimental Data Analysis*, https://doi.org/10.1007/978-981-97-6607-9_5

estimation of their uncertainty related to: the measurement principle, the measuring device, the operator, the variability of the measured material, the surrounding conditions, etc. The measurement results are used to estimate the unknown mean value of the measured quantity x and its uncertainty (uncertainty) related to the measurement model and external influences. In a statistical sense (where probability is understood as a limit of relative frequency), uncertainty is expressed as a confidence interval of parameters. In the metrological sense (when probability is expressed as a measure of confidence—belief), uncertainty is expressed more as a lack (limitation) of knowledge than as a result of repeated measurements. Expression of uncertainty published in manuals (Quantifying Uncertainty in Analytical Measurement, 1995; Taylor, 1994) is philosophically closer to the subjective definition of probability as a degree of trust (belief). This probability is then more related to a lack of knowledge than to the result of a repeated experiment.

The work of Geyser (Phillips et al., 1997) compares the ISO procedure with a statistical approach to creating confidence intervals. This chapter shows the relationships between the measurement uncertainties used in accordance with ISO and the uncertainties used in statistical data analysis, which are based on the construction of the measurement model. The approach presented in the EURACHEM manual (Quantifying Uncertainty in Analytical Measurement, 1995) is compared, respectively the NIST manual (Taylor, 1994), with a classical statistical approach based on the maximum likelihood method (Meloun & Militký, 2012; Meloun et al., 1992) and a Bayesian approach based on subjective probability (Agostini, 1995). A simplification of the ISO approach and possible sources of inaccuracies due to the invalidity of some a priori assumptions about the behavior of the data are shown. The reason is that in these materials there is a lot of information in the form of instructions without further clarification of the conditions under which the procedures can be safely used. A method of calculating uncertainties of indirect measurements based on the use of bootstrap simulation is proposed.

5.2 Uncertainties of Measurement Results

Uncertainty is a fairly well-known statistical concept related to parameter estimation. However, in engineering practice, there are a number of interpretations and definitions characterizing the uncertainty of the measurement result. Standard notation $x \pm c$ has a number of interpretations. Basic interpretations of the quantity c include:

(1) The half-width of the expected range determined by the experimenter's subjective estimate based on previous measurements.
(2) Half the distance between the scale divisions of the measuring instrument.
(3) Sampling standard deviation often used in physics (approximately 68% of all possible measurement results lie in this interval in the case of normality).

(4) Twice the sample standard deviation $c = 2\,\sigma$, often used in chemistry (approx. 95% of all possible measurement results lie in this interval in the case of normality).
(5) Standard deviation of the mean: $\sigma_m = \sigma/\sqrt{N}$, where N is the sample size and x is replaced by the mean $\overline{x}$.
(6) Statistical studies commonly determine the 95% confidence interval of the mean when x is replaced by the mean $\overline{x}$ and then $c \approx 2\sigma/\sqrt{N}$ (however, calibration certificates use the 99% confidence interval of the mean),
(7) Standard uncertainty of the measurement result u.
(8) Extended uncertainty of the measurement result u_c.
(9) Other types of statistical intervals as *tolerance intervals* (containing a portion of observations from the sampled population) and prediction intervals (containing future observations from sampled population) (Hahn & Meeker, 1991).

It is therefore evident that the determination of the parameter c is not unambiguous and can be a big problem, because based on the interval it is assessed whether the measurement results are significantly different or not, see. Figure 5.1. Here, A and B are significantly different and C, D are not significantly different, although their difference is the same in both cases.

Differences in the definition of some terms in statistics and in metrology are shown in Table 5.1

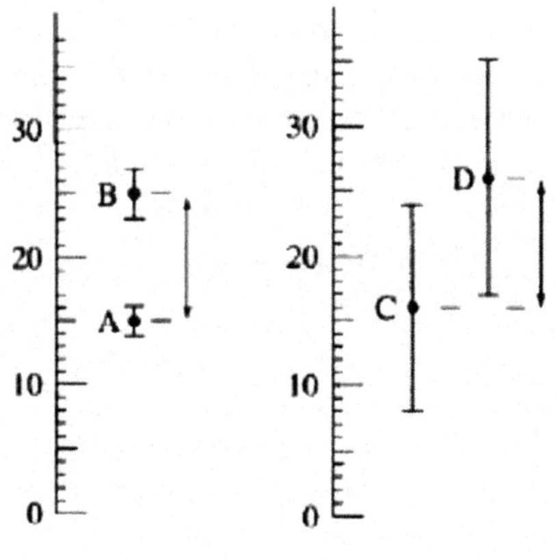

Fig. 5.1 Assessing the significance of measurement results

Table 5.1 Differences between metrology and statistics

Metrology	Statistics
Standard uncertainty A	Standard deviation of error (mastered)
Standard uncertainty B	Standard deviation of error (non-mastered)
Combain uncertainty	Standard deviation of error (combination)
Extened uncertainty	Half length of confidence interval
Cover factory	Quality of standard normal distribution

The measurement uncertainty is understood as a measure of the extent to which it is possible to estimate the true value of the measured quantity (for measurement errors, on the other hand, it is evaluated how accurately the true value of the measured quantity can be estimated). There are two basic types of uncertainties. Uncertainties of type A (marked u_{Ai}) are related to the measurement accuracy expressed by the standard deviation. They are therefore estimated based on the measurement results. Type B uncertainties (denoted u_{Bi}) are experimentally undeterminable and are related to the estimated inaccuracy caused by factors that have not been investigated (e.g. scale reading inaccuracy, container volume inaccuracy, measurement device design inaccuracy, etalon inaccuracy, etc.). For the basic additive noise model, the measurement result is in the form

$$x = \mu + \varepsilon \tag{5.1}$$

An unknown parameter μ (mean value) is a value estimated based on the results of measuring x. The random error ε has zero mean ($E(\varepsilon) = 0$) and constant variance $D(\varepsilon) = \sigma^2$. The *uncertainty of an individual measurement* (uncertainty of type A) is then expressed by the interval $x \pm u$, where u is a multiple of σ, i.e. $u = k\sigma$. When choosing $k = 1$ (see Fig. 5.2), this interval corresponds to approximately 65% of the confidence interval, and for $k = 2$, this interval corresponds to approximately 95% of the confidence interval.

The contribution of one measurement to the type A uncertainty is then $u_{Ai} = \sigma$. If averaging is performed from N measurements, the standard deviation σ is replaced by the sample standard deviation s and $u_{Ai} = s/\sqrt{N}$. In many cases the situation is more complex and some sources of error are non-random. The *extended measurement model* is better to this situation

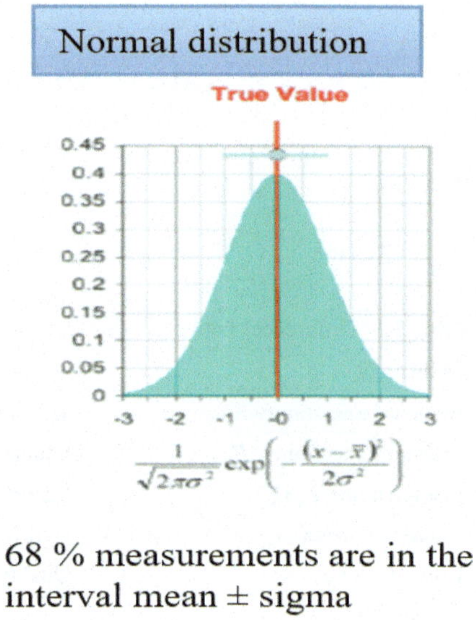

Fig. 5.2 Uncertainty of the measurement result

$$x = \mu + b + \varepsilon \tag{5.2}$$

The systematic deviation (bias) b is often only estimated based on expert estimates at a suitable interval $(a - d) \leq b \leq (a + d)$. Expert estimation of parameters a, d then enables the construction of conservative measurement uncertainties.

$$(x - a) \pm (1.96\sigma + d) \tag{5.3}$$

This so-called orthodox approach was recommended by both the American NIST and the English NPL.

ISO standards are used for evaluation of measurement uncertainties in this case the form $(x - a) \pm 1.96\sqrt{\sigma + d^2/3}$.

In metrology, B-type uncertainties are introduced. One of simplest is uncertainty due to readings on analog scales connected with the accuracy of the division (Fig. 5.3). For the fine division of the scale, the type B uncertainty corresponds to the interval between individual values. For coarse scale divisions, the type B uncertainty is half of this interval. For digital scales the type B uncertainty corresponds to the minimal difference between individual readings (Fig. 5.3).

A variety of approximate distributions are used to represent behavior of type B uncertainties. For a uniform distribution characterized by the interval $(-a,\ a)$, where errors occur with equal probability, it is $u_{Bi} = a/\sqrt{3}$ (see Fig. 5.4).

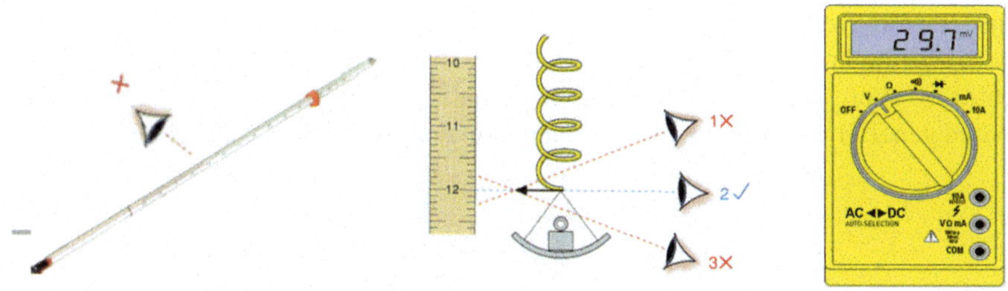

Fig. 5.3 Uncertainties in reading on scales

Fig. 5.4 Uniform distribution to calculate type B uncertainty

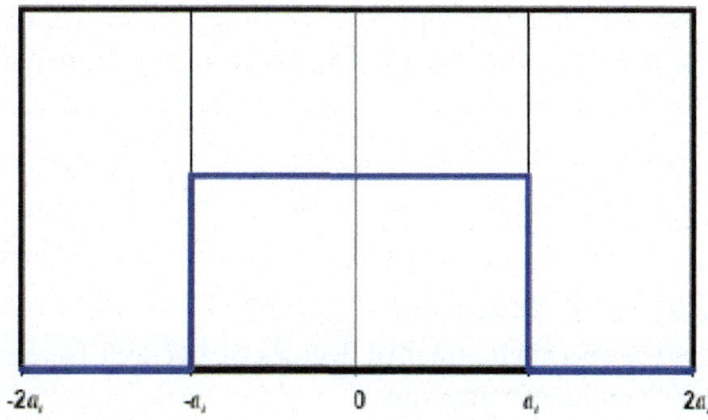

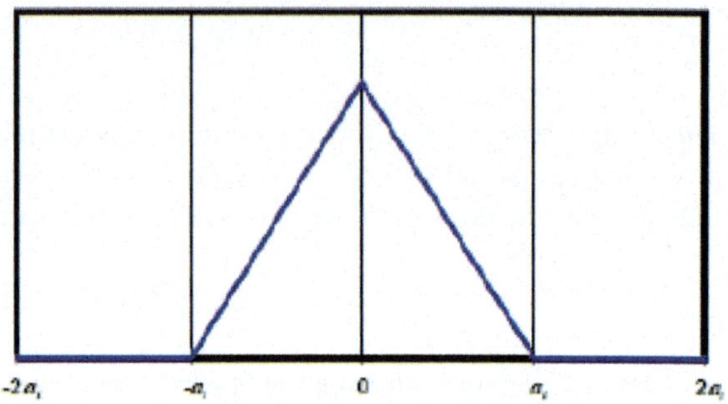

Fig. 5.5 Triangular distributions for calculating type *B* uncertainty

For the common case of a normal distribution characterized by the estimated standard deviation s is $u_{Bi} = s$.

For a triangular distribution characterized by the interval $(-a, a)$, where the probability of occurrence of errors decreases with distance from zero is $u_{Bi} = \frac{a}{\sqrt{6}}$ (see Fig. 5.5).

The uncertainty of the result, i.e. the estimate $\hat{\mu}$ of the mean value μ is expressed in statistics by the confidence interval of the mean value, for which

$$P(a^{-} \leq \mu \leq a^{+}) = 1 - \alpha \tag{5.4}$$

where calculations and interpretations of confidence limits a^{+} a^{-} is comprehensively discussed in Sect. 1.7.2.

BIMP (International Bureau of weights and measures) recommended in 1980 five rules for expressing uncertainty and based on them ISO recommended a procedure based generally on indirect measurements (in details see Sect. 5.3) when

$$\mu = f(\vec{x}, \overleftarrow{z}) \tag{5.5}$$

where $\vec{x} = (x_1, \ldots x_P)$ are measured input quantities (the corresponding uncertainties u_A of type *A* are determined by standard statistical procedures from estimates of variances $s_{x_i}^2$) and $\overleftarrow{z} = (z_1, \ldots z_R)$ are unmeasured or unmeasurable quantities. The corresponding type *B* uncertainties u_B are determined from expert estimates of variances $s_{z_i}^2$. The estimate $D(\hat{\mu})$ is calculated from the Taylor expansion to the linear terms

$$D(\hat{\mu}) = \sum_i c_i^2 s_{x_1}^2 + \sum_j e_j^2 s_{z_j}^2 \tag{5.6}$$

where $c_i = \frac{\partial f(\vec{x}, \overleftarrow{z})}{\partial x_i}$ and $e_i = \frac{\partial f(\vec{x}, \overleftarrow{z})}{\partial z_i}$.

The so-called effective degrees of freedom ν_{eff} are then used to determine $(1 - \alpha)\%$ confidence intervals

$$\nu_{\text{eff}} = \frac{D(\hat{\mu})^2}{\sum_i c_i^4 s_{\bar{x}_i}^4 / \nu_i} \tag{5.7}$$

The *uncertainty of the result* can then be determined from the relationship

$$\hat{\mu} \pm t(\nu_{\text{eff}})_{1-\alpha/2} \sqrt{D(\hat{\mu})} \tag{5.8}$$

It is useful to compare the gradual evolution of expressing the total uncertainty U as a combination of type A (statistical) and type B (non-statistical) uncertainties.

In 1985 (ASME) proposed the term

$$U = \pm\sqrt{[u_B^2 + (s_{\bar{x}}\, t_{0.95})^2]} \tag{5.9}$$

Here, point estimates and interval estimates are mixed. Finally, in 1993, ISO proposed a unified procedure leading to an interval

$$U = \pm k\sqrt{[u_B^2 + u_A^2]} \tag{5.10}$$

This definition allows the use of non-experimental information. On the other hand, the problem is how to determine the sources of uncertainty and their variability.

By default, metrology is based on the calculation of the combined uncertainty u_C as a combination of type A and B uncertainties based de facto on the assumptions of the validity of the additive measurement model and the independence of the individual uncertainties.

$$u_C = \sqrt{\sum_i u_{Ai}^2 + \sum_i u_{Bi}^2} \tag{5.11}$$

The expanded uncertainty U is then commonly defined as $U = 2\,u_C$. Formally, it is therefore the half-width of the 95% confidence interval of the mean value μ for the case of a normal distribution.

5.3 Uncertainties of Indirect Measurements

The result of the analysis in this case can be expressed as

$$y = f(\mu_1, \ldots \mu_M) \tag{5.12}$$

Here $f\ (\mu_1,\ \ldots\ \mu_M)$ is a known function of the real values of the results of direct measurements $\mu_{1,\ldots}\ \mu_M$. (For example, the radius is measured and the cross-sectional area of circular fibers is computed). The parameters estimate $(\hat{\mu}_1, \hat{\mu}_2, ...\hat{\mu}_M)$ and the corresponding estimate of the variances or squared uncertainties $D(\hat{\mu}_1), D(\hat{\mu}_2), ...D(\hat{\mu}_M)$ are available.

Standard statistical analysis is realized by the following steps (Meloun & Militký, 2012; Meloun et al., 1992).

(a) Calculation of y from estimates $\hat{\mu}_i$ $i = 1, \dots M$.
(b) Variance $D(\hat{y})$ estimation.
(c) Confidence interval of y creation.

Uncertainty analysis according to EURACHEM has the following steps (Quantifying Uncertainty in Analytical Measurement, 1995; Taylor, 1994).

(a) Calculation of y from estimates $\hat{\mu}_i$ $i = 1, \dots M$. It is not solved directly, but it is apparently assumed too approximately that $\hat{y} = f(\hat{\mu}_1, \hat{\mu}_2, ...\hat{\mu}_M)$. However, for strongly non-linear functions and higher variances, this is too approximate and it is better to use quadratic approximations (Taylor expansion into quadratic terms).
(b) Variance $D(\hat{y})$ estimate. It is actually the combined uncertainty of $u(y)$. It is based on the assumption that $f(\underline{x})$ can be replaced by a linearization based on Taylor expansion in the neighborhood of $\underline{\mu}$.

$$y = f(\underline{x}) \approx f(\underline{\mu}) + \sum_{i=1}^{M} \left(\frac{\partial f(.)}{\partial x_i} \right) (x_i - \mu_i)$$

$$u_C(y) = \sqrt{\sum_{i=1}^{M} \left(\frac{\partial y}{\partial x_i} u(x_i) \right)^2 + \sum_{i,k=1}^{M} \left(\frac{\partial y}{\partial x_i} \frac{\partial y}{\partial x_k} \operatorname{cov}(x_i, x_k) \right)^2}$$

If the uncertainty sources are linearly dependent, the correction is performed using co-variances.

Linearization can be very imprecise in many cases, especially regarding confidence intervals (expanded uncertainty).

Illustrative example of linearization inaccuracy (Agostini, 1995).

The proton energy E [GeV] can be determined from its speed c according to the relation $E = m c^2/\sqrt{1 - (v/c)^2}$. For the case that the velocity is measured with a relative accuracy of 0.2% and $v/c = 0.9971$, a 95% confidence interval is estimated.

Linearization	Correct solution
$0.7 \leq E \leq 24$	$\leq E \leq \infty$

(c) Confidence interval estimate of y: Almost always incorrectly approximate normality is assumed. (The non-linear function of normally distributed random variables no longer a normal distribution!!). Half of the 95% confidence interval, or the expanded uncertainty is then approximately $U = 2\,u(y)$. Here 2, (or more precisely 1.98) is the quantile of the normalized normal distribution. However, for a non-linear transformation, non-symmetric distributions result, leading to a non-symmetric confidence interval. In special cases (e.g. trace analysis) this can significantly affect the conclusions (for positively skewed distributions, the correct interval will be narrower in the direction of lower values and wider in the direction of higher values). Complications caused by non-linearity are mainly related to the fact that the non-linear transformation of a random variable generally changes its distribution and changes the dependence of the variance on the mean value (for a normal distribution it is independence). If the measured variable x has a constant variance $\sigma^2(x)$, the result of processing (analysis) $y = f(x)$ has a *non-constant variance*,

$\sigma^2(y) \approx \left(\frac{\mathrm{d}f(x)}{\mathrm{d}x}\right)^2 \sigma^2(x).$

Also, the mean $\overline{y}$ cannot be estimated by directly substituting the mean $\overline{x}$ into the function $y = f(x)$, i.e. generally $\overline{y} \neq f(\overline{x})$. Approximation by Taylor expansion

$$f(x) \approx f(\overline{x}) + \sum_{i=1}^{n} \frac{1}{i!} \frac{\mathrm{d}^i f(x)}{\mathrm{d}x^i} (x - \overline{x})^i$$

is valid only in a limited interval. This can be seen, for example, in Fig. 5.6, where the approximation of the function $f(x) = \exp(x)$ by Taylor expansion to degrees 0, 1, and 2 is shown.

In metrology, the product power function is commonly found: $y = (x_1^{a1}\ x_2^{a2} \ldots x_M^{aM})$, where a_1 to a_M are small integers (– 1, – 0.5, 0.5, 2, 3). For this case is a

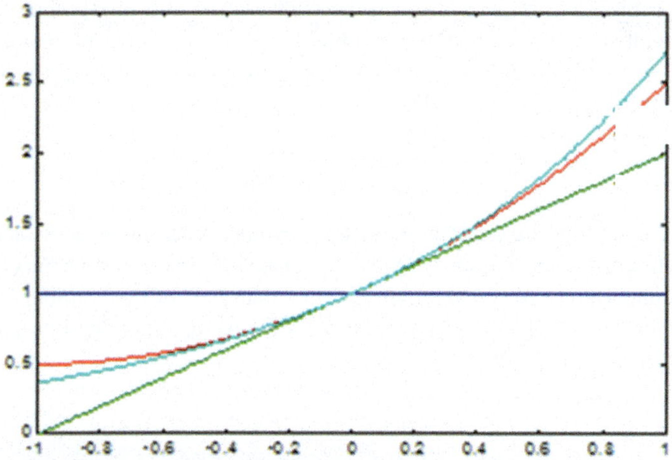

Fig. 5.6 Taylor expansion of the function $y = \exp(x)$ in the vicinity of the origin ($y = 1$ is dark blue line, $y = 1 + x$ is green line, $y = 1 + x + 0.5x^2$ is red curve, and $y = \exp(x)$ is pale blue curve)

combined uncertainty equal to

$$\mathrm{u_C}(y) = y(\overline{\mathbf{x}})\sqrt{\sum_{i=1}^{M}\left(a_i\frac{u(x_i)}{\overline{x}_i}\right)^2 + 2\sum_{i=1}^{M-1}\sum_{k>i}^{M}\left(a_i\, a_k\frac{u(x_i)}{\overline{x}_i}\frac{u(x_k)}{\overline{x}_k}\,\mathrm{r}(x_i, x_k)\right)^2}$$

where $r(x_i, x_k)$ is correlation coefficient between input quantities x_i and x_k. For the case of a linear function: $y = (x_1 + x_2 + \cdots + x_M)$ and a linearly dependent all uncertainties, the combined uncertainty is equal to

$$u_C(y) = \sum_i u(x_i).$$

5.4 Uncertainty and Statistics

Standard evaluation of uncertainties is based on the following strict assumptions without verification:

(1) validity of additive measurement model, or the effect of noise components (sources of uncertainty);
(2) constant variance of measurements (or sources of uncertainty);
(3) approximate normality of a non-linear function of normally distributed variables (to determine the extended uncertainty, or confidence interval—CI);
(4) no-correlation of measurements;
(5) small non-linearity $f(\underline{x})$ allowing the use of linearization.

Another problem is the inaccuracy in the construction and interpretation of *U* (or *CI*). Classical statistics leads to the fact that for number of measurements $N \to \infty$ the $100(1 - \alpha)\%$th confidence interval of the parameter μ is equal to (Meloun & Militký, 2012; Meloun et al., 1992):

$$\hat{\mu} \pm u_{1-\alpha/2}\sqrt{D(\hat{\mu})}$$

When calculating the uncertainties, the combined uncertainty u_c^2 is actually not just an estimate of the variance $D(\hat{\mu})$, but contains other components. Then the expanded uncertainty is *systematically higher* than half of the confidence interval, a value of 2 does not ensure approximately 95% coverage, and the interpretation of such an interval is difficult.

5.5 Bayesian Approach to Uncertainties

This approach is based on the Bayesian definition of conditional probability and uses a priori information (marked by the index o). The likelihood function is generally the joint probability density, where $\underline{x} = (x_1, \ldots x_n)$ are the measurements and $\underline{h} = (h_1, \ldots h_n)$ are the values of the influencing variables (sources of uncertainty—i.e. external unobserved parameters, systematic deviations, calibration constants, etc.). A priori information about μ is expressed by the probability density $f_o(\mu)$. According to the Bayesian formula, the posterior probability density function for μ, i.e. $f(\mu/\underline{x})$ can be expressed as

$$f(\mu/\underline{x}) = \frac{f(\underline{x}/\mu, h_o) f_o(\mu)}{\int f(\underline{x}, h_o) f_o(\mu) \mathrm{d}\mu} \tag{5.13}$$

This equation is based on the assumption that all influencing variables have values h_o. Knowing the posterior probability density $f(\mu/\underline{x})$ (which is a combination of the a priori probability $f_o(\mu)$ and the information hidden in the experiment) $f(\underline{x}/\mu, \underline{h})$, the estimates of mean are easily determined from the definition

$$\hat{\mu} = \int \mu f(\mu/\underline{x}) \mathrm{d}\mu \tag{5.14}$$

and for variance it is valid

$$D(\hat{\mu}) = \int (\mu - \hat{\mu})^2 f(\mu/\underline{x}) \mathrm{d}\mu \tag{5.15}$$

For large sample sizes, the a priori information is of little importance and is then directly proportional to the likelihood function. Maximum likelihood estimates then correspond to Bayesian estimates.

Bayesian formulas can also be used in cases where there are systematic errors as well (noise components do not have a zero mean value). Suppose that the quantity h, affecting the quantity μ has the distribution $h \sim N(h_o, \sigma_h^2)$ and the measurements x_i have the distribution $x_i \sim N(\mu, \sigma_M^2)$. Then, based on the Bayes relation, it is valid that $\hat{\mu} \approx N(\overline{x} - h_o, \sqrt{\sigma_M^2 + \sigma_h^2})$. This means that the result is corrected by the value of the systematic error and the total variance is the sum of the variance of the measurement (determinable from repeated measurements) and the variance of the influencing quantity (can be determined by other factors as device design, or simulation, etc.). Here σ_M is the type A uncertainty and σ_h is the type B uncertainty.

5.6 Simulation for Calculating Uncertainties of Indirect Measurements

As it was shown above, the accuracy of the calculation of uncertainties of indirect measurements depends on the non-linearity of the function $\mu(\mathbf{x})$ [see Eq. (5.5)]. For strongly non-linear functions $\mu = f(\overrightarrow{x}, \overleftarrow{z})$, the calculation of the corresponding variance $D(\hat{\mu}) = \sum_i c_i^2 s_{x_1}^2 + \sum_j e_j^2 s_{z_j}^2$ is strongly biased. Advantageously, in these situations, a simulation based on the ideas of Monte Carlo methods can be used (see Fig. 5.7).

If only estimates of mean values $(\hat{\mu}_1, \hat{\mu}_2, \ldots \hat{\mu}_M)$ and variances $D(\hat{\mu}_1), D(\hat{\mu}_2), \ldots D(\hat{\mu}_M)$ are known for uncertainties of type A or approximate distribution of type B uncertainties are available, the Monte Carlo method described in Sect. 3.51 can be used. It is also possible to choose more complicated schemes of action of individual uncertainties and their possible combinations (Evaluation of Measurement Data—Supplement 1 to the "Guide to the Expression of Uncertainty in Measurement"—Propagation of Distributions Using a Monte Carlo Method, 2008; Meloun et al., 1992; Ogilvie, 1984; Phillips et al., 1997).

In the case that samples from which estimates of mean values and variances have been made are available for all uncertainties of type A, the Bootstrap principle can be used to determine the empirical distribution (i.e. a special "resampling" implemented using "artificial samples", the elements of which are obtained from the original sample by random selection with return).

5.6.1 Example

Aim is to estimate the measurement uncertainty of a mercury thermometer. Sources of uncertainty—all are type B, i.e. not experimentally measured:

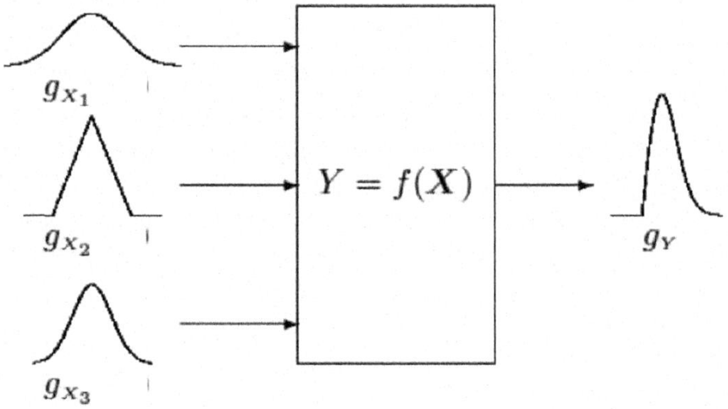

Fig. 5.7 Principle of Monte Carlo simulation

x_1 calibration inaccuracy $a = \pm 1$ °C.
x_2 design inaccuracy $a = \pm 0.1$ °C.
x temperature reading inaccuracy $a = \pm 0.25$ °C.

The first two sources are systematic for a given thermometer and cannot be reduced by repeating measurements. Let all three sources have rectangular distribution, where $u_B = 0.5774a$.

Uncertainty $u_{B1} = 0.5774 \times 1 = 0.5774$ °C.
Uncertainty $u_{B2} = 0.5774 \times 0.1 = 0.05774$ °C.
Uncertainty $u_{B3} = 0.5774 \times 0.25 = 0.14435$ °C.

Combined Uncertainty—uncorrelated sources: $u_c = \sqrt{u_{B1}^2 + u_{B2}^2 + u_{B3}^2} = 0.59796$ °C.

Combined Uncertainty—linearly dependent sources: $u_c = u_{B1} + \sigma_{B2} + \sigma_{B3} = 0.77949$ °C.

Thus, expanded uncertainties are determined for the two extreme cases of the correlation relationship between sources of measurement uncertainties.

5.7 Conclusion

It is evident that the calculation of uncertainties as proposed by ISO and EURACHEM is applicable only under special assumptions about the action of errors, the type of modeled function, and the sources of uncertainties. For more complex situations, it is always better to first find a suitable measurement model and then determine the uncertainty interval within it. Also, the problem of random and systematic non-experimental errors has not yet been satisfactorily resolved.

References

Agostini, G. D. (1995). Probability and measurement uncertainty in physics. Rept. DESY, 95–242.

Evaluation of measurement data—Supplement 1 to the "Guide to the expression of uncertainty in measurement"—Propagation of distributions using a Monte Carlo method (2008).

Hahn, G. J., & Meeker, W. Q. (1991). *Statistical intervals.* Wiley. https://doi.org/10.1002/9780470316771

Meloun, M., & Militký, J. (2012). *Interactive statistical data analysis.* Carolinum. (in Czech)

Meloun, M., Militký, J., & Forina, M. (1992). *Chemometrics for analytical chemistry* (Vol. 1). Ellis Horwood.

Ogilvie, J. F. (1984). A Monte-Carlo approach to error propagation. *Computers & Chemistry, 8*(3), 205–207. https://doi.org/10.1016/0097-8485(84)80007-8

Phillips, S. D., Eberhardt, K. R., & Parry, B. (1997). Guidelines for expressing the uncertainty of measurement results containing uncorrected bias. *Journal of Research of the National Institute of Standards and Technology, 102*(5), 577. https://doi.org/10.6028/jres.102.039

Quantifying Uncertainty in Analytical Measurement. (1995).

Taylor, B. N. (1994). *Guidelines for evaluating and expressing the uncertainty of NIST measurement results*. https://doi.org/10.6028/NIST.TN.1297

Chapter 6
Treatment of Measurement Results

Faith moves mountains, but it is impossible to determine where.

6.1 Introduction

The purpose of data analysis is to obtain relevant information from experiments and observations. The processing of measurement results is a frequent case of technical data analysis. When processing the results of routine measurements, an additive measurement model is commonly assumed. The data are a priori judged to be independent equally distributed quantities coming from a normal distribution. These assumptions are the basis of practically all classical methods of experimental data analysis. In many cases where a constant parameter is repeatedly measured under the same conditions, this approach is sufficient if a sufficient number of repetitions is ensured. For smaller samples and imprecise measurements, simple robust techniques can be used that work well if the data distribution is symmetric. It should be kept in mind that a small violation of the assumption of normality may not be catastrophic with respect to the outcome of the statistical analysis. On the other hand, however, it is bad when estimates and tests depend on other factors than on the behavior of the bulk of the data (on sample size, order of results of unobserved variables, etc.). A frequent requirement is to use measurements ($x_1 \ldots x_N$) forming in a certain sense a selection of size N to determine a suitable estimate of the mean value of μ and compare it with the specified value of μ_o. Considering the variability of the measurement, it is also advisable to check whether μ_o falls within the confidence interval CI of the μ parameter or not. Standard analysis based on the assumption of an additive measurement model and the normality of measurement errors often leads to overly optimistic conclusions. For positively skewed distributions, even though the arithmetic mean is an asymptotically unbiased estimate, it is very likely to be less than the true value of the location parameter μ. Thus, the estimate of the mean value is underestimated. Moreover, the construction of the confidence interval is only very approximate and its symmetric limits are often unusable. When analyzing special types of data, where measurement errors are negligible compared to the variability of

J. Militký et al., *Textile Material Research—Statistical Tools of Experimental Data Analysis*, https://doi.org/10.1007/978-981-97-6607-9_6

the measured material, the analyzed samples are strongly different in terms of investigated properties and the distribution of results is significantly asymmetric (usually skewed toward higher values). Then both standard and robust analysis often lead to incorrect conclusions, or exclusion of data that do not correspond to the assumption of symmetry, but are really “acceptable”. In such cases, regardless of the quality of the measurement, the quality of the data processing determines the result. If the data do not meet the assumption of normality, in many cases, it is possible to improve their distribution by a suitable transformation. It turns out that the vast majority of data analysis time is spent preparing it and verifying assumptions about its behavior (estimated at 60–90% of the time). These activities may seem less important, but as a rule, they are not so trivial and without them, it is not possible to carry out a correct statistical analysis. Problems with data acquisition, data integrity, and data quality, etc., often need to be addressed. With the increase in the availability of powerful personal computers, there is decentralization and interactivity in the processing of experimental data and the interpretation of results. This places greater demands on practitioners, who find it difficult to defend simple data evaluation procedures often based on simplified or even incorrect assumptions. The offer and possibilities of computer-oriented statistical data processing force the experimenter to a deeper analysis, which usually leads to a radical change of perspective on routinely performed research work. The tasks of evaluating experimental data in technical practice have some common features:

(a) the ranges of processed data are either small or extremely large,
(b) the distribution of the data rarely conforms to the normality assumed frequently in standard statistical analysis,
(c) outlying measurements and various heterogeneities occur in the data,
(d) statistical models are often created on the basis of preliminary information from data (data-oriented approaches),
(e) there is some uncertainty in choosing a model describing the behavior of the data,
(f) the parameters of the model describing the behavior of the data often have a physical meaning, which limits their signs, respectively range.

If the statistical analysis is not to be a mere numerical calculation without a deeper meaning, it is of course necessary that all the assumptions that led to the design of the given analysis procedure be verified. An interesting option for using standard data analysis when the basic assumptions are not met is their transformation leading to an improvement in the shape of the distribution of the sample and an approach to normality.

A problem with data analysis in general is dealing with a single finite range sample. This has the consequence that the results of the analyses are indeterminate (with respect to samples repetition) and the samples are limitedly informative (with respect to sample size). According to the approach to solving this problem, there are two basic variants:

- Data-oriented approach (faith in data)
- Model-oriented approach (faith in the model).

Advantageously, the principles of Bootstrap methods can be used to analyze the effect of imperfect selection on the results of data processing. The data-oriented approach results in the generation of simulated selections based on the classic non-parametric Bootstrap. The model-oriented approach uses the parametric Bootstrap. This chapter discusses routine data analysis procedures, exploratory analysis, data transformation, and computer-intensive methods for identifying and limiting subproblems associated with the distribution of measurement results.

6.2 Methods of Data Processing

Routine processing of experimental data usually involves:

A. descriptive analysis (estimation of location, dispersion, and shape parameters),
B. determination of confidence intervals,
C. parameter significance testing.

The classic solution is given, for example, in (Meloun & Militký, 2012).

Two techniques are commonly used in descriptive data analysis:

(a) *classic*, when the arithmetic mean, variance, skewness, and kurtosis are calculated according to known relationships. Moments (analogy of defining relations) are used to determine selection characteristics.
(b) *exploratory*, when robust characteristics of the position (median) of dispersion, skewness, and kurtosis are used (see Sect. 6.5). Individual techniques usually assume only the independence of the sample elements and use *quantile characteristics*.

In classic descriptive data analysis, it is assumed that:

- the individual sample elements are mutually independent and have the same probability of being selected,
- the selection is homogeneous, i.e. they all come from the same distribution $f(x)$,
- the distribution of $f(x)$ is Gaussian-normal (which is especially advantageous for further stages of data processing).

When processing the results of routine measurements, an additive measurement model is commonly assumed. The data (x_i), $i = 1, \dots N$ are often a priori judged to be independent equally distributed quantities coming from the normal distribution $N(\mu, \sigma^2)$. These assumptions are the basis of practically all classical methods of descriptive analysis of experimental data.

If the data do not meet the assumption of normality, it is possible to improve their distribution by a suitable transformation. Finally, it is also possible to construct empirical probabilistic models that allow data processing according to their behavior and not a priori assumptions.

In a number of cases, the range of analyzed data varies within several orders of magnitude, which limits the use of standard statistical methods based on the assumption of constant variance or additive measurement model (Meloun & Militký, 2012).

In the work (Militký & Meloun, 1999), the possibilities of using a transformation stabilizing the variance or a multiplicative measurement model were discussed. This leads to a logarithmic transformation of the data (Meloun & Militký, 2012). A procedure combining both measurement models and removing their disadvantages was proposed (Militký & Meloun, 1999). Although the multiplicative measurement model leads to the use of an asymmetric lognormal distribution, it is far from universal.

One of the more general procedures for eliminating asymmetry is a suitable ordinary power transformation of the data (Meloun & Militký, 2012). Even here, however, problems arise, especially with reverse transformation and applicability only for some tasks. It can be deduced that for small variances σ^2, the estimate of the power-law transformation parameter is difficult to identify (Bickel & Doksum, 1981).

In the case that data from different samples are analyzed, it is commonly assumed that the measurement errors are negligible in comparison with the variability of the measured material.

A significant bias of the sample mean–variance can occur when the data are not independent. This can be a situation where samples for analysis are taken from different locations that are somehow related to each other (spatial or temporal autocorrelation). In the case of the simplest first-order autocorrelation expressed by the autocorrelation coefficient ρ, the variance of the sample mean will increase.

$$D(\overline{x}) = \frac{\sigma^2}{N\left(1-\rho^2\right)} \tag{6.1}$$

For more complicated situations (long-range spatial dependence), the distortion caused by the dependence in sampling positions can be disproportionately high. Routine statistical analysis is based on the estimates of $\overline{x}$, s^2 calculated from the relationships given in Sect. 1.4

The basic role in the interpretation of measurement results is played by the $100(1 - \alpha)\%$ confidence interval of the mean value, described in Sect. 1.7.2. By default, a 95% confidence interval is constructed, which has, in the case of a normal distribution of errors or measurements, form

$$\overline{x} - t_{1-\alpha/2}(N-1)\frac{s}{\sqrt{N}} \leq \mu \leq \overline{x} + t_{1-\alpha/2}(N-1)\frac{s}{\sqrt{N}} \tag{6.2}$$

where $t_{1-\alpha/2}(N - 1)$ is the quantile of the Student distribution with $N - 1$ degrees of freedom. For larger samples, this quantile is commonly replaced by the quantile of the standard normal distribution. For non-normal distributions, the estimates of $\overline{x}$, s^2 no longer have optimal statistical properties, and the confidence interval defined by Eq. (6.2) is not reasonably applicable. For asymmetric distributions of data, the interval (6.2) is inappropriate because it is symmetric. In addition, it will no longer be the case that it is $100(1 - \alpha)\%$.

When the above standard procedure cannot be used for real data, there are basically three ways:

I. Finding a suitable distribution for original data and constructing special confidence intervals.
II. Improving the distribution of the original data so that standard analysis can be applied to the transformed data
III. The use of computer-oriented procedures where "artificial samples" with the same statistical behavior are generated, and then based on the central limit theorem, the average of these samples is worked with.

Each way has its advantages and disadvantages. It is logical to use the possible improvement of data distribution by a suitable transformation, especially in cases where the goal is only to determine the confidence interval of the location parameter and not to construct probabilistic models. Moreover, it is very easy to determine whether this improvement is statistically significant or not. On the other hand, however, one transformation may not be suitable for all data and problems arise if it is necessary to implement a reverse transformation. Advantageously, the generation of "artificial samples" (Bootstrap) can be used for a deeper assessment of the quality of the transformation.

6.3 Data Quality Validation

The extent of the sample size, the independence of the elements of the sample, the normality, and the homogeneity of the sample are commonly assessed to evaluate data quality. A more detailed overview of suitable algorithms and methods can be found in (Meloun & Militký, 2012). When analyzing data on a computer, the output reports should include information on test results from the above groups. During interactive analysis, based on partial tests, either data adjustments can be made, or ideas about their distribution can be revised (modification of the statistical model).

6.3.1 Determination of the Minimum Sample Size

The sample size N primarily affects the accuracy of location and dispersion estimates (the variances of these estimates are a function of N^{-1}). Indirectly, the sample size is reflected in the construction of confidence intervals (the larger N, the narrower the confidence interval) and in hypothesis testing (the larger N, the lower the probability of a type II error and the greater the power of the test—see Sect. 1.7.3). For extremely small samples, it even happens that the width of the confidence interval and the results of the tests are more affected by the value of N than by the variability of the data. When determining a sufficient sample size, various criteria related to requirements for confidence intervals, hypothesis tests, or other statistical characteristics are used. If it is verified that the sample comes from a normal distribution, it is possible to require that the arithmetic mean lies with probability $(1 - \alpha)$ in the range $[\mu - d, \mu + d]$.

The following relationship can then be used to determine the minimum sample size N_m

$$N_m = s^2 \left(t_{1-\alpha/2}(N_1 - 1)/d\right)^2 \tag{6.3}$$

Here $t_{1-\alpha/2}(N_1 - 1)$ is the quantile of the Student distribution with $N_1 - 1$ degrees of freedom, and s^2 is the variance estimate obtained from N_1 preliminary experiments. Another option is to require that the mean length of the confidence interval be $E(\mathrm{CI}) = k\sigma$, where k is the chosen number. A rather complicated expression then applies to the selection of sample size N_m

$$N_m = \frac{2\sqrt{2}}{\sqrt{k(N_1 - 1)}} \frac{\Gamma(0.5\,k)}{\Gamma(0.5(N_1 - 1))} t_{1-\alpha/2}(N_1 - 1) \tag{6.4}$$

where $\Gamma(x)$ is the gamma function. If the variance s^2 was estimated from a larger number of N_1 measurements (i.e. $N_1 - 1$ is large), the following approximations can be used.

$$N_m = \left[\frac{2}{k}\left(1 - \frac{(N_1 - 1)}{4}\right) t_{1-\alpha/2}(N_1 - 1)\right]^2 \tag{6.5}$$

If the purpose is to perform significance tests, it is possible to choose the minimum sample size N_m so that the size of the error of the first type α and also the size of the error of the second type β are preserved. Thus, if the data come from a normal distribution and the hypothesis H_0: $\mu = \mu_0$ is tested against the hypothesis H_A: $\mu > \mu_0$ (where $\mu_0 = u_p$ is expressed as the 100 P% quantile), the minimum sample size can be determined from the relation

$$N_m = \left(\frac{u_\alpha + u_\beta}{u_P}\right)^2 + 0.5u_\alpha^2 \tag{6.6}$$

where u_z is 100 z% quantile of the standard normal distribution. Thus, if $\alpha = 0.05$, $\beta = 0.1$ and $P = 0.3$, Eq. (6.6) yields $N_m = 33$. Equation (6.6) can also be used for other distributions if the quantiles are replaced by the standard quantiles of the respective distribution. In many cases, a minimum sample size of N_m is determined so that the coefficient of variation of the standard deviation $v(s)$ has a prescribed value. Then it is valid.

$$N_m = \frac{g_2 - 1}{4\,v^2(s)} + 1 \tag{6.7}$$

Here, g_2 is the kurtosis of the population from which the selection comes (for a normal distribution, $g_2 = 3$). The coefficient of variation $v(s) = 0.1$ is usually chosen. Then from Eq. (6.7) $N_m = 50$ is obtained for samples from the normal distribution. For the Laplace distribution ($g_2 = 6$), the minimum sample size is $N_m = 125$. For a uniform distribution with extremely short ends ($g_2 = 1.8$), $N_m = 21$. This is apparently the cutoff size for samples from symmetric distributions.

It follows from the above that the selection sample sizes from (3–5), 10 to 15, which are common in the processing of experimental data in material science, are statistically insufficient for the construction of confidence intervals and the implementation of significance tests.

The results here will therefore depend on the scope of the selection. This is why it is recommended to always use appropriate robust estimation methods for these extremely small samples (which in these cases also have the greatest effect in eliminating the effect of biased measurements).

6.3.2 Verification of Independence of Sample Elements

The basic assumption of quality measurements is that individual results are mutually independent (measurement errors are independent random variables). The dependence of the measurement results (and thus the selection of elements) can be caused mainly by the following factors:

- instability of the measuring device (e.g. time drift caused by a change in the state of the measuring device during longer operation),
- non-constant measurement conditions (which can change either in leaps or gradually),
- neglecting factors that significantly influence the results of experiments (e.g. sample volume),
- incorrect selection of samples for measurement (non-random selection).

It is obvious that it is not always possible to identify the dependence of measurement results, or heterogeneity of the selection (caused, for example, by a sudden change in the conditions of the experiments). An indication of the inhomogeneity of the selection is usually a higher variance. If it is suspected that the experimental results are dependent on some specific factor (time, sample volume, etc.), it is often possible to estimate this dependence directly from the scatter diagram (the y-axis shows the measurement results x_i and the corresponding x-axis values of the considered factor, such as time, sample volume, humidity, etc.). The appearance of a non-random pattern is an indication that the selection is not independent. To process such data, either regression methods are used (if the sample size N is sufficient and the factor on which the measurement result x_i depends varies within a sufficient range), or new experiments are performed. A common type of dependence is *time dependence* (usually another factor, such as temperature or humidity, can change systematically with time). If the data are recorded in the order in which they were measured, it is possible to test the dependence of successive sample elements using the autocorrelation coefficient. It is usually based on von Neumann statistics (Meloun & Militký, 2012).

$$T = \frac{\sum_{i=1}^{N-1} (x_{i+1} - x_i)^2}{\sum_{i=1}^{N} (x_i - \overline{x})^2} \tag{6.8}$$

It is valid that the quantity $T^* = 1 - T/2$ is closely related to the first-order autocorrelation coefficient ρ_1 (so the hypothesis H_0: $\rho_1 = 0$ can be tested against the alternative H_A: ρ_1 # 0). It is easy to find that $-1 \leq T^* \leq 1$. For large samples, asymptotic relations hold for the mean $E(T^*)$ and variance $D(T^*)$.

$$E(T^*) = 0 \quad D(T^*) = (N-2)/(N^2-1) \tag{6.9}$$

Thus, for larger samples, the quantity $q = T^* \sqrt{(N^2 - 1)/(N-2)}$ has a standard normal distribution. If $|q| > u_{1-\alpha/2}$, where $u_{1-\alpha/2}$ is the quantile of the standard normal distribution, H_0 is rejected, and the sample is not considered to be composed of independent elements. For practical purposes, modified statistics are used.

$$t_N = \frac{T_1 \sqrt{(N+1)}}{\sqrt{(1 - T_1^2)}} \quad \text{kde} \quad T_1 = T^* \sqrt{\frac{(N^2 - 1)}{(N^2 - 4)}} \tag{6.10}$$

It is true that, under the validity of the H_0 hypothesis, t_N has an approximately Student's distribution with $N + 1$ degrees of freedom. If it will be $|t_N| > t_{1-\alpha/2}(N+1)$ (where $t_{1-\alpha/2}(N + 1)$ is the quantile of the Student's distribution with $N + 1$ degrees of freedom), the hypothesis of independence of the sample elements is rejected at level α. A number of other tests for verifying the randomness of the data series (x_i), $i = 1 \ldots N$, (against the hypothesis of autocorrelation of different orders) are based directly on the autocorrelation coefficients.

$$\rho_k = \frac{\sum_{i=1}^{N-k} (x_i - \overline{x})(x_{i+k} - \overline{x})}{\sum_{i=1}^{N} (x_i - \overline{x})^2} \tag{6.11}$$

For $1 \leq k \leq N - 1$, if the sample elements are independently identically distributed random variables with finite variance, the asymptotic distribution of the variable $\rho_k \sqrt{N}$ is standard normal $N(0, 1)$. For sufficiently large samples, the significance of individual autocorrelation coefficients can therefore be tested very easily. If $\left|\rho_k \sqrt{N}\right| \leq u_{1-\alpha/2}$, the kth-order autocorrelation is negligible at the α significance level.

For the mean value $E(\rho_k)$, it is true that

$$E(\rho_k) = -\frac{N-k}{N(N-1)} \tag{6.12}$$

Assuming normality, the variance $D(\rho_k)$ can be expressed in the form

$$D(\rho_k) = \frac{N^4 - (k+3)N^3 + 3kN^2 + 2k(k+1)N - 4k^2}{(N+1)(N-1)^2N^2} \tag{6.13}$$

A more accurate significance test of the H_0: $\rho_k = 0$ is then based on the assumption that the random variable

$$u_k = \frac{\rho_k - E(\rho_k)}{\sqrt{D(\rho_k)}} \tag{6.14}$$

has a standard normal distribution $N(0,1)$. To verify the randomness of the selection, it is necessary to test the significance of all ρ_k $(k = 1, \ldots N-1)$. It is more advantageous to use summary statistics, which are a combination of all autocorrelation coefficients. A simple statistic of this type has the form

$$Q_1 = N \sum_{k=1}^{m} \rho_k^2 \tag{6.15}$$

where m is the number of autocorrelation coefficients that are simultaneously tested.

Assuming randomness of selection, the quantity Q_1 has an asymptotically $\chi^2(m)$ distribution. However, for medium sample sizes, this approximation is quite imprecise. Another option is to use normalization transformations and statistics.

$$Q_2 = \sum_{k=1}^{m} u_k^2 = \sum_{k=1}^{m} \frac{(\rho_k - E(\rho_k))^2}{D(\rho_k)} \tag{6.16}$$

This statistic also has an asymptotically $\chi^2(m)$ distribution. However, it is preferable to Q_1 for smaller samples. Both of these statistics assume of an approximately normal distribution of the data.

To ensure the robustness of the Q_2 statistic, it is advantageous to replace the quantities x_i with their ranks $R(x_i)$, which are actually the indices (i) of the corresponding order statistics $\mathbf{x}(i)$. Analogous to the original sample elements, ordinal *rank autocorrelation coefficients* can be defined by relation.

$$P_k = \frac{\sum_{i=1}^{N-k} \left(R(x_i) - \overline{R}\right)\left(R(x_{i+k}) - \overline{R}\right)}{\sum_{i=1}^{N} \left(R(x_i) - \overline{R}\right)^2} \tag{6.17}$$

where

$$\overline{R} = \frac{\sum_{i=1}^{N} R(x_i)}{N} = \frac{N+1}{2} \tag{6.18}$$

It is true, that

$$\sum_{i=1}^{N} \left(R(x_i) - \overline{R}\right)^2 = \frac{N\left(N^2 - 1\right)}{12} \tag{6.19}$$

For the mean $E(P_k)$, Eq. (6.18) holds, but for the variance $D(P_k)$, the relation (6.20) must be used.

$$D(P_k) = \frac{5N^4 - (5k+9)N^3 + 9(k-2)N^2 + 2k(5k+8)N + 16k^2}{5(N-1)^2 N^2 (N+1)} \tag{6.20}$$

A robust test statistic based on rank autocorrelation coefficients then has the form

$$Q_3 = \sum_{k=1}^{m} \frac{[P_k - E(P_k)]^2}{D(P_k)} \tag{6.21}$$

The asymptotic distribution of the Q_3 statistic is again $\chi^2(m)$.
Instead of statistics in the form of sum of squares, the Q_4 statistic can be used

$$Q_4 = \sum_{k=1}^{m} (\rho_k / k) \tag{6.22}$$

for which the mean value of $E(Q_4)$ is in the same form as Eq. (6.12). For the variance $D(Q_4)$, it is valid.

$$D(Q_4) = \sum_{j=1}^{m-1} \sum_{k=1}^{m-1} \left[\mathrm{cov}\left(\rho_j, \rho_k\right) / (j\,k)\right] \tag{6.23}$$

Assuming the normality of the sample, it is

$$\mathrm{cov}\left(\rho_j, \rho_k\right) = \frac{2\left[kj(N-1) - (N-k)\left(N^2 - j\right)\right]}{(N-1)^2 N^2 (N+1)} \tag{6.24}$$

where, $1 \leq k \leq j \leq N - 1$. Again, $(Q_4 - E(Q_4))/D(Q_4)$ has an approximately standard normal distribution. Thus, autocorrelations of arbitrary orders can be tested simultaneously using these statistics, and then more thoroughly examine the randomness of the sample.

6.3.3 Normality Verification

Approximate normality of the sample distribution or the normality of the population from which the sample was made is one of the most important assumptions about the data. It is base of a number of significance tests, outlier tests, and tests of sample randomness. So-called *directional tests* are used to test different types of deviations from normality (which are pre-specified by the alternative hypothesis H_A). These are always optimal with respect to the chosen alternative (e.g. H_A: "the distribution is skewed" or H_A: the distribution is more peaked").

If the type of deviations from normality cannot be specified a priori, so-called *omnibus tests* are used. For these tests, the alternative hypothesis is H_A: "the distribution of the data cannot be assumed to be normal". An overview of selected tests of both groups is presented in the paper (Meloun & Militký, 2012). Only some simpler or generally more suitable tests are described here.

It should be noted that in many cases, it is more advantageous to use special types of graphs to verify normality, which also allow the identification of outliers, or assessment of data behavior in the tail region. Standard for checking normality are special *Q-Q* graphs for normal distribution named *rankit graphs.*

Rankit graphs often better reveal what causes deviations from normality and are sufficiently sensitive (see Sect. 6.7). A disadvantage of *Q-Q* graphs, especially for small samples, is the wavy character caused by autocorrelation between order statistics.

A. Tests based on descriptive statistics

Since the skewness g_1 of the normal distribution is known to be zero and the kurtosis g_2 is equal to 3, it is possible to test how much the sample skewness $\hat{g}_1$ and kurtosis $\hat{g}_2$ differ from these values. Assuming normality, the first four moments of the sample skewness $\hat{g}_1$ are

$$E(\hat{g}_1) = 0 \quad D(\hat{g}_1) = \frac{6}{N}\left(1 - \frac{3}{N} + \frac{6}{N^2} - \frac{15}{N^3} + \cdots\right) \tag{6.25}$$

$$g_1(\hat{g}_1) = 0 \quad g_2(\hat{g}_1) = 3 + \frac{36}{N} - \frac{864}{N^2} + \frac{12096}{N^3} \cdots \tag{6.26}$$

Similarly, the moments for sample kurtosis $\hat{g}_2$ are

$$E(\hat{g}_2) = \frac{3(N-1)}{(N+1)} \quad D(\hat{g}_2) = \frac{24}{N}\left(1 - \frac{15}{2N} + \frac{271}{8N^2} - \frac{2319}{16N^3} + \cdots\right) \tag{6.27}$$

$$g_1(\hat{g}_2) = \frac{216}{N}\left(1 - \frac{29}{N} + \frac{519}{N^2} - \frac{7637}{N^3} + \cdots\right)$$
$$g_2(\hat{g}_2) = 3 + \frac{540}{N} - \frac{20196}{N^2} + \frac{470412}{N^3} \cdots \tag{6.28}$$

Here $g_1(.)$ is the skewness and $g_2(.)$ is the kurtosis of the given estimate. It follows from these relationships that the sample skewness has a symmetric distribution, which is already close to normal for smaller samples.

The distribution of sampling kurtosis is asymmetric and approaches the normal distribution very slowly (up to $N \geq 1000$). Any system of empirical densities can be used to approximate the distribution of $\hat{g}_1$ and $\hat{g}_2$. A combination of skewness and kurtosis can be used as a suitable omnibus test. Relatively simple statistics have the form

$$C_1 = N\,\hat{g}_1^2/6 + N(\hat{g}_2 - 3)^2/24 \tag{6.29}$$

which has a roughly $\chi^2(2)$ distribution.

Other tests from this group include the ratio of absolute deviation and root mean square deviation. The test statistic has the form

$$C_2 = \frac{\sum_{i=1}^{N} |x_i - \overline{x}|}{\sum_{i=1}^{N} (x_i - \overline{x})^2} \tag{6.30}$$

Assuming normality, these relationships can be derived for the moments of C_2 statistic

$$E(C_2) = \sqrt{2/\pi} + 0.1995/N + 0.0249/N^2 - 0.0312/N^3 - \cdots \tag{6.31}$$

$$D(C_2) = 0.0451/N - 0.1246/N^2 + 0.0849/N^3 + 0.0063212/N^4 + \cdots \tag{6.32}$$

$$g_1(C_2) = \left(-1.7618/\sqrt{N}\right)\left(1 - 2.3681/N - 8.8646/N^2 + \cdots\right) \tag{6.33}$$

$$g_2(C_2) = (3 + 5.441/N)(1 - 7.628/N + \cdots) \tag{6.34}$$

For testing purposes, it is commonly assumed that C_2 has an asymptotically normal distribution $N\left(\sqrt{2\pi},\ 0.0451/N\right)$.

B. Tests based on *Q*-Q graphs.

Among the basic ones from this group is the Shapiro-Wilk *W* test, which is also specified in Czech standard ČSN 01 0255. The test statistic of the *W* test has the form

$$W = \frac{\left[\sum_{i=1}^{N} a_i x_{(i)}\right]^2}{\sum_{i=1}^{N} (x_i - \overline{x})^2} \tag{6.35}$$

The values a_i are the normalized coefficients related to the expected values of the standardized order statistics $E(x_{(i)}) \approx u_{P_i}$, where $P_i = (i - 0,375)/(N + 0,25)$ and

u_{P_i} are the quantiles of the standard normal distribution. Knowing u_{P_i} the coefficients a_i(i = 1, … N) can be determined approximately by this procedure.

1. Calculation of non-normalized coefficients $\hat{a}_i^*$ defined by relations

$$\hat{a}_i^* = \sqrt{\frac{\hat{a}_1^2}{1 - 2\hat{a}_1^2} \sum_{i=2}^{N-1} \hat{a}_i^2} \quad i = 1, \dots N \tag{6.36}$$

where is

$$\hat{a}_i = 2\, u_{P_i} \quad i = 2, \dots N - 1 \tag{6.37}$$

elsewhere

$$\hat{a}_1^2 = \hat{a}_N^2 = g(N-1) \text{ pro N} \leq 20 \text{ a } \hat{a}_1^2 = \hat{a}_N^2 = g(N) \text{ jinde} \tag{6.38}$$

Function g (N) is defined as

$$g(N) = \frac{\Gamma(0,5\ (N+1))}{\sqrt{2\,\Gamma(0,5N+1)}} \tag{6.39}$$

In Eq. (6.39), $\Gamma(x)$ is a gamma function. In the range $7 \leq N \leq 2000le2000$, it can be sufficiently accurately approximated by Stirling's formula, which, after substituting into Eq. (6.39), leads to the relation

$$g(N) = \frac{6N+7}{6N+13} \sqrt{\frac{\exp(1)}{N+2} \left[\frac{N+1}{N+2}\right]^{N-2}} \tag{6.40}$$

2. Normalization, when it is valid

$$a_i = \frac{\hat{a}_i^*}{\sqrt{\sum_{i=1}^N \hat{a}_i^{*2}}} \tag{6.41}$$

The critical values of W statistics are tabulated in e.g. in Czech standard ČSN 01 0225 for $3 \leq N \leq 50$.

However, from a computational point of view, it is more advantageous to perform a normalization transformation

$$y = (1 - W)^\lambda \qquad Z = \left(y - \mu_y\right)/\sigma_y \tag{6.42}$$

Using this transformation, the domain of applicability of the W test can be extended to $7 \leq N \leq 2000$, where the random variable Z (assuming the data come from a normal distribution) has a standard normal distribution $N(0, 1)$. Therefore, if

$|Z| > u_{1-\alpha/2}$, the hypothesis H_0 is rejected at the significance level α and the sample is not considered to come from a normal distribution.

Approximate formulas were found to determine the parameters $\lambda,\ \mu_y, \sigma_y$, which are a function of sample size [47]

$$\lambda = \sum_{i=0}^{6} C_i(\ln N - d)^i \ \ln(\mu_y) = \sum_{i=0}^{6} C_i'(\ln N - d)^i \ \ln(\sigma_y) = \sum_{i=0}^{6} C_i''(\ln N - d)^i \tag{6.43}$$

where for $N \leq 20$ is $d = 3$ and for higher *íN* is $d = 5$ (Royston, 1982).

Using a computer, it is therefore very simple to calculate the quantity Z and compare it with the corresponding quantile of the standardized normal distribution.

A simplified version of the W test does not require any special coefficients and has the form

$$D = \frac{\sum_{i=1}^{N} (i - 0.5(N+1))\, x_{(i)}}{N \sum_{i=1}^{N} (x_i - \overline{x})^2} \tag{6.44}$$

Assuming normality, the mean value of the D statistic is equal to

$$E(D) = \frac{(N-1)\,\Gamma(0.5\,(N-1))}{2\sqrt{2N\pi}\ \Gamma(N/2)} \approx \frac{1}{2\sqrt{\pi}} \tag{6.45}$$

The asymptotic variance of the D statistic has the form

$$D(D) = \frac{12\sqrt{3} - 37 + 2\pi}{24\pi N} \approx 8.99110^{-4}/N \tag{6.46}$$

Again, standardization can be used for larger sample sizes

$$C_3 = (D - E(D))/\sqrt{D(D)} \approx^{\circ} 33,349(D - 0.282)\sqrt{N} \tag{6.47}$$

The statistics C_3 has an asymptotically standard normal distribution $N(0, 1)$. If the distribution from which the sample comes has long tails, the quantity C_3 is negative (and, conversely, is strongly positive for distributions with short tails).

A measure of linearity in rankit *Q-Q* plots is the Shapiro-Francis statistic W_s. It is actually the correlation coefficient between $x(_i)$ and the expected values $E(x(_i))$ of the *i*-th order statistic of the standard normal distribution. If u_{P_i} is used (as with the W test), it results in the Weisberg and Bingham statistic, which has the same properties as W.

Normalization transformations of the type (6.42) can also be used for the W test. It is also possible to make a direct comparison of the sample quantiles with the quantiles of the standard normal distribution $\Phi^{-1}(P)$. For the test of hypothesis: $F_e \approx^{\circ} N\left(\overline{x}, s^2\right)$ two test statistics are calculated.

$$T_1 = \left(\sum_{i=1}^{N} \left| Z_i - \Phi^{-1}\left(\frac{i}{N+1}\right) \right| \right) / N \tag{6.48}$$

and

$$T_2 = \left(\sum_{i=1}^{N} \left(Z_{(i)} - \Phi^{-1}\left(\frac{i}{N+1}\right) \right)^2 \right) / N \tag{6.49}$$

where

$$Z_i = (x_i - \bar{x})/s \ \text{ a } \ Z_{(i)} = \left(x_{(i)} - \bar{x}\right)/s$$

Hypothesis H_0 is rejected if it is $T_1 \geq K_{1-\alpha}$ resp. $T_2 \geq Q_{1-\alpha}$
For the K_p quantiles, the following relation can be used

$$K_p = a_p + b_p/N + c_p/N^2 \tag{6.50}$$

where for $P = 0.95$, $a_p = 0.7105$, $b_p = -0.1751$, and $c_p = 0.0108$.
Similarly, an approximate relationship holds for Q_p

$$Q_p = a_p + b_p \ln N + c_p (\ln N)^2 \tag{6.51}$$

where for $P = 0.95$, $a_p = 0.0178$, $b_p = 2.8736$, and $c_p = -8.2894$.

C. Tests based on the empirical distribution function

These tests generally compare the degree of agreement between the empirical distribution function of F_e (sample) and the selected theoretical distribution function F.

Of the tests in this group, the *Anderson–Darling* test is the best for testing normality. The null hypothesis is again of the form H_0: $F_e \approx^\circ N\left(\bar{x}, s^2\right)$. The test statistic can be expressed in the form.

$$\mathrm{AD} = \left[\sum_{i=1}^{N} (2i - 1)\left(\ln Z_{(i)} + \ln\left(1 - Z_{(N-i+1)}\right)\right) \right] / N \tag{6.52}$$

Here

$$Z_{(i)} = \Phi\left(\frac{x_{(i)} - \bar{x}}{s}\right) \tag{6.53}$$

where $\Phi(x)$ is the distribution function of the standard normal distribution.

The null hypothesis is rejected if $\mathrm{AD} > D_{1-\alpha}$, where the approximate relation holds for the quantiles $D_{1-\alpha}$

$$D_p = a_p\left(1 + b_p/N + c_p/N^2\right) \tag{6.54}$$

where for $P = 0.95$ $a_p = 1.0348$, $b_p = -\,1.013$, $c_p = -\,0.93$ (valid for large N).

A number of other tests of this type are described in the book (Mardia, 1980).

D. Tests using transformations

Procedures using Box-Cox transformations can be included in this group. The verification of normality is that the most plausible estimate of the transformation parameter λ is equals to one.

The test proposed by Thomas and Pierce is based on the fact that a random variable transformed by its own distribution function has a uniform distribution. The null hypothesis is again $H0$: $F_e \approx^\circ N\left(\overline{x}, s^2\right)$. The test statistic has the form

$$W_k = po + 27.3809\left[\sum_{i=1}^{N}\left(Z_i^2 - 1/3\right) - \sum_{i=1}^{N}\left(Z_i - 0.5\right)\right]^2 /N \tag{6.55}$$

where

$$po = 16.3172^2\left[\sum_{i=1}^{N}\left(Z_i - 0.5\right)\right]^2 \tag{6.56}$$

The W_k statistic has a $\chi^2(2)$ distribution when H_0 valid.

The comparison of the appropriateness of individual tests shows that the best omnibus tests include Shapiro–Wilk W, Anderson–Darling AD, and the combined skewness and kurtosis test. In summary, information on various simulation experiments for comparing normality tests is given in the book (Mardia, 1980).

6.3.4 Verification of Sample Homogeneity

If the data were obtained by measurement, inhomogeneities in the sample arise as a result of:

- non-compliance with the same measurement conditions
- unsuitability of the measuring device
- incorrect selection of samples.

If the accuracy of the measurement is known, expressed by e.g. the value of the variance, a significantly greater variance of the measured results is always a sign of the heterogeneity of the sample. The processing of inhomogeneous samples is very common in practice, especially as a consequence of the inhomogeneity of the measured material. (A compromise must be chosen when specifying the concept of a *population* so that the inhomogeneity in the sample is not too high).

In many cases, on the basis of a logical analysis of the problem, different *subgroups* of data can be specified (usually according to some objective criteria such as the source of obtaining the sample, their size, etc.), which are assumed to be already homogeneous. Then it is possible to test whether individual subgroups differ in mean values and variances. If significant deviations are found, individual subgroups should be treated separately.

Consider the case where we assume that the data is apparently to be divided into two groups

$$J_1 \sim^{\circ} N(\overline{x}_1, s_1^2) \text{ and } J_2 \sim^{\circ} N(\overline{x}_2, s_2^2).$$

There is a total of N_1 data in J_1 group and $N_2 = N - N_1$ data in J_2 group. A test statistic Z can be used to test the significance of dividing the selection elements into groups.

$$Z = N \ln s^2 - N_1 \ln s_1^2 - N_2 \ln s_2^2 \tag{6.57}$$

Here, $N = N_1 + N_2$ is the total sample size and s^2 is the sample variance corresponding to the entire N-data sample. The Z statistic has a $\chi^2(2)$ distribution.

If $Z > \chi^2_{1-\alpha}(2)$, the division into groups is considered statistically significant. It is obvious that the division into groups should have a logical justification.

In some cases, the exact division of data into groups is not possible. Then a method based on a likelihood function can be used (again assuming normality of the data in both groups).

The whole procedure consists of these steps.

1. Performing a preliminary division of data into groups J_1 and J_2.
2. Determination of averages $\overline{x}_1, \overline{x}_2$ and variances s_1^2, s_2^2 for data from group J_1 (N_1 elements) and J_2 (N_2 elements).
3. Calculation of the quantity Z_i according to the relationship.

$$Z_i = \frac{(x_i - \overline{x}_2)^2}{s_2^2} - \frac{(x_i - \overline{x}_1)^2}{s_1^2} + \ln\left(s_2^2/s_1^2\right) \tag{6.58}$$

If $Z_i > 1$, the element x_i is placed in the group J_1, if $Z_i < 1$, the element x_i is placed in the group J_2, and if $Z_i = 1$, it remains where it is. This procedure can be repeated as long as all the elements are correctly included.

Attention, the division into groups obtained in this way must be interpreted factually, or assess its meaning.

The homogeneity of the selection is often disturbed by outliers, which usually significantly deviate from the given sample in terms of magnitude. They can often be identified very easily on exploratory analysis graphs (they lie outside the pattern formed by the other points). Outliers strongly distort both estimates of $\overline{x}$, and (even more significantly) s^2 and thus invalidate further statistical analysis.

The problem of deviating measurements (obviously outliers) is very complicated. To identify them (if there are more), it is necessary to know their distribution as well as the distribution of the remaining elements of the sample. In addition, "masking" of outliers and specification of the model, according to which the outliers behave, play an important role here.

It is evident that the analysis of outliers is based on a number of strict assumptions. In practice, it is often (without additional information) not very reliable. An alternative to the identification and subsequent elimination of outlier measurements is the use of robust statistical methods. These provide estimates of location and variance parameters that are little affected by outliers. On the other hand, where it is not outlier, but a sample has skewed distribution, robust methods may be inappropriate. Their effectiveness often decreases in cases where the data come from a normal or weakly contaminated distribution, which will be reflected in the growth of the variance of the robust estimates.

Here are some simple rules of thumb to identify outliers. The simplest is the well-known 3σ rule, where all values for which it is $|x_i - \overline{x}| \geq 3\sigma$ are considered as outliers (assuming that σ is known). It is recommended to calculate $\overline{x}$ and estimate σ, already without the assumed outliers. The threshold values of x_H for which the deviation from the mean is just at the acceptance limit (with a larger deviation, the given point is considered as outlier) depend on the sample size.

Unfortunately, the 3σ rule, or its modifications, can only be used for a strictly normal distribution. 3σ is too high for a uniform distribution and too low for a Laplace distribution.

In the paper (Novitskii & Zograf, 1985) was shown how to find the constant K_c in the expression $|x_H| = K_c\,\sigma$ so that for a number of symmetric distributions (differing in kurtosis g_2), it is possible to determine the limit beyond which all points are considered to be outliers. It was found that

$$K_c = 1.55 + 0.8\sqrt{(g_2 - 1)}\,\log(N/10) \tag{6.59}$$

In Eq. (6.59), the kurtosis g_2 is replaced by an estimate computed from all points except those a priori assumed to be possible outliers. For small N, the variances of the estimates s and $\hat{g}_2$ are quite high, so that the calculated K_c lies in a rather wide interval and the evaluation becomes complicated.

In exploratory data analysis (Meloun & Militký, 2012), so-called inner walls, which are a function of the interquartile range $R_F = \tilde{x}_{0.75} - \tilde{x}_{0.25}$, are used to identify outlying measurements. All elements that lie outside the inner walls end points $H_D = \tilde{x}_{0.25} - KR_F$ and $H_H = \tilde{x}_{0.75} + KR_F$ are considered to be outliers. By default, $K = 1.5$ is chosen and quartiles are replaced by order statistics $\tilde{x}_{0.25} = x_{(f)}$ and $\tilde{x}_{0.75} = x_{(N-f+1)}$. Usually, the index f is determined from the relation

$$f = 0.5\,\mathrm{int}[(N + 3)/2] \tag{6.60}$$

Table 6.1 Values of K ensuring that $P(N, K) = 0.9$ or 0.95 (when choosing $f =$ fI)

N	$P(N, K) = 0.9$	$P(N, K) = 0.95$
7	2.3	1.7
8	2.1	1.6
9	2.1	1.7
10	2.2	1.8
12	2.1	1.8
15	2.2	1.8
20	2.2	1.8
40	2.2	2.0
100	2.2	2.0
200	2.4	2.2

With regard to the smoothest determination of quartiles, it is preferable to use linear interpolation in order to find the interpolated value of $x(_{\mathrm{fI}})$ for generally non-integer fI $= N/4 + 5/12$. The probability $P(N, K)$ that there will be no value out of inner walls $[H_D, H_H]$ in a sample of size N drawn from a normal distribution depends on the size of K and N. For a given K, the $P(N, K)$ approaches to zero as N grows.

To identify outliers, it is convenient for $P(N, K)$ to be constant (e.g. 0.95). However, this means varying K depending on the sample size (Hoaglin et al., 1986). In Table 6.1, the values of K for different N are given, ensuring that $P(N, K) = 0.95$ or 0.9 (again assuming that the sample comes from a normal distribution). For a rough approximation of K in the range $8 \leq N \leq 52$ at $P(N, K) = 0.9$, the following relation can be used

$$K = 2.04 - 5.1\,/N \tag{6.61}$$

Similarly, for $P(N, K) = 0.95$ and range $8 \leq N \leq 100$, this approximation can be used

$$K = 2.25 - 3.6\,/N \tag{6.62}$$

The advantage of this procedure, consisting in the calculation of the limits for a given sample size, is primarily robustness (outliers or their groups practically do not affect the size of the quartiles). Moreover, there is no masking effect and it is not necessary to know a priori the number (or distribution) of outliers. The crucial assumption is that the "correct" data come from a normal distribution.

Another elementary option for identifying the highest element of the sample $x_{(N)}$ as a possible outlier is the use of the 100 α% quantile of the distribution of the largest order statistic $x_{(N)}(\alpha)$ generally

$$x_{(N)}(\alpha) = F_e^{-1}\left(\alpha^{1/N}\right) \tag{6.63}$$

where, F_e is the sample distribution. This procedure requires the determination of the distribution of sample and the estimation of its parameters (again, the suspect values $x_{(N)}$ must be omitted in the calculations).

If the assumption of normality of data distribution can be accepted, there are a number of tests for identifying a single outlier (see example in (Militký & Militká, 1985).

In the case of multiple measurements that appear to be outliers, it is necessary to first estimate their number and then carry out tests. An overview of the basic tests of outliers measurements and their groups can be found in work (Meloun & Militký, 2012).

6.4 Visualization of Data Distribution

In exploratory data analysis, it is common to use techniques that are simple and little influenced by possible outlier measurements (data heterogeneity). Classical moment characteristics (calculated from all the data) are highly unstable and, in addition, more time-consuming during manual calculations. Quantile characteristics are also suitable for smaller samples and are robust and easy to be determined. The procedures of exploratory data analysis are described in detail in the book (Meloun & Militký, 2012).

The basic stage of the preliminary analysis is the ascending sorting of the data. Data sorted in ascending order $x_{(i)} \leq x_{(i+1)}$ (where $x_{(i)}$ is the ith smallest element of the sample) are called *order statistics*. It forms a total of $N + 1$ intervals on the real axis.

$$\left(-\infty, x_{(1)}\right), \left(x_{(1)}, x_{(2)}\right), \ldots \left(x_{(i)}, x_{(i+1)}\right), \ldots \left(x_{(N-1)}, x_{(N)}\right), \left(x_{(N)}, \infty\right)$$

Order statistics $x_{(i)}$ are the values of the sample elements sorted in ascending order. Let $F_e(x)$ be the distribution function from which the sample elements come. It is known that a transformed random variable

$$z_{(i)} = F_e\left(x_{(i)}\right) \tag{6.64}$$

has a Beta distribution Be[i, $N - i + 1$] independently of the F_e. From the well-known properties of the beta distribution, it follows that the mean value is given by the relation

$$E\left(z_{(i)}\right) = \frac{i}{N + 1} \tag{6.65}$$

and for the variance, it true that

$$D\left(z_{(i)}\right) = \frac{[i/(N-i+1)]}{(N+1)^2(N+2)} \tag{6.66}$$

Moreover, the $z_{(i)}$, $z_{(j)}$ are mutually correlated. It can be shown that the elements V_{ij} of the corresponding covariance matrix V for all pairs $i, j = 1, \ldots N$ are functions of only i, j, and N. If $E[z(_i)]$ is back-transformed to the original order statistic, the following relation results

$$E\left(x_{(i)}\right) = F_e^{-1}\left(z_{(i)}\right) = Q_e(P_i) \tag{6.67}$$

In Eq. (6.67) is $Q_e(P_i)$ the quantile function and

$$P_i = \frac{i}{N+1} \tag{6.68}$$

is the rank probability. Equation (6.67) directly implies that the mean value of the order statistic $E(x_{(i)})$ is 100 P_i% quantile of the distribution from which the sample originates. The order statistic $x_{(i)}$ itself is then understood as an estimate of this quantile. For grouped data expressed by class values x_i^* and corresponding frequencies, f_{i} is x_i^* actually 100 P_i% quantile for

$$P_i = \sum_{j=1}^{i} f_j - 0.5 f_i \tag{6.69}$$

A more accurate approximation to the theoretical quantile function (especially for an approximately normal distribution) can be achieved by selection.

$$P_i = \frac{i - 3/8}{N + 0.25} \quad \text{nebo} \quad P_i = \frac{i - 1/3}{N + 1/3} \tag{6.70}$$

Given that order statistics divide the real axis into $N + 1$ interval that have the same probability $p = 1/(N + 1)$ that the realization of the random variable X occurs in them, it will not be possible to estimate the quantiles for arbitrary $0 \leq p \leq 1$. If the intervals $\left(-\infty, x_{(1)}\right)$ and $\left(x_{(N)}, \infty\right)$, about which nothing is known, are removed, the maximum possible rank probability will be $P_{\mathrm{Ma}} \leq (N - 1)/(N + 1)$ and the minimum possible rank probability will remain $P_{\mathrm{Mi}} \geq 1/(N + 1)$. To reduce the sensitivity of quantile estimates to outlier observations, n_V order statistics are usually omitted at each end. Then for P_{Ma}, it is true that

$$P_{\mathrm{Ma}} \leq (N - 1 - 2n_V)/(N + 1) \tag{6.71}$$

From there, it is possible to determine the number of measurements N_o, for which it will be possible to obtain a 100 P_{Ma}% quantile (when n_V extreme order statistics are omitted). Equation (6.71) directly implies that

$$N_o \geq (1 + P_{\mathrm{Ma}} + 2n_V)/(1 - P_{\mathrm{Ma}}) \tag{6.72}$$

For $n_V = 2$ and $P_{\mathrm{Ma}} = 0.75$ is $N_o = 23$. Thus, to determine the upper quartile when dropping the two highest and two lowest ordinal statistics, it is necessary to have at least 23 measurements. If will be excluded only the highest and lowest order statistics, $N_o = 15$ measurements will be necessary. It follows that for $n_V = 1$ the median $\tilde{x}_{0.5}$ can be determined from at least $N_o = 7$ measurements. The $N_o = 80$ measurements will be needed to calculate the 95% quantile (at $n_V = 1$) and 400 measurements will be needed for the 99% quantile (at $n_V = 1$). It is therefore evident that quantiles close to one or to zero will not be possible to obtain from smaller samples. If the value of the quantile $\tilde{x}_\alpha$ for $i/(N + 1) \leq \alpha \leq (i + 1)/(N + 1)$ is needed, it is possible to use piecewise linear approximation.

$$\hat{Q}(\alpha) = \tilde{x}_\alpha = (N + 1)[\alpha - i/(N + 1)]\left(x_{(i+1)} - x_{(i)}\right) + x_{(i)} \tag{6.73}$$

For the variance of a quantile $\hat{Q}(\alpha)$ from a sample of size N, it is valid that

$$D\left(\hat{Q}(\alpha)\right) = \frac{\alpha(1 - \alpha)}{N\left(f\left[Q(\alpha)\right]\right)^2} \tag{6.74}$$

where $f\left[Q(\alpha)\right]$ is the probability density value of the distribution of the data at $Q(\alpha)$. Quantile estimates can also be calculated as a weighted average of all order statistics $x_{(i)}$ with special weights (Harrell & Davis, 1982). For the purposes of exploratory analysis, a simple procedure for determining quantiles for selected rank probabilities is developed (Meloun & Militký, 2012). Various graphical procedures are advantageous for identifying data peculiarities, which allow a more comprehensive assessment of the behavior of individual data and their influence on conclusions about the behavior of the entire sample. By combining these procedures, it is possible to both verify the validity of the assumptions and find alternative methods of data processing (e.g. division into groups).

For the graphical visualization of data distributions, there are a number of simple techniques for estimating the shape of the distribution and features such as bimodality, outlier measurement, etc. The basic ones include: histogram, quantile plot, *Q-Q* plot, and kernel probability density estimation (Meloun & Militký, 2012). These techniques are demonstrated on 100 intentionally mixed data characterizing the strengths [*N*/tex] of 7.4 tex fineness cotton yarns obtained by compact and ring spinning technology. It is therefore a mixture of two types of data that should be identified at the exploratory analysis stage. In Table 6.2, the basic statistical characteristics of both the entire mixed sample ($N = 100$) and both samples for individual technologies separately (always $N = 50$) are shown. Also, individual graphs are always constructed both for original samples (individual technologies) and for a mixture of both original samples.

Table 6.2 Statistical characteristics of the strength of cotton yarns

Sample	N	$\bar{x}$ [N/tex]	s [N/tex]	95% CI
Mixture	100	0.1782	0.0304	0.172–0.184
Compact	50	0.2013	0.0195	0.197–0.205
Ring	50	0.1552	0.0199	0.153–0.159

The simplest characterization of the data distribution is an estimate of the quantile function, which is based on order statistics $x_{(i)}$, especially for small samples. The properties of the quantile function and its advantages for the construction of the sampling distribution function are presented in the book (Meloun & Militký, 2012). The quantile function is generally the inverse of the distribution function. If $F(x) = P$ is a continuous distribution function and $f(x)$ is the corresponding probability density function, $Q(P) = x$ is a quantile function. It can be easily verified that $F(Q(P)) = P$ for all $0 \leq P \leq 1$ and $f(Q(P))\, q(P) = 1$. The quantity $q(P) = \mathrm{d}Q(P)/\mathrm{d}P$, which is denoted as the quantile density function and $f(Q(P) = 1/q(P)$, is a density quantile function. From Eq. (6.67), it can be seen that the order statistics $x_{(i)}$ are rough estimates of the quantile function $Q_e(P_i)$ at P_i locations. The asymptotic distribution of quantile $\tilde{x}_P$ is normal with mean $Q(P)$ and variance defined by Eq. (6.74).

Piecewise linear interpolation can be used to estimate special sample quantiles $\tilde{x}_{P_i}$, and $\tilde{x}_{1-P_i}$ for $P_i = 2^{-i}$ $(i = 1, \ldots N)$. These quantiles are referred to as letter values (Meloun & Militký, 2012). All letter values except the median $(i = 1)$ are in pairs. For example, it is possible to determine the lower quartile $(P_i = 0.25)$ and the upper quartile $(P_i = 0.75)$, etc. Possibilities of a more accurate estimate of P_i are presented in the paper (Meloun & Militký, 2012). The median is then used to characterize the location, and the difference between the upper $(P_i = 0.75)$ and lower $(P_i = 0.25)$ quartiles is used to characterize the variability. The quantile deviation DQ is then twice this difference.

$$\mathrm{DQ} = 2(\tilde{x}_{0.75} - \tilde{x}_{0.25}) \tag{6.75}$$

DQ is equal to the numerical derivative of $Q(P)$ at $P = 0.5$, which roughly approximates the density quantile deviation $q(0.5)$.

$$\mathrm{DfQ} = 1/f(Q(0.5)) = 1/f(\tilde{x}_{0.5}) = q(0.5) \tag{6.76}$$

Using these location and variance estimates, it is possible to standardize the sample probability density so that DQ = 1 and median = 0. The standardized quantile function is referred to as the form identifying the quantile function QI(P). The sample function QI$_e$(P) is determined by the relation (Meloun & Militký, 2012).

$$\mathrm{QI}_e(P) = \frac{\left(x_{(i)} - \tilde{x}_{0.5}\right)}{2(\tilde{x}_{0.75} - \tilde{x}_{0.25})} \tag{6.77}$$

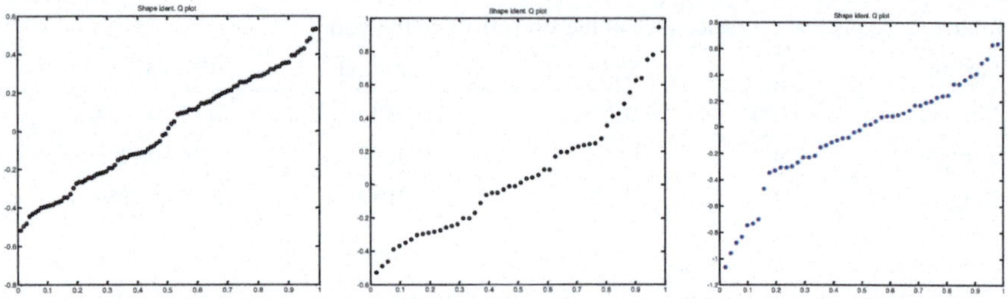

a) mixture b) ring c) compact

Fig. 6.1 $QI_e(P)$ graph for cotton yarn strength

The graph of *shape identifying quantile function* $QI_e(P)$ is the dependence of $QI_e(P_i)$ on P_i. The shape identifying quantile functions $QI_e(P)$ for the cotton yarn strengths is shown in Fig. 6.1. The values of $|QI_e(P)| \geq 1$ are identified outliers (for a normal distribution) or to indicate a distribution with long tails.

In Fig. 6.1a shows a small bend in the central region, which should identify bimodality. The bend is also visible in Fig. 6.1c, where it is probably due to grouping of data. Therefore, $QI_e(P)$ is here non-suitable for the purpose of indicating the distribution of the mixture of yarn strengths. The classical quantile graph is somewhat better (see Fig. 6.12). $QI_e(P)$ values can easily be used to specify the skewness and tail length of sampling distributions. The quantity $SQ = QI_e(0.25) + QI_e(0.75)$ is used as a measure of skewness (it is equal to zero for symmetric distributions). The tail length measures are $QI_e(0.05)$ and $QI_e(0.95)$. It is valid, that:

for distribution with short tails is $QI_e(0.95) < 0.5$,
for distribution with long tails is $QI_e(0.95) > 1$.
for distribution with moderate tails is $0.5, < QI_e(0.95) < 1$.

The empirical quantile graph QP is a dependence of $x_{(i)}$ on P_i. Special features of the data such asymmetry, local concentration, and approximate normality can be easily identified on this graph. A detailed analysis of QP is described in the book (Meloun & Militký, 2012). To improve interpretation, it is convenient to add a course of the normal distribution function to this graph

$$Q_N(P) = \mu + \sigma\, u_P \tag{6.78}$$

where u_p are the quantiles of the standard normal distribution $N(0, 1)$. The parameters μ and σ are location and scale estimates. Since there is no inverse function to the probability density function of the normal distribution in compact form, various approximations are used, the simple is

$$u_P = \frac{9.4 \ln(1/P - 1)}{|\ln(1/P - 1)| + 14}$$

By default, two types of quantile functions for normal distribution can be displayed. The first is based on moment estimates, i.e. the sample mean $\bar{x}$ and the sample standard deviation s. The second uses a robust quantile estimate of location, i.e. the median $\tilde{x}_{0.5}$, and the standard deviation calculated from the quartiles $s_M = (\tilde{x}_{0.75} - \tilde{x}_{0.25})/1.349$. This variant of QP allows better visualization of deviations from normality. If the data can be assumed a priori to come from some known distribution, the quantile function of that distribution can be used instead of the quantile function of the normal distribution. From the quantile graph QP, the following peculiarities of the sample can be identified:

(a) symmetry, or asymmetry of sample distribution. If the quantile function has a concave-convex shape and its inflexion is at $P_i \approx^\circ 0, 5$, it is a symmetric distribution. When sample distribution is skewed toward higher values, the quantile function has a convex increasing course, and when is skewed toward lower values, a concavely increasing course is appeared,
(b) local concentration of data (multimodality), when in the mode area the local slopes of the quantile graph are close to zero,
(c) the presence of outliers (if they strongly differ from the trend of other data).

In the preliminary analysis, the normal distribution is often considered as the base model and it is observed how much the sampling distribution differs from this model. Then it is advantageous to "extract" the shape of the normal distribution for all letter values and quantiles. This can be achieved by subtracting the value of the corresponding quantile of the normal distribution. The adjusted quantiles $d_{(i)}$ then have the form $d_{(i)} = x_{(i)} - s\, u_{P_i}$ for $P_i = (i - 1/3)/(N + 1/3)$. Here, s is the estimate of the standard deviation. To verify normality, it is then plotted $d_{(i)}$ on the y-axis and u_{P_i} on the x-axis. A horizontal line indicates normality. (The area below the median and the area above the median can be monitored separately, and the sample behavior in both areas can be evaluated separately). However, it is often difficult to determine whether it is a skewed measurement or a strongly skewed distribution (for this, e.g. letter characteristics to express symmetry must be used).

A simple estimate of the *distributionfunction* is the local sum of the order statistics.

$$\text{cdf}(x) = \frac{\sum_{i=1}^{i} x_{(i)}}{\sum_{i=1}^{N} x_{(i)}}, \quad \text{for} \quad x_{(i)} \leq x \leq x_{(i+1)}$$

This estimate is shown for example of yarn strength data in Fig. 6.2. The same figure shows the distribution function curves of the normal and uniform distributions. A typical "jump" on the empirical distribution function generally indicates bimodality.

The graphs in Fig. 6.2 indicates the approximation of the empirical distribution function to normality is evident, especially for strengths from ring and compact technology. The mixture is significantly different from normal and is approaching to rectangular one, the sudden jump is not significant.

To compare the empirical distribution of the sample with the selected theoretical distribution, it is possible to simply use different variants of the *Q-Q graph*. The

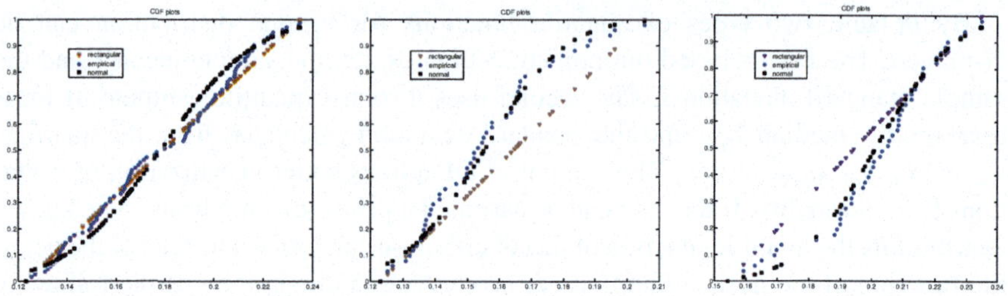

a) mixture b) ring c) compact

Fig. 6.2 Estimation of the distribution function for yarn strengths (points lying on straight line) are for rectangular distribution, blue points are based on data, and black points are for normal distribution

classical *Q-Q* graph is based on the comparison of the empirical quantile function $Q_e(P_i) \cong x_{(i)}$ with the selected theoretical quantile function $Q_T(P_i)$. If the theoretical distribution function is of the type $F_T((x - T)/S)$, it is preferable to use the standardized quantile function $Q_{TS}(P_i)$ (Meloun & Militký, 2012), or the form identifying quantile function $Q_{IT}(P_i)$. If the sample distribution is identical to the theoretical one, a straight line will appear.

$$x_{(i)} = T + SQ_{TS}(P_i) \tag{6.79}$$

The parameter T characterizes the location and the parameter S the dispersion. For some three-parameter distributions, the shape characteristic is used as a plot parameter. The disadvantage of the *Q-Q* graph is the frequent non-linearity and local fluctuations caused by the dependence between the order statistics and their non-constant variance. The *Q-Q* graph for the case of a normal theoretical distribution (rankit graph) is for yarn strengths of cotton yarns samples shown in Fig. 6.3.

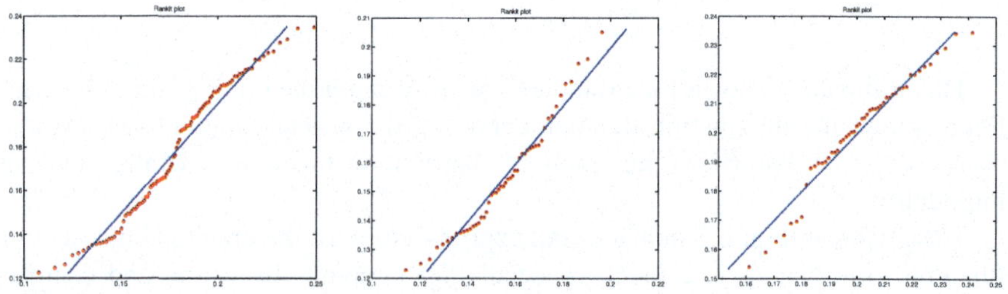

a) mixture b) ring c) compact

Fig. 6.3 *Q-Q* Rankit graph for samples of cotton yarn strengths (blue line is course for normal distribution)

It can be seen from Fig. 6.3 that the strength distribution of mixture deviates more significantly from the straight line and indicates a long-tailed distribution or bimodality. For the distribution of strength from individual technologies, the approach to normality is somewhat better.

When testing normality, it is often more convenient to use *conditional Q-Q graphs*. When constructing conditional rankit graphs, order statistics $x_{(i)}$ against $\Phi^{-1}\left[0.5\left(U_{(i-1)}+U_{(i+1)}\right)\right]$ are plotted. Here $\Phi^{-1}[u]$ is the standardized quantile function of the normed normal distribution (a suitable approximation is given in (Meloun & Militký, 2012)). The order statistics $U(_i)$ (where $U(_0)=0$, $U(_{N+1})=1$) are ordered random variables U_i with a uniform distribution defined by

$$U_i=\Phi\left(\frac{x_i-\hat{\mu}_R}{\hat{\sigma}_R}\right)$$

Here $\Phi(x)$ is the distribution function of the standard normal distribution (its approximation is given in (Meloun & Militký, 2012). The robust estimate of the location parameter is $\hat{\mu}_R=\tilde{x}_{0.5}$, i.e. equal to the median, and the robust standard deviation $\hat{\sigma}_R=0.75(\tilde{x}_{0.75}-\tilde{x}_{0.25})$ is proportional to the interquartile range ($\tilde{x}_\alpha$ is, the 100 $\alpha\%$ quantile). The construction of conditional rankit graphs is relatively simple. If the points in this graph form an (approximately) straight line, the assumption of normality can be accepted.

The *P-P* (or *Probability-Probability*) graph enables a direct comparison of distribution functions. In a *P-P* plot, the empirical distribution function (estimated as the rank probability P_i) is plotted against the theoretical distribution function $F_T(x_{(i)})$. If this dependence forms a straight line (with a zero intercept and unit slope), the empirical distribution is identical to the theoretical one. The disadvantage of this graph is the necessity of a complete specification of the theoretical distribution function, including its parameters. This usually requires estimation of the parameters of the given distribution, e.g. by the maximum likelihood method.

Parzen (1979) suggested use of the so-called *comparing distribution function* defined by the expression

$$\text{CDD}_i=F_T(Q_e(P_i)) \tag{6.80}$$

for these purposes.

The CDD function is approximately equal to the theoretical distribution function at $x_{(i)}$. Thus, the dependence of CDD_i on P_i is approximately equal to the *P-P* plot.

To indicate bimodality, it is more advantageous to use the so-called *comparative P-P graph.* In this graph, CDD_i is replaced by the difference CDD_i-P_i. In the case of a normal distribution, the corresponding comparative *P-P* graph is a horizontal line with zero intercept. Comparative *P-P* graph for cotton yarn strengths did not indicate bimodality and were not informative for this type of data.

The combination of *Q-Q* and *P-P* graphs is therefore not very advantageous for indicating bimodality in this case. A simple probability density estimate (Pdf) is a histogram with constant column width classes (the number of classes is M). A histogram is a piecewise constant probability density estimate. The height of the column in j-th class bounded by the values (t_{j-1}, t_j) is determined from the relation.

$$f_H(x) = \frac{C_N\left(t_{j-1}, t_j\right)}{N h_{\mathrm{j}}} \tag{6.81}$$

Here, the function $C_N(a, b)$ denotes the number of sample values in the interval $\prec a, b \succ$ and $h_j = t_j - t_{j-1}$ is the length of the jth class (interval). When constructing a histogram, the boundary values $\{t_{\mathrm{j}}\}\, j = 1, \ldots M$, the number of intervals M and their lengths h_j must be determined with regard to its quality. For approximately normal data, the constant class length is given by the relation.

$$h = 3.49\,(\min(s, Dq/2)/1.34)/N^{1/3} \tag{6.82}$$

Because the standard deviation estimate is sensitive to outliers, it is preferable to use the interquartile range $R_F = (\tilde{x}_{0.75} - \tilde{x}_{0.25})$. For approximately normally distributed data, it is then possible to use the relation

$$h = 2R_F/N^{1/3} \tag{6.83}$$

In the case of non-constant class length, a simple two-step method can be used. In the first stage, the number of classes M is determined

$$M = \mathrm{int}\left[2.46\,(N - 1)^{0.4}\right] \tag{6.84}$$

Here int[x] is integer part of number x. The optimal number of classes determined from Eq. (6.84) for the sample of size $N = 100$ containing cotton yarn strength is equal to $M = 15$.

In the second phase, individual class lengths are calculated based on the principle of equal probability in all classes. For this purpose, it is convenient to use the empirical quantile function $Q(P_i)$ using order statistics $x_{(i)}$. Practically, this means that the x-axis is divided into identical intervals of size $1/M$. Corresponding quantile estimates are determined for these intervals. Practical experience shows that this procedure is advantageous even for strongly skewed distributions (Meloun & Militký, 2012). The histogram for the strength of cotton yarns and non-constant length classes (Meloun & Militký, 2012) is in Fig. 6.4.

It is obvious that the non-constant length of the classes did not help much here, especially with the strength data from individual technologies. It is also due to relatively small samples ($N = 50$). A natural generalization of the histogram is the kernel estimate of the probability density $f(x)$ (a smooth function, dependent on the smoothing parameter h). This estimate has for the case of a constant smoothing parameter h the form.

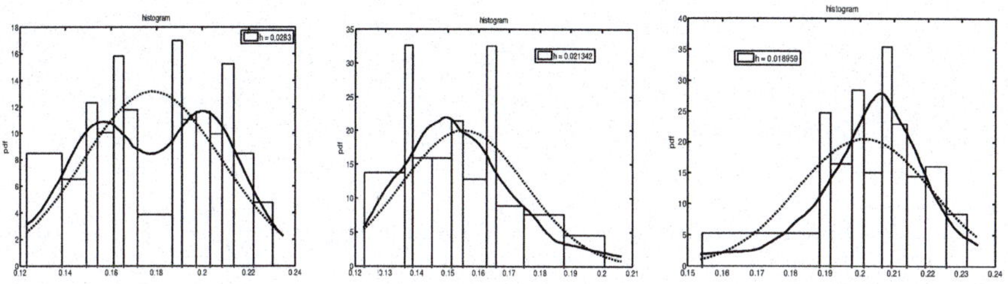

a) mixture b) ring c) compact

Fig. 6.4 Histogram for cotton yarn strengths along with kernel estimate (solid line) and probability density function of normal distribution (dotted)

$$\hat{f}(x) = \frac{1}{N}\sum_{i=1}^{N} K\left[\frac{x - x_i}{h}\right] \tag{6.85}$$

The selection of the kernel function $K[x]$ and the calculation of the smoothing parameter h are given in the book (Meloun & Militký, 2012). A simple biquadratic kernel function $K(x)$ has the form.

$$K(x) = 0.9375\left(1 - x^2\right)^2 \quad \text{for} \quad -1 \leq x \leq 1 \tag{6.86}$$

The kernel probability density estimate defined by Eq. (6.85) and the kernel function expressed by Eq. (6.86) for cotton yarn strengths are indicated by a solid black curve in Fig. 6.5. Typical bimodality for mixture strengths, unimodality for strengths from individual technologies, and moderate skewness are visible. It can be seen that the bimodality created by mixing the two types of data was best reflected in the probability density estimates.

If the data can be assumed to be a sample from either one or a mixture of two normal distributions, it is possible to compare the unimodal model

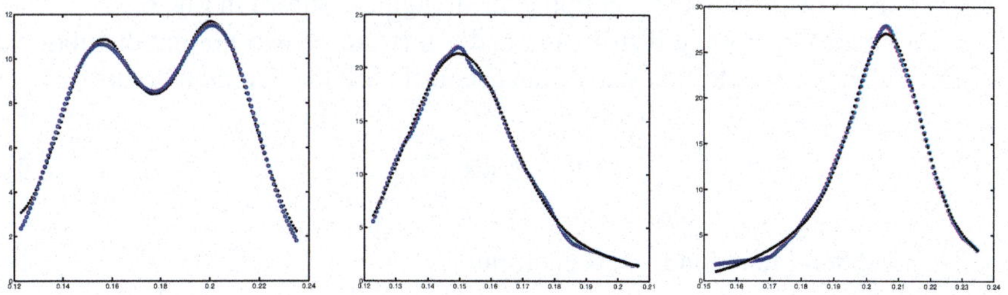

a) mixture b) ring c) compact

Fig. 6.5 Empirical probability density function approximation (blue circles) by a mixture of two normal distributions (black points in **a**) and one normal distribution (black points in **b**) and **c** for cotton yarn strengths

$$f_u(x_i) = A_0 \exp\left(-\frac{(x_i - B_1)^2}{2\,C_1^2}\right) \tag{6.87}$$

and a bimodal model (mixture of two normal distributions)

$$f_B(x_i) = A_1 \exp\left(-\frac{(x_i - B_1)^2}{2\,C_1^2}\right) + A_2 \exp\left(-\frac{(x_i - B_2)^2}{2\,C_2^2}\right) \tag{6.88}$$

Here A_1, A_2 are the proportions of the first normal distribution (index *1*) and the second normal distribution (index *2*). Parameters B_1 and B_2 are mean values of individual distributions and parameters C_1, C_2 are standard deviations.

The mixture of two normal distributions can be unimodal or bimodal (see Chap. 1). To estimate the parameters of a mixture of two normal distributions (A_1, A_2, B_1, B_2, C_1 and C_2), it is possible to use non-linear least squares methods, where the data is obtained from a histogram. The residual r_i for *i*th value is the difference between the empirical probability density function [see Eq. (6.85)] and the model probability density function [see Eq. (6.88)]

$$r_i = \hat{f}(x_i) - f_B(x_i) \tag{6.89}$$

The residual sum of squares is then equal to

$$S = \sum_{i=1}^{N} r_i^2 \tag{6.90}$$

where N is the number of classes. The assumptions leading to the use of the least squares method (i.e. minimization of S) are given in the book (Meloun & Militký, 2012). The BIMODAL program in MATLAB uses least squares minimization algorithms based on the Levenberg–Marquardt method and its improvement (Trust-region). The description of these algorithms is given in the book (Meloun & Militký, 2012). An approximation of the empirical probability density function of a mixture of two normal distributions for cotton yarn strengths is shown in Fig. 6.5

To compare the normal distribution and a mixture of two normal distributions (unimodal and bimodal), it is also advantageous to use the likelihood ratio

$$\mathrm{LR} = 2\,\ln\left(\frac{L_B}{L_U}\right) \tag{6.91}$$

The likelihood function L_U has the form

$$L_U = \prod_{i=1}^{N} \frac{1}{\sqrt{2\pi\,C_1^2}} \exp\left(-\frac{(x_i - B_1)^2}{2C_1^2}\right) \tag{6.92}$$

and for the likelihood function L_B is valid

$$L_B = \prod_{i=1}^{N} A_1 \exp\left(-\frac{(x_i - B_1)^2}{2\,C_1^2}\right) + A_2 \exp\left(-\frac{(x_i - B_2)^2}{2\,C_2^2}\right) \tag{6.93}$$

The LR statistic has an approximate $\chi^2(4)$ distribution, i.e. a simple unimodal distribution can be accepted for LR ≤ 9.

6.5 Transformation of Data

Data transformations are encountered very often when processing experiments. According to the causes, it is possible to divide the transformation into two basic groups:

A. Transformation improving data distribution. Here, the transformation is *requested* and contributes to the improvement of data distribution (simplifies their processing).
B. Transformation as a result of mathematical operations (usually functions) with measured quantities. This is the case when, for complicated systems, the input random variables are known, and the properties of the output random variable need to be determined. This includes all transformations where other quantities are calculated on the basis of experimental results (e.g. from radius values, the area of circular elements). Here, the transformation is actually undesirable because it distorts the original distribution of the data.

In the case of (A) a suitable transformation is searched. In the case of (B), suitable data processing procedures are selected that limit the negative influence of the transformation. This duality often leads to a situation where formally identical (mathematically correct) methods give vastly different results. It is clear from the above that the transformation can be either a "useful tool" or a "fundamental obstacle" in the statistical analysis of data.

As was shown in Sect. 6.2, it is ideal for statistical data analysis if the sample elements are mutually independent random variables with the same normal distribution. Real samples differ more or less from this state. In the simpler case, there are sample distributions with longer tails and higher kurtosis than corresponds to a normal distribution. This is often a consequence of the presence of symmetrical outliers in both ends of sample or non-constant sample elements variance. Here, in statistical analysis, the center of symmetry is still at the mode, which is identical to the median and mean value. The effective estimate of location is the median (the mean of x has approximately twice the variance). Common statistical tests are relatively robust to higher kurtosis of the data (this applies in particular to the t-test of significance). Also, the vast majority of robust methods for estimating location and dispersion parameters are based on the notion of a symmetric distribution of data contaminated by a certain proportion of symmetrically placed outliers.

More complicated is the case where the sampling distribution is skewed (usually toward higher values). Then the mode is no longer identical to the median or the mean value, and the actual interpretation of the location parameter is more difficult. Effective estimation of the location parameter is only possible if the law of probability distribution is known (which, however, is not known a priori during data analysis). Common statistical tests are generally not robust to skewed data distributions. Also, the basic robust methods for estimating location and dispersion parameters do not work well here. Standard statistical analysis often leads to overly optimistic conclusions here. For positively skewed distributions, even though the arithmetic mean is an asymptotically unbiased estimate, it is very likely to be less than the true value of the location parameter μ.

In Fig. 6.6, the results of a simple simulation consisting of repeated generation ($n = 100$) of samples of size 30 from the lognormal distribution with parameters $\mu = 0.8$ and $\sigma = 1.4$ are shown. For each selection, the arithmetic mean was determined and the deviation $ro = (\overline{x} - \exp(\mu + 0, 5\sigma^2))$ was calculated. From all 100 results, the percentage of cases roz where ro is negative (arithmetic mean is lower than the true mean value) was determined. This procedure was repeated 50 times and the median difference was determined (it is plotted as a horizontal line). It can be seen that the median line is at the 60% level and then most of the points lie above the 50% level. Thus, the arithmetic mean is systematically lower than the true value $\mu = \exp(\mu + 0.5\sigma^2)$.

Analogous results were obtained for other sample sizes. Thus, for positively skewed distributions, the estimate of the mean value is underestimated, and it is clear that the symmetrization transformation will be very useful for data analysis.

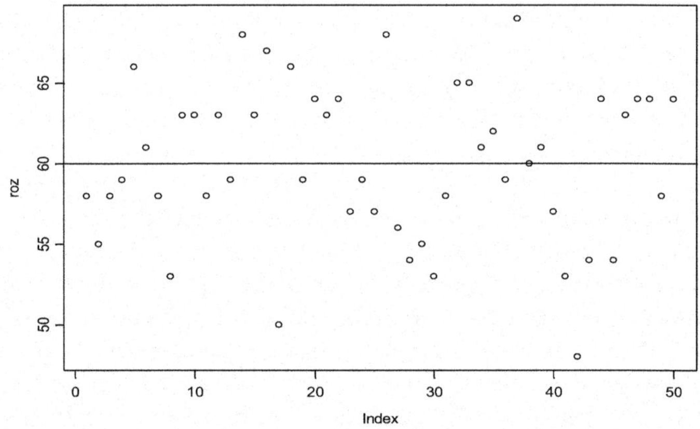

Fig. 6.6 Percentage of cases where the arithmetic mean is less than the mean (lognormal distribution)

It is known that if the sample of size N is from a non-normal distribution with mean μ and variance σ^2, the random variable $Z = \sqrt{N}\,(\bar{x} - \mu)/\sigma$ has an asymptotically normal distribution. If σ is not known, it is replaced by the sample standard deviation s. Then the so-called Student's random variable $t = \sqrt{N}\,(\bar{x} - \mu)/s$ has a Student's distribution with $(N - 1)$ degrees of freedom. The asymptotic normality of the quantity Z, or Student's distribution of the quantity t, allows the construction of a confidence interval of the mean value μ. The size of N sufficient for asymptotic normality depends strongly on the skewness $g_1(x)$ of the original distribution (from which the data originate) (Boos & Hughes-Oliver, 2000). To quantify the effect of skewness on the distribution of the random variable Z, it is possible to use the first term of the Edgeworth expansion which is valid.

$$P(Z \leq x) = F_n(x) - \frac{g_1(x)\left(x^2 - 1\right)}{6\sqrt{N}} f_n(x) \tag{6.94}$$

Here $F_n(x)$ is the distribution function of the standard normal distribution and $f_n(x)$ is the corresponding probability density function. The skewness of the random variable Z is given by the relation

$$g_1(Z) = g_1(x)/\sqrt{N} \tag{6.95}$$

The closer $g_1(Z)$ to zero, the closer the distribution of the quantity Z is to normal one. It can be seen that for a data distribution that is skewed toward higher values (i.e. $g_1(x)$ positive), the distribution of the random variable Z is also skewed toward higher values (i.e. $g_1(Z)$ positive). The confidence interval of the mean value then has a higher upper limit and a higher lower limit that corresponds to the real distribution of the Z statistic. For example, for sample of size $N = 10$ from the standard exponential distribution, when $g_1(x) = 2$, the 97.5% quantile of the distribution of the quantity Z is equal to 2.24, and the corresponding quantile of the standard normal distribution is only 1.96. Similarly, it can be determined that the 2.5% quantile of Z is only $-$ 1.65 compared to the corresponding quantile of the standard normal distribution $-$ 1.96. The confidence interval is thus entirely shifted to the right compared to the actual one (Boos & Hughes-Oliver, 2000). The first term of the Edgeworth expansion can also be used to quantify the effect of skewness on the distribution of the random variable t

$$P(t \leq x) = F_n(x) + \frac{g_1(x)\left(2x^2 + 1\right)}{6\sqrt{N}} f_n(x) \tag{6.96}$$

Here again $F_n(x)$ is the distribution function of the standard normal distribution and $f_n(x)$ is the corresponding probability density function. The opposite sign of the correction term is apparent, meaning that for a data distribution that is skewed toward higher values (i.e. $g_1(x)$ positive), the distribution of the random variable t is skewed toward lower values (i.e. $g_1(t)$ negative). The confidence interval of the mean value then has a lower upper limit and a lower limit than corresponds to the real distribution of the t statistic and is therefore completely shifted to the left compared to the real one (Boos & Hughes-Oliver, 2000). This is particularly annoying for data heavily skewed to the right. The reason for the difference between the behavior of the random variable Z and t is the correlation between the estimates $\overline{x}$ and s. The asymptotic correlation coefficient is equal to (Boos & Hughes-Oliver, 2000)

$$\rho(\overline{x}, s) = \frac{g_1(x)}{\sqrt{(g_2(x) - 1}} \tag{6.97}$$

An accompanying phenomenon for a number of "non-normally" distributed samples is also a non-constant variance (only for a normal distribution is the mean value independent of the variance). The variance stabilizing transformation is therefore also a transformation approximating to normality. Questions related to the existence of a transformation leading to normality are theoretically solved in the paper (Efron, 1992).

6.5.1 *Variance Stabilizing Transformation*

No constancy of variance is an accompanying phenomenon in a number of measurements. It indicates either the invalidity of the additive measurement model of Eq. (4. 1) or the non-normality of the distribution of the random variable from which the selection was made. Here, it is discussed the case where the variance $D(x)$ is a certain known function of the mean value μ, which it is possible to formally express by the relation

$$D(x) = g(\mu) \tag{6.98}$$

In the known (presumed) shape of $g(x)$, the stabilizing transformation $h(x)$ is then searched. Since the requirement is to choose such a function $h(x)$ that $D(h(x)) =$ const. and $D(x) = g(x)$, it is easily found from Eq. (6.98) that

$$h(x) \approx \text{const} \int \frac{\mathrm{d}x}{\sqrt{g(x)}} \tag{6.99}$$

By solving this integral (the constant const. is not important for the shape of the transformation), it is then possible to determine the transformation stabilizing the variance.

In many cases, the measurement is carried out under the condition of constancy of the relative error, i.e. constancy of the coefficient of variation $\text{CV} = (s_x/\overline{x})100$. The variance σ_x^2 at location x is then apparently $\sigma_x^2 = [\text{CV}/100]^2 x^2$ and the function (6.98) is therefore $g(x) = x^2$. After substitution and analytical integration of Eq. (6.99) is the result $h(x) = \ln(x)$. Using a logarithmic transformation here eliminates the no constancy of the variance (in general, this transformation is always advantageous if

Table 6.3 Effect of origin shift O on skewness $g_1(x)$ for various stabilizing transformations

O	0	1	2	3	10	100
$\sqrt{x}$	1.58	0.94	1.11	1.21	1.42	1.56
$\log(x)$	1.58	0.44	0.72	0.88	1.07	1.54
$1/x$	1.58	0.12	0.18	0.39	0.67	1.50

the individual elements of the sample change within several orders of magnitude). A common case is that $g(x) = x^P$ (general power dependence of the variance). Then, knowing the specific P, the stabilizing transformation $h(x) = x^{1-P/2}$ can easily be found from Eq. (6.99).

For a discrete distribution (Poisson or Binomial), the variance-stabilizing transformation is square root. Better is $\sqrt{x+0.5}$ or respectively $\sqrt{x} + \sqrt{x+1}$ (Bartlett, 1947). For more complicated relations between the variance and the mean value, the transformation $1/x \sinh^{-1}(x)$, $\sinh^{-1}(\sqrt{x})$ etc., (Bartlett, 1947) is used. For highly skewed data (such as the $\chi 2$ distribution), a square root transformation is recommended. For the gamma distribution, the stabilizing transformation is the third root $h(x) = x^{1/3}$. The transformation generally depends on the position of the data minimum $x_{(1)}$ (Osborne, 2002).

Shifting the data by a constant $c = 1 - x_{(1)}$ will result in a significant reduction in skewness $g_1(x)$ when using variance-stabilizing transformations (root, logarithmic, reciprocal). However, if the origin is already shifted by $c_O = O - x_{(1)}$ for $O = 2$ or more, the skewness does not change much. The larger O, the less efficient the transformation. This is clearly visible in Table 6.3 (modified for the sample given in (Osborne, 2002)).

The logarithmic transformation also depends on the base of the logarithms, as is visible in Fig. 6.7.

Cleveland recommended comparing at least $\log(x)$, $\ln(x)$ and $\log_2(x)$. When using a logarithmic transformation.

6.5.2 *Power Data Transformation*

Power transformation is quite widely applicable for solving a huge range of problems. It is known that both additive and multiplicative models can be expressed as special cases of the power-law class of measurement models, which is characterized by the fact that the transformation of both sides using the function $h(.)$ results in an additive model.

$$h(x) = h(\mu) + \varepsilon \tag{6.100}$$

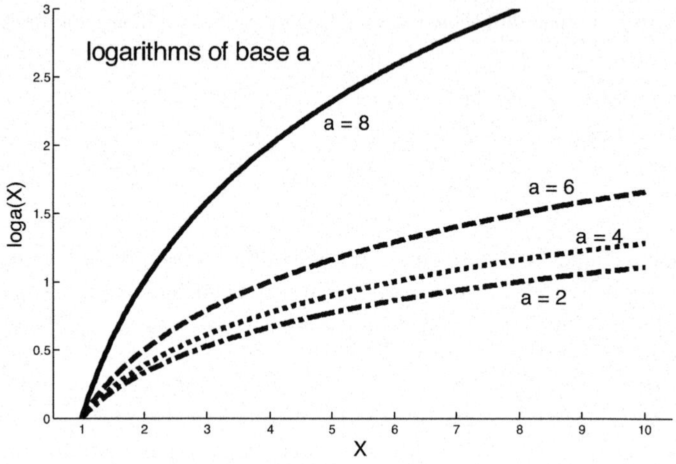

Fig. 6.7 The influence of the base of logarithms on the effect of the logarithmic transformation

For the probabilistic model (6.1), a suitable data transformation can stabilize the variance, bring the skewness of the distribution closer to zero and the shape of the distribution to a normal distribution. The goal is to find a suitable power, or appropriate member (if the entire family of transformations is used) from behavior of sample data x_i, $i = 1, \ldots N$.

The simplest is a *simple power transformation*

$$\begin{aligned} y &= \mathrm{hp}(x) = \mathrm{sign}(x)\mathrm{abs}(x)^P \quad \text{pro}\, P \neq 0 \\ y &= \mathrm{hp}(x) = \ln(x) \qquad\qquad\quad \text{pro}\, P = 0 \end{aligned} \tag{6.101}$$

where abs(x) is the absolute value and sign(x) is the sign function

$$\mathrm{sign}(x) = 1 \;\text{for}\, x > 0,\; \mathrm{sign}(x) = -1 \;\text{for}\, x < 0,\; \mathrm{sign}(x) = 0 \;\text{for}\, x = 0$$

This transformation is not scale preserving, nor is it continuous everywhere with respect to P. However, it preserves the order of the data in the sample (like as all power transformations) and is applicable to any x. It is used as a simple symmetrizing transformation and, therefore, the optimal power of P is selected to minimize the appropriate symmetry measures of the sample. It is possible to directly use the sample skewness $g_1(y)$, or its robust version $g_R(y)$, see

$$g_R = \frac{(\tilde{y}_{0.75} - \tilde{y}_{0.5}) - (\tilde{y}_{0.5} - \tilde{y}_{0.25})}{\tilde{y}_{0.75} - \tilde{y}_{0.25}} \tag{6.102}$$

where $\tilde{y}_P$ is the P% quantile of the transformed sample. It is just as simple to observe the difference between the mean $\bar{y}$ and the median $\tilde{y}_{0.5}$ in the transformation.

$$g_P = \frac{\bar{y} - \tilde{y}_{0.5}}{\sqrt{\sum_{i=1}^{N} (y_i - \bar{y})^2 / (N - 1)}} \quad (6.103)$$

To assess the quality of the transformation or find the optimal P, it is also possible to use the scatterplot with quantiles (GRK), quantile graphs, or Q-Q graphs, the construction of which is described in the book (Meloun & Militký, 2012) and in Sect. 6.5. The disadvantages of the simple power transformation (especially the discontinuity around zero and the changes of scales in the transformation) are removed by the family of Box-Cox transformations $h\ (x)$, which is a linear transformation of the simple power transformation hp(x).

The Box-Cox class of polynomial transformations is applicable only for positive x and has the form (Box & Cox, 1964; Draper & Cox, 1969)

$$\begin{aligned} y &= h(x) = \frac{x^\lambda - 1}{\lambda} \quad \lambda \neq 0 \\ y &= h(x) = \ln(x) \quad \lambda = 0 \end{aligned} \quad (6.104)$$

where λ is the transformation parameter. For $\lambda = 1$ results an additive measurement model and for $\lambda = 0$, it is a multiplicative measurement model (see Chap. 4). Using the Taylor expansion, it can be deduced that in this case, it is

$$x \approx \mu + \varepsilon / \mu^{1-\lambda} \quad (6.105)$$

In case when the variance $D(\varepsilon) = \sigma^2$ is small, it is multiplicative model of measurements equal to an additive model with non-constant errors, for which a weighted arithmetic mean with weights proportional to $\mu^{-(1-\lambda)/2}$ can be used as an estimate. It can be shown that a suitable estimator of the parameter (the location parameter) is the sample median, which is invariant to a monotonic transformation. Since $h(x)$ is a linear transformation of hp(x), it is valid for the re-transformed mean values.

$$h^{-1}[E(h(x))] = \mathrm{hp}^{-1}\left[E(\mathrm{hp}(x))\right] \quad (6.106)$$

For both transformations, the estimate of the re-transformed mean is the *generalized mean*

$$M = \left(\frac{1}{N} \sum_{i=1}^{N} x_i^\lambda \right)^{1/\lambda} \quad \text{for } \lambda \neq 0 \quad (6.107)$$

or

$$M = \left(\prod_{i=1}^{N} x_i \right)^{1/N} \quad \text{for } \lambda = 0 \quad (6.108)$$

If the power transformation is applied to the additive measurement model, the output is $h(x) = h(\mu + \varepsilon)$. The Taylor expansion shows bias due to this inaccuracy.

$$B = E(h(x)) - h(\mu) \approx \frac{\sigma^2}{2!} \frac{\mathrm{d}^2 h(x)}{\mathrm{d}x^2} \downarrow_{(x=\mu)} \tag{6.109}$$

For example, for a logarithmic transformation, $B = -\ 0.5\mathrm{CV}^2$, where CV is the coefficient of variation. The simple power transformation is invariant to scaling and the Box-Cox transformation is not invariant to scaling. Details can be found in (Emerson & Stoto, 1982). To eliminate this disadvantage, modified transformations can be used

$$h(x, p) = \frac{x^\lambda - m^\lambda}{\lambda} \quad \lambda \neq 0 \tag{6.110}$$

$$h(x, p) = \ln(x/m) \quad \lambda = 0 \tag{6.111}$$

where the parameter m is chosen as an arithmetic mean, a geometric mean, or the median of the original data. It also follows directly from the above that both transformations are dependent on data shift. Thus, transforming to the power of $(x + a)$ will give different results than transforming to the power of x only.

It is easy to see that:

- The family of transformations defined by Eq. (6.104) is continuous with respect to the power. In the neighborhood to zero, $\lim\left(x^\lambda - 1\right)/\lambda = \lim x^\lambda \ln(x) = \ln(x)$ is used.
- All transformations $h(x)$ are passing through one point with coordinates $y = 0$, $x = 1$ and have a common slope at this point (they are identical in shape here).
- At $x = 1$, the curvature of the function $h(x)$ is given by the relation

$$K(1) = |\lambda - 1|\sqrt{2^3} \tag{6.112}$$

For $\lambda < 0$ is then $K(1) \approx^\circ 1 - \lambda$

- Power transformations with exponents $-\ 2$, $-\ 3/2$, $-\ 1$, $-\ 0.5$, 0, 0.5, 1, 3/2, 2 are evenly spaced in terms of curvature.
- Due to the transformation (6.104), however, the location and dispersion characteristics generally change, which complicate the comparison of differently transformed samples (of course, this does not affect the approach to normality, or the improvement of the symmetry of the sample). The shape of the transformation (6.104) for different values λ is shown in Fig. 6.8.

Additional linear transformation can be applied to ensure that the transformed data has approximately the same location and dispersion as the untransformed data (Militký & Meloun, 1999).

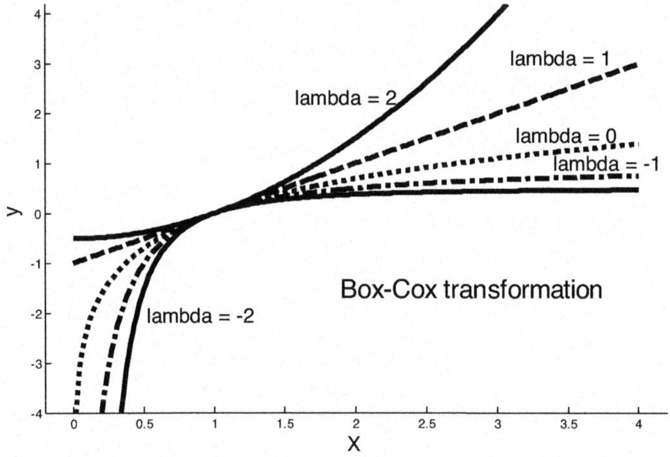

Fig. 6.8 Box-Cox transformation for different λ

From a data analysis point of view, transformation is always desirable if $x_{(N)}/x_{(1)} \geq 20$ (assuming positive data). Equation (6.104) is applicable only for positive data. If another data origin x_0 is known (below which the data cannot be occurred) the *generalized Box-Cox power transformation* is chosen.

$$y = h(x) = \frac{(x + c)^\lambda - 1}{\lambda} \quad \lambda \neq 0 \tag{6.113}$$

$$y = h(x) = \ln(x + c) \quad \lambda = 0 \tag{6.114}$$

In general, two parameters c and λ are searched for this transformation (usually c is chosen as little lower than x_0).

Considering that these transformations are used to a data bounded from below, it is apparently not possible for their distribution to be strictly normal. To eliminate this (practically not very important) disadvantage, Bickel and Doksum (Bickel & Doksum, 1981) recommend an extended Box-Cox transformation (for parameter λ > 0), which covers the entire real axis.

$$h(x) = \frac{\text{sign}(x)\ \text{abs}(x)^\lambda - 1}{\lambda} \quad \lambda \neq 0 \tag{6.115}$$

The disadvantage is that this transformation does not include a logarithmic transformation. This transformation is already scale independent. Close to this transformation is *the modular transformation* (John & Draper, 1980).

$$\begin{aligned} y = h(x) &= \operatorname{sign}(x)\left[\frac{(|x|+1)^{\lambda}-1}{\lambda}\right] \quad \lambda \neq 0 \\ y = h(x) &= \operatorname{sign}(x)\ln[|x|+1] \quad \lambda = 0 \end{aligned} \tag{6.116}$$

The Yeo Johnson transformation (Yeo, 2000) was designed to be defined on the entire real axis and satisfy the condition of continuity of the second derivative around the origin. For this transformation, it is valid that

$$\begin{aligned} y = h(x) &= \frac{(x+1)^{\lambda}-1}{\lambda} \quad \lambda \neq 0 \text{ and } x \geq 0 \\ y = h(x) &= \ln(x+1) \quad \lambda = 0 \text{ and } x \geq 0 \\ y = h(x) &= \frac{-(-x+1)^{2-\lambda}-1}{2-\lambda} \quad \lambda \neq 2 \text{ and } x<0 \\ y = h(x) &= -\ln(-x+1) \quad \lambda = 2 \text{ and } x<0 \end{aligned} \tag{6.117}$$

It can be seen that for $\lambda < 1$, the function $h(x)$ is concave and for $\lambda > 1$, the function $h(x)$ is convex. It can also be seen that for positive x, $h(x)$ is also positive, and for negative x, $h(x)$ is also negative. For positive data, the Yeo Johnson transformation is equivalent to the generalized Box-Cox transformation defined by Eq. (6.113) for $c = 1$.

Problems associated with limiting the range of values in the Box-Cox transformation can also be eliminated by using a family of exponential transformations (Manly, 1976).

$$\begin{aligned} y = h(x) &= \left[\frac{\exp(\lambda x)-1}{\lambda}\right] \quad \lambda \neq 0 \\ y = h(x) &= x \quad \lambda = 0 \end{aligned} \tag{6.118}$$

The shape of the exponential transformation (6.118) for different values λ is shown in Fig. 6.9.

It can be shown that the transformation defined by Eq. (6.118) is a limiting case of the generalized Box-Cox transformation (6.113). Again, different skewness and kurtosis characteristics can be used for parameter estimates in these families of transformations. In the case of one-parameter families of transformations, only one shape characteristic (usually skewness) can be focused on.

It is more advantageous to use tests of normality of the data in a power transformation (i.e. use the quantity y). The well-known Shapiro–Wilk test is proportional to the significance test of the slope in a *Q-Q* graph, so linearity in *Q-Q* graph can also be assessed.

Considering the requirement that the sampling distribution in the transformation is as close as possible to the normal distribution, the maximum likelihood method can be used to estimate the optimal parameter of transformation. If the transformation

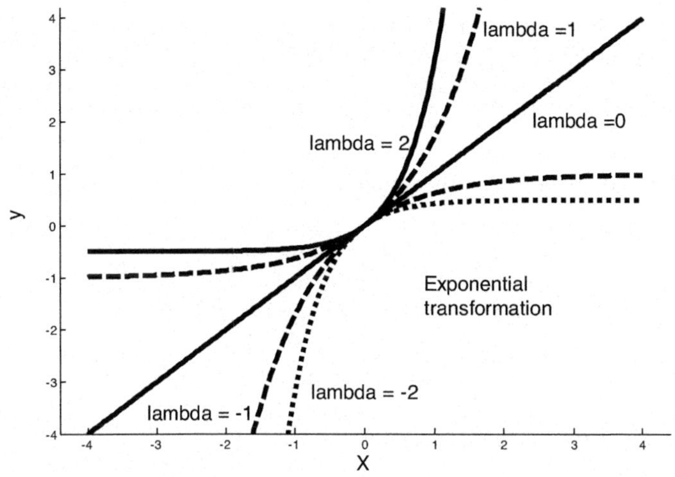

Fig. 6.9 Exponential transformation for different λ

$y = h(x)$ leads to a normal distribution $N\left(\mu_y, \sigma_y^2\right)$, the probability density of the original variables can be expressed in the form

$$f(x_i) = \frac{1}{\sigma_y\sqrt{2\pi}} \exp\left(-\frac{\left(x_i - \mu_y\right)^2}{2\sigma_y^2}\right) |J_i| \tag{6.119}$$

Here $|J_i| = \mathrm{d}y/\mathrm{d}x \downarrow_{x_i}$ is a Jacobian of transformation.

If the assumptions of the additive measurement model (normality and independence) are valid, the logarithm of the likelihood function has the form

$$\ln L(\lambda) = -\frac{N}{2}\ln(2\pi) - N\ln\left(\sigma_y^2\right) - \frac{1}{2\sigma_y^2}\sum \left(y_i - \mu_y\right)^2 + \sum \ln|J_i| \tag{6.120}$$

By replacing the parameters μ_y, σ_y^2 by maximum likelihood estimates for the selected λ,

$$\overline{y} = \left(\sum_{i=1}^{N} y_i\right)/N \qquad s_y^2 = \sum_{i=1}^{N} (y_i - \overline{y})^2/(N-1) \tag{6.121}$$

a concentrated likelihood function (for the parameter of the power transformation λ) is created

$$\ln L(\lambda) = -\frac{N}{2}\ln\left(s_y^2\right) + \sum \ln|J_i| \tag{6.122}$$

Thus, the different transformations differ only in the second term in Eq. (6.122). For the classical Box-Cox transformation defined by Eq. (6.113) is

$$\ln|J_i| = (\lambda - 1)\ln|x_i| \tag{6.123}$$

The modular transformation defined by Eq. (6.116) is

$$\ln|J_i| = (\lambda - 1)\ln(|x_i| + 1) \tag{6.124}$$

The Yeo Johnson transformation defined by Eq. (6.117) is

$$\ln|J_i| = (\lambda - 1)\mathrm{sign}(x_i)\ln(|x_i| + 1) \tag{6.125}$$

And finally, for exponential transformation, it is

$$\ln|J_i| = \lambda\, x_i \tag{6.126}$$

Therefore, it is easy to determine $\ln L(\lambda)$ for the chosen family of transformations and by one-dimensional search, usually in the range $-3 \le \lambda \le 3$, a maximal likelihood estimate $\hat{\lambda}$ (for which is $\ln L(\hat{\lambda})$ maximal) can be found.

Thus, for the generalized Box-Cox transformation, it is possible for each chosen λ to find a suitable $c < x_{(1)}$ by a one-dimensional search that minimizes the function

$$\ln L(\lambda) = \sum (\lambda - 1)\ \ln(x_i + c) - \frac{N\ \ln s_c^2}{2} \tag{6.127}$$

Here s_c^2 is calculated from the relation $s_c^2 = \sum_{i=1}^{N} (y_i - \overline{y})^2/(N - 1)$ where y_i is determined from the relation (6.118). It can be seen that the maximization $\ln L(\lambda)$ is equivalent to the minimization of the variance of the transformed variables s_y^2. Based on the Taylor expansion of the function $h(x)$ for a fixed λ, an approximate expression is valid

$$D\left(\frac{x^\lambda - 1}{\lambda}\right) = \frac{1}{\lambda^2} D(x^\lambda) \approx E(x)^{2\lambda - 2} D(x) = E(x)^{2\lambda}\delta^2 \tag{6.128}$$

where δ is the coefficient of variation. It is obvious that for fixed λ, the variance in the transformation will be higher, for the greater dispersion of the data. This will allow identification of the extreme (minima). For small dispersion of data, the variance in the transformation will be small and identifying the extreme will be more difficult. In work (Bickel & Doksum, 1981), it was shown that for $D(x) \to 0$, the variance of $\hat{\lambda}$ is $D(\hat{\lambda}) \to \infty$ and similarly the variance of the generalized average grows beyond all limits. For easy identification of the transformation, it is therefore advantageous to have a greater dispersion of the data, as is, for example, common in samples with asymmetric distributions. Formally, the maximization of $\ln L(\lambda)$ for the Box-Cox transformation can be expressed in the form

$$\frac{\mathrm{d}\ln(L)}{\mathrm{d}\lambda} = \sum_i \ln(x_i) - \frac{1}{\sigma^2}\sum_i \left(h(x_i) - \frac{1}{N}\sum_i h(x_i)\right)\frac{\mathrm{d}h(x_i)}{\mathrm{d}\lambda} = 0 \tag{6.129}$$

where

$$\frac{\mathrm{d}h(x_i)}{\mathrm{d}\lambda} = \frac{(1+\lambda\,x_i)\ln(1+\lambda\,x_i) - \lambda\,x_i}{\lambda^2} \tag{6.130}$$

From the second derivative of the likelihood function, the variance of the maximum likelihood estimate of the power transformation can be derived (Draper & Cox, 1969).

$$D(\hat{\lambda}) = \frac{2\left(1 - 0.333\,g_1^2 + 0.388\,g_2\right)}{3\,Nw} \tag{6.131}$$

where $w = \sigma^2(1+\lambda)$. Here, σ^2, g_1 and g_2 are the variance, skewness, and kurtosis of the original data. It can be seen that for $\sigma^2 \to 0$, the variance of the estimate of the power transformation increases above all limits. An inequality can be established for the asymptotic $(1-\alpha)$ confidence interval of the parameter λ of the power transformation.

$$\ln L(\lambda) \geq \ln L(\hat{\lambda}) - 0,5\,\chi^2_{1-\alpha}(1) \tag{6.132}$$

Here, $\chi^2_{1-\alpha}(1)$ is the quantile of χ^2 distribution with 1 degree of freedom. All λ satisfying this inequality lie within the confidence interval and are therefore acceptable. This can easily be used to distinguish between additive and multiplicative measurement models.

It is valid, that:

- If the 95% confidence interval also contains one, the additive model of measurements is chosen.
- If the 95% confidence interval contains zero and not one, the multiplicative model of measurements is chosen.
- In other cases, it is possible to choose the probabilistic model (6.1) and use the procedure proposed in (Meloun & Militký, 2012) for further analysis.

It is advantageous to use a graph of the dependence of $\ln L(\lambda)$ on λ with a drawn (usually 95%) confidence interval. From this graph, it is easy to estimate both the quality of the transformation and assess the limits within which the value λ can appear. It is true that the narrower these limits are, the higher the quality of the transformation, if $\lambda = 1$ does not lie within them.

The power transformation parameter λ is apparently related to the skewness of the data distribution. For the quantification of this relationship, its expansion into the Taylor series can be substituted into the condition (6.128) instead of $h(x)$, and the maximum likelihood estimate can be determined analytically. This derivation is carried out in work (Box & Cox, 1964). The result can be written in the form

$$\lambda \approx 1 - \frac{E(x)\,\sigma\,g_1}{6} \tag{6.133}$$

where g_1 is the skewness of the original data. It can be seen that for data skewed to higher values, the transformation parameter is significantly less than one. Using relation (6.133), it is possible to easily assess the effect of data shift on the power transformation parameter. For example, in case when data are shifted by a constant a, i.e. $y = a - x$ it is valid, that

$$\lambda_y = \lambda_x - (a\,\sigma\,g_1)/6$$

As can be seen, any linear transformations of the data and their range must also be considered when using the power transformation procedure. From a numerical point of view, it is more advantageous to transform the data $u_i = y_i/C$ so that the Jacobian of the transformation is equal to one. It is then necessary to search C for which it is

$$\sum_{i=1}^{N} \ln|CJ_i| = 0 \tag{6.134}$$

After rearrangements, the Eq. (6.134) has the form

$$C = \exp\left(\frac{1}{N}\sum_{i=1}^{N} \ln|J_i|\right) \tag{6.135}$$

For the classical Box-Cox transformation, after substituting from Eq. (6.113) into Eq. (6.135), it follows that

$$C_B = \exp\left(\frac{\lambda - 1}{N}\sum_{i=1}^{N} \ln x_i\right) = \exp\left(\ln \overline{x}_g^{\lambda-1}\right) = \overline{x}_g^{\lambda-1} \tag{6.136}$$

where $\overline{x}_g$ is the geometric mean of the original sample (Hinkley & Runger, 1984). For the modular transformation, it is

$$C_M = \exp\left(\frac{\lambda - 1}{N}\sum_{i=1}^{N} \ln(|x_i|) + 1\right) = \overline{H}_g^{\lambda-1} \tag{6.137}$$

where $\overline{H}_g$ is the geometric mean of the quantities $H_i = |x_i| + 1$. Finally, for exponential transformation is

$$C_E = \exp\left(\frac{\lambda}{N}\sum_{i=1}^{N} x_i\right) = \exp(\lambda\,\overline{x}) \tag{6.138}$$

where $\overline{x}$ is the arithmetic mean of the original data.

When using variables $u_i = y_i/C$, the second term in Eq. (6.122) disappears, so it is sufficient to calculate only the variance s_u^2 for each λ. Instead of dividing all y_i elements by the number C, modified transformation formulas $h(x)/C$ can be used.

Especially for the purposes of exploratory data analysis (Meloun & Militký, 2012), a procedure was proposed that enables a graphical assessment of the appropriateness of a power transformation. A simple class of transformations is used.

$$h(x) = ax^{\lambda} + \mathrm{b} \quad \lambda \neq 0$$
$$h(x) = c\ \ln(x) + d \quad \lambda = 0 \tag{6.139}$$

The parameters a, b, c, d are chosen by Emerson and Stotto (1982) so that the approximate linearity of the transformation around the median is preserved, i.e.

$$h(\tilde{x}_{0,5}) \approx \tilde{x}_{0,5} \quad \frac{\mathrm{d}}{\mathrm{d}x}(h(x) \downarrow_{x=\tilde{x}_{0,5}} \approx 1 \tag{6.140}$$

Substituting from Eq. (6.139) into Eq. (6.140), it will be obtained estimates

$$a = \tilde{x}_{0.5}^{1-\lambda}/\lambda \quad b = \tilde{x}_{0.5}(\lambda - 1)/\lambda \quad c = \tilde{x}_{0.5} \quad d = \tilde{x}_{0.5}[1 - \ln(\tilde{x}_{0.5})]$$

where $\tilde{x}_{0.5}$ is the sample median of the original data. When searching for a suitable transformation, the fact that for symmetric distributions all half-sums are equal to the median is used. It means

$$(\tilde{y}_P + \tilde{y}_{1-P})/2 = \tilde{y}_{0.5} \tag{6.141}$$

Equation (6.141) indicates that for any P from the interval $0<P<0.5$, of symmetric distributions the mean value of the 100 P% and 100 $(1 - P)$% quantiles of the transformed data is exactly equal to the median. Emerson and Stotto (1982) performed an expansion of the quantiles and into a Taylor series around the median $\tilde{y}_{0.5}$.

$$\tilde{y}_P = \tilde{y}_{0.5} + \lambda\,\tilde{y}_{0.5}^{\lambda-1}(\tilde{x}_P - \tilde{x}_{0.5}) + \frac{\lambda(\lambda - 1)}{2\,\tilde{y}_{0.5}^{\lambda-2}\,(\tilde{x}_P - \tilde{x}_{0.5})^2} \tag{6.142}$$

and an analogous relationship also applies to $\tilde{y}_{1-P}$. After substituting into Eq. (6.141), the following relation is obtained

$$\frac{(\tilde{x}_P + \tilde{x}_{1-P})}{2} \simeq (1 - \lambda)\frac{(\tilde{x}_{1-P} - \tilde{x}_{0.5})^2(\tilde{x}_{0.5} - \tilde{x}_P)^2}{4\tilde{x}_{0.5}} \tag{6.143}$$

The sample quantiles $\tilde{x}_{P_i}$ and the median $\tilde{x}_{0.5}$ are then used to determine the appropriate transformation. By plotting the points

$$y_i^* = \frac{\left(\tilde{x}_{p_i} + \tilde{x}_{1-p_i}\right)}{2} \quad x_i^* = \frac{\left[\left(\tilde{x}_{1-p_i} - \tilde{x}_{0.5}\right)^2 + \left(\tilde{x}_{0.5} - \tilde{x}_{p_i}\right)^2\right]}{4\tilde{x}_{0.5}} \tag{6.144}$$

Therefore, in the case of suitability of the symmetrization transformation, a linear dependence passing through the beginning results

$$y^* = (1 - \lambda)\, x^* \tag{6.145}$$

Thus, an estimate of the transformation parameter λ can be directly found from the slope of this dependence. In the practical application of this procedure, individual letter values are chosen (Meloun & Militký, 2012) for which $P_i = 2^{-(i+1)}$ $i = 1, \ldots$. To obtain a robust estimate of the slope (1 − λ), it is recommended to calculate the local slope $k_i = y_i * / x_i *$ for all points and then take the median of all k_i.

This procedure is suitable for slightly and moderately skewed distributions. Cameron (Cameron, 1984) showed that for strongly skewed distributions and quantiles $\tilde{x}_{p_i}$ far from the median, the graph of y^* versus x^* has a systematic curvature. This is clearly evident if the factor τ is introduced

$$\tau = (\tilde{x}_{1-P} - \tilde{x}_{0.5}) / (\tilde{x}_{0.5} - \tilde{x}_P) \tag{6.146}$$

After substituting it into Eq. (6.143) and rearrangements, the following inequality results

$$|1 - \lambda| < \frac{2\,\tilde{x}_{0.5}}{(\tilde{x}_{1-P} - \tilde{x}_{0.5})}\left(1 - \tau^{-1}\right) \tag{6.147}$$

For data strongly skewed to higher values, τ is large and $(\tilde{x}_{1-P} - \tilde{x}_{0.5})/\tilde{x}_{0.5}$ is also large. Then from Eq. (6.147) is λ → 1. However, for suitable transformation, it is necessary to have values of λ significantly smaller than 1, or even negative. This practically leads to a situation where, for significantly skewed distributions, the graphical dependence of y^* on x^* has systematic curvature. Then it is appropriate to carry out an iterative search of the optimum, when the result from the first determination of the slope of (y^* vs. x^*) is inserted into Eq. (6.100) and in subsequent graphs, the transformed quantiles $h(x)$ (for λ determined from the previous graph) are used instead of the quantiles of the variable x. Also, it is advantageous in the first stages to use slope determined from quartiles ($P_i = 0.25$). It follows that the Emerson-Stott procedure is not completely automatic and often requires an iterative search for a suitable λ, where in each iteration, a graph of type y^* on x^* is constructed. On the other hand, this procedure is very simple and enables the assessment of the influence of possible influential points on the transformation result. It is easy to see that all the given types of transformation are members of the general *Johnson family* of transformations $J(x)$, when (Johnson, 1949)

$$J(x) = \alpha + \beta h(\delta x + \gamma) \tag{6.148}$$

where α, β, δ and γ are the parameters and $h(x)$ is the appropriate monotonic function. For the power, transformation is $h(x) = x^{\lambda}$ and for the exponential transformation is $h(x) = \exp(x)$. It can be shown that there exist functions $h(x)$ for which the transformation $J(x)$ covers the entire range of skewness and kurtosis (arcsine hyperbolic, logarithm) (Johnson, 1949). Using these functions and estimating the parameters α, β, δ and γ, it is possible to find the optimal normalization transformation (Slifker & Shapiro, 1980). In work (Kaskey et al., 1980), a suitable empirical system of probability densities is used to achieve the normalization transformation of data. The classical Pearson system of probability density functions is chosen and the transformation $y = h(x)$ is selected, for which probability density function, of the Pearson system is transformed to other probability density function (which is chosen as a normal distribution).

6.5.3 *Treatment of Transformed Data*

If the transformation of the data led to approximate normality, a confidence interval of the mean value $\overline{y} \pm t_{1-\alpha/2}(N-1)\, s_y/\sqrt{N}$ can be determined for the quantities $y = h(x)$ and estimates $\overline{y}$, s_y^2. If it is necessary, significance tests can be also performed. In many cases, adequate results are achieved with this procedure (although it is often better to use t-tests based on the α-truncated mean (Doksum & Wong, 1983). In spite of some theoretical problems (Hinkley & Runger, 1984) and the related discussion), basic statistical data analysis can be performed very easily in the correct transformation case. The problem, however, is that it is often required to determine both statistical characteristics and confidence intervals in the original variables. The *naïve* approach is to use simply reverse transform

$$\overline{x}_R = h^{-1}(\overline{y}) \tag{6.149}$$

to obtain the re-transformed average $\overline{x}_R$ of the original data and to calculate the H_D and H_H limits of the confidence interval of the mean value

$$H_D = h^{-1}\left(\overline{y} - t_{1-\alpha/2}(N-1)\, s_y/\sqrt{N}\right) H_H = h^{-1}\left(\overline{y} + t_{1-\alpha/2}(N-1)\, s_y/\sqrt{N}\right) \tag{6.150}$$

Unfortunately, this naive re-transformation is only valid if $h(x)$ is a linear transformation. It is therefore commonly happens that the original interval (in y variable units) covers a mean $\overline{y}$, but the re-transformed interval (in x variable units) no longer contains $h^{-1}(\overline{y})$. A simple improvement option is to use the Taylor expansion of the function $h\ (x)$ around the value of $\overline{y}$. After adjustments, an approximate expression for the re-transformed average can be obtained.

$$\overline{x}_R = h^{-1}\left[\overline{y} - \frac{1}{2}\frac{\mathrm{d}^2h(x)}{\mathrm{d}x^2}\left(\frac{\mathrm{d}h(x)}{\mathrm{d}x}\right)^{-2} s_y^2\right] \tag{6.151}$$

For the corresponding variance $s_{x_R}^2$, it is valid.

$$s_{x_R}^2 = \left(\frac{\mathrm{d}h(x)}{\mathrm{d}x}\right)^{-2} s_y^2 \tag{6.152}$$

For the general transformation, it must be considered that both derivatives are functions of $\overline{x}_R$, so after modifying Eq. (6.151) non-linear equation for calculation $\overline{x}_R$ is obtained. They may not always have an analytical solution. Its solvability will generally depend on the magnitude of $\overline{y}$ and s_y^2. When using Eq. (6.151), it is possible to “refine” the boundary points of the confidence interval of the mean value

$$H_D = h^{-1}\left(\overline{y} + G - t_{1-\alpha/2}(N-1)\, s_y/\sqrt{N}\right) H_H = h^{-1}\left(\overline{y} + G + t_{1-\alpha/2}(N-1)\, s_y/\sqrt{N}\right) \tag{6.153}$$

where

$$G = -\frac{1}{2}\frac{\mathrm{d}^2h(x)}{\mathrm{d}x^2}\left(\frac{\mathrm{d}h(x)}{\mathrm{d}x}\right)^{-2} s_y^2 \tag{6.154}$$

The quality of these approximations will depend on how much the Taylor expansion into quadratic terms is sufficient to approximate $h(x)$ around the mean value.

For the case of the classical Box-Cox transformation expressed by Eq. (6.113) and $\lambda \neq 0$, it is possible to substitute into Eq. (6.151) and calculate an approximate expression for $\overline{x}_R$.

$$\overline{x}_R \approx \left[1 + \lambda(\overline{y} - 0,5(\lambda - 1))\, \overline{y}^{\lambda} s_y^2\right]^{1/\lambda} \tag{6.155}$$

This non-linear equation can be modified to the form

$$a^2 - (1 + \lambda\, \overline{y})\, a + 0,5\, \lambda(\lambda - 1)\, s_y^2 = 0 \tag{6.156}$$

where $a = \overline{x}_R^{\lambda}$. It follows from the solvability condition of the quadratic Eq. (6.156) that

$$|(1 + \lambda\, \overline{y})| > s_y\sqrt{2\,\lambda\,(\lambda - 1)} \tag{6.157}$$

This is a constraint on the interrelationship between $\overline{y}$ and s_y^2. Solving Eq. (6.156) can be expressed simply in the known form

$$a_{1,2} = 0,5(1 + \lambda\overline{y}) \pm 0,5\sqrt{1 + 2\lambda\left(\overline{y} + s_y^2\right) + \lambda^2\left(\overline{y} - 2s_y^2\right)} \tag{6.158}$$

After substituting for a, the following solution is obtained

$$\overline{x}_{iR} = a_i^{1/\lambda} \quad i = 1, 2 \tag{6.159}$$

The $\overline{x}_{iR}$ that is closer to the median $\tilde{x}_{0.5}$ is used as the solution. With knowledge of $\overline{x}_R$, it will be possible to calculate the estimate s_{xR}^2.

$$s_{xR}^2 = \overline{x}_R^{2-2\lambda} s_y^2 \tag{6.160}$$

For the case of the classical Box-Cox transformation expressed by Eq. (6.113) and $\lambda = 0$, i.e. logarithmic transformation, simplification occurs and from Eq. (6.151), it is directly valid.

$$\overline{x}_R = \exp\left[\overline{y} + 0.5s_y^2\right] \tag{6.161}$$

Equation (6.152) can also be used directly for estimation of s_{xR}^2. when substituting $\lambda = 0$.

In the case where the determination of the median is required as a location parameter (which is a robust estimate, but the Box-Cox transformation itself is quite not robust), the situation is greatly simplified. As is well known, the median of transformation is a transformation of the median, so for the re-transformed median, it is valid.

$$\tilde{x}_{R0.5} = h^{-1}(\tilde{y}_{0.5}) \tag{6.162}$$

Comparison of Eqs. (6.151) and (6.162) for determining location parameters by re-transformation after Box-Cox transformation is discussed in work (Taylor, 1986). It was found that for sufficient sample sizes and in cases where the transformed distribution does not have very long tails, it is sufficient to use Eq. (6.151).

For a more accurate estimate of the re-transformed mean $\overline{x}_R$, in cases where λ it is not too close to zero (and the classical Box-Cox transformation is applied), the following relation can be used.

$$\overline{x}_R = \left(1 + \hat{\lambda}\,\overline{y}\right)^{1/\hat{\lambda}} \frac{1 + s_y^2\left(1 - \hat{\lambda}\right)}{2\left(1 + \hat{\lambda}\,\overline{y}\right)^2} \tag{6.163}$$

Here, $\hat{\lambda}$ is the maximum likelihood estimate obtained by maximizing the likelihood function in $L(\lambda)$. The simpler case is when the back transformation $g(y) = h^{-1}(y)$ is of the recursive type (Neyman & Scott, 1960), i.e. it corresponds to the differential equation $g''(y) = A + Bg(y)$. It can be easily proved that the logarithmic transformation $h(x) = \ln(x)$, where $g(y) = \exp(y)$ ($A = 0, B = 1$) belongs to this class. The re-transformed value of the mean is then

$$\overline{x}_R = \exp(\overline{y}) \sum_{i=1}^{\infty} \frac{\Gamma(0.5(N-1))}{\Gamma(0.5(N-1)+i)} \frac{1}{i!} \left[(1 - 1/N) s_y^2 (N-1) \right]^2 \tag{6.164}$$

It is clear that this reverse transformation is a rather complicated problem. Most estimates $\overline{x}_R$ are biased and also have larger variances. Therefore, it is always advantageous to work only with transformed values (unless it is absolutely necessary to know the characteristics of the original quantities). This approach is particularly suitable for the implementation of significance tests, where the entire analysis can be transformed. Also, with probabilistic reasoning, it is possible to only work in transformation.

6.6 Empirical Probability Models

Knowing the distribution of the data, it is quite routine to construct various probabilistic considerations such as:

- what is probability that the observed random variable (measurement result) lies in the given interval,
- what is probability that the random variable does not exceed the given value,
- what is probability that the observed random variable lies in the given interval, etc.

These probabilistic considerations make it possible to answer a number of practical questions about the observed random variable. Knowledge of the distribution also enables a more correct implementation of the calculation of basic statistical characteristics, etc. Unfortunately, when solving practical problems, the probability distribution of the experimentally measured quantity can only be determined in special cases. Usually, some distributions are then used a priori (e.g. normal) and the whole further analysis is based on them. In Sect. 6.6, the possibilities of data transformation enabling the approximation of the data distribution to the normal distribution are described. An alternative possibility is the construction of empirical probability models expressed, for example, by the empirical probability density $f(x)$. Non-parametric methods of empirical probability density function estimation are discussed in Sect. 6.5. Another option is to use a selected system of parametric probability densities (probability distribution laws) that are suitable for data analysis. It is necessary to state that the empirical densities or their systems have quite wide application in practice. They can be used to solve the following basic tasks:

A. Approximation of known (sampling) distributions, which are difficult to quantify in practice (examples are normal, Student and Fisher distributions),

B. the construction of an empirical probability model to express the distribution of data,
C. modeling of special types of distribution to monitor robustness or sensitivity of statistical procedures,
D. Simulation of inputs and outputs in the statistical analysis of complex systems.

A number of different empirical density systems are described, the vast majority of which are four-parameter (they can be ensured to have the same mean, variance, skewness, and kurtosis equal to the sampling distribution). These systems are designed with several basic requirements including:

1. The ability to identify a number of forms of distribution occurring in practice,
2. simplicity of choosing a specific distribution from the given system,
3. ease of data distribution approximation,
4. easy handling, especially when expressing quantiles.

It is known that every probability distribution is uniquely determined by all its general moments or central moments. Classification of the distribution is therefore possible based on the analysis of moments. The first general and second central moments, i.e. the mean $E(x)$ and the variance $D(x)$, determine only the position and dispersion, but not the shape of the distribution. The third standardized moment g_1 and the fourth standardized moment g_2 are commonly used to characterize the shape. It must be noted that these four moments are not even sufficient to accurately identify the normal distribution, as there are a number of probability densities with zero skewness $g_1 = 0$ and kurtosis $g_2 = 3$.

Knowing the skewness and kurtosis of the data, the methods of statistical processing of sample distributions strongly different from the normal can be selected. Most empirical density systems are unimodal. The Johnson system (Johnson, 1949) and the system of the generalized lambda distribution (Karian et al., 1996) are particularly advantageous for processing experimental data (task ad C) and for various simulation studies (tasks A, B, D). The Johnson distribution system contains three groups of distributions resulting from the transformation of the standard normal distribution and covers the entire acceptable range of skewness and kurtosis.

A quantity x having a Johnson probability distribution is a function of a quantity z with a standard normal distribution in the form.

$$x = g\left(\exp\left(\frac{z-\mu}{\sigma}\right)\right) \tag{6.165}$$

The function $g(z)$ is monotonous and has three basic forms.

For the class of bounded distributions SB $(0 < x < 1)$ is

$$g(z) = \frac{z}{1+z} \tag{6.166}$$

For the class of SL of lognormal distributions $(0 < x < 1)$ is

$$g(z) = z \tag{6.167}$$

For the class of unbounded distributions SB ($-\infty < x < \infty$) is

$$g(z) = 0,5(z - 1/z) \tag{6.168}$$

Details about the processing of this system of densities are given in (Slifker & Shapiro, 1980).

The lambda distribution system is based on the well-known fact that the basic task in many applications is to determine the quantile function. Unlike the Johnson system, this function is determined here as a non-linear transformation of the variable q, which has a rectangular distribution on the interval [0, 1].

The variable x has a lambda distribution if it is valid

$$x = \lambda_1 + \left[q^{\lambda_3} - (1 - q)^{\lambda_4}\right]/\lambda_2 \tag{6.169}$$

Obviously, λ_1 is a location parameter and λ_2 is a dispersion parameter. The parameters λ_3 and λ_4 are related to skewness and kurtosis. If $\lambda_3 = \lambda_4$, it is possible to use Eq. (6.169) for a wide class of symmetric distributions differing in kurtosis. This distribution system allows the description of a number of unimodal symmetric and asymmetric distributions for which $g_2 \geq 2.75 + 1.6625\, g_1^2$. Details on the processing of this system of densities are presented in the work (Karian et al., 1996).

It is often beneficial to choose a system of densities with shape parameters related to the quantile characteristics, which allows easy diagnosis and quantification of the shape of the distribution. One such system is referred to as *g-h*.

The *g-h* system of empirical distributions is based on a monotonic transformation of a standardized quantity with a normal distribution and uses appropriate quantile functions to define the shape. It is one of the translational distribution systems, where the original random variables (quantiles) x are first standardized to normalized quantiles.

$$Y = (x - \tilde{x}_{0.5})/R \tag{6.170}$$

where $\tilde{x}_{0.5}$ is the location parameter (median) and R is the appropriate scale parameter [e.g. DQ or the interquartile estimate of the standard deviation from Eq. (6.193)]. The variable y has the same distribution shape as the variable x and is a monotone transformation of a standard random variable z having a normal distribution. Formally, then is $Y = Y(z)$. For the g-distribution class (skewed only), a general type transformation is chosen

$$Y = Q_{g,0}(z) = G(z) \cdot z \tag{6.171}$$

where $G(z)$ is an odd function for which it must hold

$$G_{g,0}(0) = 0 \text{ and } \lim_{z \to 0} Q_{g,0}(z) \approx z \tag{6.172}$$

$G(z)$ is actually a skewness operator that depends on the appropriate *skewness parameter* g. It is appropriate that for $g > 0$, the distributions are skewed to the right, for $g < 0$, the distributions are skewed to the left, and for $g = 0$, a distributions are symmetrical. A simple function which meets all these requirements has the form

$$G(z) = \left[\exp(gz) - 1\right]/(gz) \tag{6.173}$$

The probability density function of g-distribution $f_g(x)$ has the form

$$f_g(x) = 1/\left[\sqrt{2\pi}\,|(x - \tilde{x}_{0.5})\,g + R|\right]\,\exp\left(-\left\{\ln\left[(x - \tilde{x}_{0.5})\,g/R + 1\right]\right\}^2/(2g^2)\right) \tag{6.174}$$

It is therefore a lognormal distribution (SL type distribution in the Johnson system of densities).

For the *h-distribution* class (symmetric with different kurtosis), this type of transformation is selected.

$$Y = Q_{0,h}(z) = H(z)z \tag{6.175}$$

The kurtosis operator $H(z)$ must be an increasing positive even function dependent on a suitable kurtosis parameter. For $h = 0$, it is a normal distribution and for the higher $h > 0$, the corresponding distribution has longer tails. Proper function has the form

$$H(z) = \exp\left(h z^2/2\right) \tag{6.176}$$

It follows from this equation that the parameter h can formally be negative. However, the monotonicity condition of the function Q_g, 0 is violated if $z^2 > -1/h$. The h-distribution class defined by Eq. (6.176) behaves like a Pareto distribution in the region of tails. The choice of factor 1/2 in Eq. (6.176) also ensures that for $h \approx 1$ the Q_0, h distribution is close to the Cauchy distribution (Thiele et al., 1987). However, the probability density function of h distribution can only be obtained numerically. A two-parameter transformation is chosen for the general class of *g-h distributions*.

$$Y = Q_{g,h}(z) = G(z)\,H(z)\,z \tag{6.177}$$

where $Q(z)$ is chosen according to Eq. (6.171) and $H(z)$ is defined by Eq. (6.176). In this case, the corresponding probability density function must be calculated numerically. For the generation of random numbers from the g-h distribution with the selected parameters g and h, it is sufficient to generate the quantiles z_p of the standard normal distribution (here p determines 100 p% quantile) and substitute into

Eq. (6.177) for $z = z_p$. In order to use this system in the processing of asymmetric sampling distributions, it is necessary to know the expressions for the first four moments. These are functions of the parameters g and h. For a standard random variable $Y = Q_{g,h}(z)$ from a g-h distribution, the mean value is defined by

$$E(Y) = \frac{1}{g\sqrt{h-1}}\left[\exp\left(\frac{g^2}{2(1-h)}\right) - 1\right]$$
$$\leq 0 \leq h \leq 1 \tag{6.178}$$

and for variance is valid

$$D(Y) = \frac{1}{g^2\sqrt{1-2h}}\left[\exp(2\omega) - 2\exp\left(\frac{\omega}{2}\right) + 1\right] - \frac{1}{g^2(1-h)}\left[\exp\left(\frac{\omega}{2}\right) - 1\right]^2$$
$$0 \leq h < 0.5 \tag{6.179}$$

where $\omega = g^2/(1-2h)$. For the g-distribution case, $h = 0$ and then from Eq. (6.178), it is

$$E(Y) = \left[\exp\left(\frac{g^2}{2}\right) - 1\right]/g \tag{6.180}$$

and from Eq. (6.179), it is

$$D(Y) = \exp(g^2)\left[\exp(g^2) - 1\right]/g^2 \tag{6.181}$$

Skewness g_1 of the g-distribution can be expressed in the form

$$g_1 = \left[\exp(3g^2) - 3\exp(g^2) + 2\right]/\sqrt{(\exp(g^2) - 1)^3} \tag{6.182}$$

And kurtosis g_2 has the form

$$g_2 = \left[\exp(6g^2) - 4\exp(3g^2) + 6\exp(g^2) - 3\right]/(\exp(g^2) - 1)^2 \tag{6.183}$$

The third central moment, which is needed in some cases, is given by the relation

$$E[(Y - E(Y))]^3 = \exp(3g^2/2)\left[\exp(3g^2) - 3\exp(g^2) + 2\right]/g^3 \tag{6.184}$$

For the case of the *h-distribution* is $g = 0$, the odd central moments are zero and thus the moment $E(Y) = g_1 = 0$. For the variance is then

$$D(Y) = \frac{1}{\sqrt{(1-2h)^3}} \quad \text{pro} \quad 0 \leq \text{h} \leq 0,5 \tag{6.185}$$

and kurtosis has form

$$g_2 = 3(1-2h)^3/\sqrt{(1-4h)^5} \tag{6.186}$$

When calculating the kurtosis of the higher moments of the *g-h distribution*, it is possible to use the Martinez relation, which is valid for $g \neq 0$ and $0 \leq h \leq 1/n$ (Thiele et al., 1987).

$$E(Y^n) = \frac{1}{g^n \sqrt{1-nh}} \sum_{i=0}^{n} (-1)^i \binom{n}{i} \exp\left\{ \frac{[(n-i)g]^2}{2(1-nh)} \right\} \tag{6.187}$$

Advantageously, to estimate the parameters g, h, or both, the sample median $\tilde{x}_{0.5}$ and pairs of quantiles $\tilde{x}_p$, $\tilde{x}_{1-p}$ for suitable $0 < p < 0.5$ can be used (*this range of p is used in other relations*). For small samples, it is possible to use all order statistics $x_{(i)}$. For larger samples, letter values are used, where $p = 2^{-i}$ $i = 2, 3$ (corresponding to quartiles, octiles, sedeciles, etc.). In the case of the g-distribution, it is possible to determine the parameter g (and also R) from two conditions for the selected probability p

$$\tilde{x}_p = \tilde{x}_{0.5} + R\,Y = \tilde{x}_{0.5} + R\left[\frac{\exp(g\,z_p) - 1}{g}\right]$$

$$\tilde{x}_{1-p} = \tilde{x}_{0.5} + R\left[\frac{\exp(-g\,z_p) - 1}{g}\right] \tag{6.188}$$

where z_p is the standard quantiles of the normal distribution (due to symmetry, it is $z_p = -z_{1-p}$). After simple rearrangements, the skewness parameter g_p has the form

$$g_p = -\frac{1}{z_p} \ln\left[\frac{\tilde{x}_{1-p} - \tilde{x}_{0.5}}{\tilde{x}_{0.5} - \tilde{x}_p}\right] \tag{6.189}$$

The subscript p here indicates that the skewness parameter in general may depend on the position of the quantiles relative to the median. If it is g_p approximately constant (or not a function of p), the procedure is to determine the median g of all g_p. It is then simply possible to estimate the R parameter as a slope in the Q-Q graph, where x_p values are plotted against $Q_{g,o}(z_p)$. The straight line in this plot indicates that the g-distribution with a constant value of g can be used. For the same purpose, it is also possible to use a symmetry graph, when half-sums $0.5(\tilde{x}_p + \tilde{x}_{1-p})$ are plotted against $z_p^2/2$. Using Eq. (6.188), the theoretical dependence is easily determined.

$$\begin{aligned} 0,5(x_p + x_{1-p}) &= x_{0,5} + 0.5\frac{R}{g}[\exp(g\,z_p) + \exp(-g\,z_p) - 2] \\ &\approx x_{0.5} + R\,g\,z_p^2/2 \end{aligned} \tag{6.190}$$

Therefore, if the assumption of constancy of the skewness parameter g is valid, an approximately linear dependence will appear in the symmetry graph. However, this graph can only be used for small g. Non-linearity becomes apparent at higher g. If g_p, it is not constant; it is usually assumed that it is a simple polynomial for the normalized quantiles z_p, in respective of their squares. Simple is generalization of type

$$g_p(z_p) = a_o + a_1 z_p^2 \tag{6.191}$$

Equation (6.191) can be easily verified on the basis of the *skewness graph*, when g_p is plotted against z_p^2. If an approximately linear dependence appears in this graph, a more general distribution with a non-constant g_p expressed by Eq. (6.191) can be used. Parameters a_o and a_1 in Eq. (6.191) can be easily determined from the intercept and slope of the skewness plot (use of robust regression is recommended). For the h-distribution case, knowing the location parameter $\tilde{x}_{0.5}$ and the scale parameter R, the values of h_p can be determined directly from the relation (6.176) for each $\tilde{x}_p$. It will be in form

$$h_p = \left[2 \ln \left(\frac{(\tilde{x}_p - \tilde{x}_{0.5})/R}{z_p} \right) \right] / z_p^2 \tag{6.192}$$

In the case that the sampling distribution is symmetric, it must of course be $h_p = h_{1-p}$. An interquartile estimate of the standard deviation of the median is recommended as a suitable scaling parameter.

$$R = 0.926\,(\tilde{x}_{0.75} - \tilde{x}_{0.25})/\sqrt{N} \tag{6.193}$$

where N is the sample size. Under the assumption of symmetric distribution, it is possible to directly use relations

$$\begin{aligned} \tilde{x}_p &= \tilde{x}_{0.5} + R\, z_p \exp\left(h\, z_p^2/2 \right) \\ \tilde{x}_{1-p} &= \tilde{x}_{0.5} - R\, z_p \exp\left(h\, z_p^2/2 \right) \end{aligned} \tag{6.194}$$

(here $z_p < 0$ and $z_{1-p} = -z_p$) for evaluating linear function

$$\ln \left[\frac{\tilde{x}_{1-p} - \tilde{x}_p}{-2 z_p} \right] = \ln R + h\, z_p^2/2 \tag{6.195}$$

From Eq. (6.195), it can be seen that by plotting $\ln\left[(\tilde{x}_{1-p} - \tilde{x}_p)/(-2 z_p) \right]$ against $z_p^2/2$, in the case of the validity of the h-distribution with a constant kurtosis parameter h, an approximately linear dependence appeared. This dependence is referred to as a *pseudo-sigma graph* and allows estimation of h and R from the slope and intercept of the regression line. For a *normal distribution*, the pseudosigma plot is practically a horizontal straight line with zero slope and intercept ln (R). For non-symmetric distributions, pseudo-sigma plots are non-linear. If h_p is not constant (but $h_p \approx h_{1-p}$), it is assumed that this is again a simple type dependence

$$h_p = b_0 + b_1\, z_p^2 \tag{6.196}$$

Equation (6.196) can be easily verified on the basis of the kurtosis plot, plotting h_p as a function of z_p^2. If an approximately linear dependence appears in this graph, it means that a more general distribution with a non-constant kurtosis parameter is expressed by Eq. (6.196). Another possibility is the direct use of the h distribution for the construction of asymmetric distributions, when two branches with different h are considered. This *HH* distribution can be used if graph $\ln[(\tilde{x}_p - \tilde{x}_{0.5})/(z_p)]$ against $z_p^2/2$ is composed from two straight lines intersecting at the point $z_p = 0$. Other simple generalizations of the h distribution also satisfy this condition [*HQ*, *HHH*, and *HR* see (Thiele et al., 1987)].

The very common is the case when the distribution of the data is not symmetrical and even after a symmetrizing (logarithmic) transformation, the length of the tails does not correspond to a normal distribution. For this case, it is necessary to use the g-h distribution. Given the choice of $Q_{g,h}(z)$ in the form (6.175) and the functions $G(z)$, $H(z)$, it can be easily determined that Eq. (6.191) holds regardless of the size of h. This means that it will be possible to estimate of the kurtosis parameter g (as in the case of a "pure" g-distribution) and then apply a skewness correction before estimating h. Knowing g, it can be easily determined quantile range.

$$\tilde{x}_{1-p} - \tilde{x}_{0,5} = \frac{R}{g}\left(\exp(-g\, z_p) - 1\right) \cdot \exp\left(h\, z_p^2/2\right) \tag{6.197}$$

After rearrangements, a linear dependence appears

$$y^* = \ln\left[\frac{g\left(\tilde{x}_{1-p} - \tilde{x}_{0,5}\right)}{\exp(-g\, z_p) - 1}\right] = \ln R + h\, z_p^2/2 \tag{6.198}$$

By plotting y^* against $z_p^2/2$ (*modified pseudo-sigma graph*) in the case of the validity of the g-h distribution with constant g, h, an approximately linear dependence appears. It is also easy to proceed in the case that g_p is expressed by Eq. (6.191) and h_p is expressed by Eq. (6.192). Since the estimate of g_p is independent of h_p, as with the "pure" g-distribution, the parameters a_0, a_1 can be found. However, Eq. (6.196), can no longer be used for parameter b_0, b_1 estimation. But a skewness correction must be made [as in Eq. (6.197)]. After rearrangements, the following relationship is valid.

$$y_p^* = \ln R + \frac{h_0}{2} z_p^2 + \frac{h_1}{2} z_p^4 \tag{6.199}$$

Firstly, the skewness correction is again performed and following relation is determined.

$$y_p^* = \ln\left[\frac{(\tilde{x}_{1-p} - \tilde{x}_{0.5})g_p(z_p)}{\exp(-z_p\, g_p(z_p)) - 1}\right] \tag{6.200}$$

knowing a_0, a_1 for each p. The dependence of y_p^* against $z_p^2/2$ is known as a *generalized kurtosis plot.* If this graph is parabolic, it means that it is necessary to use a generalized g-h distribution with non-constant g, h parameters.

To verify the validity of different types of g-h distribution, it is of course appropriate to use the formal tools of linear regression and test the significance of the slope, respectively intercept. However, for the purposes of data analysis, it is usually sufficient to assess only in the case of constant parameters g, h the corresponding skewness and kurtosis for further analysis.

To assess the shape of the distribution with regard to skewness, it is possible to construct graphs:

- Symmetry i.e. dependence of $0.5(\tilde{x}_p + \tilde{x}_{1-p})$ on $z_p^2/2$
- Skewness, i.e. dependence of g_p on $z_p^2/2$

To assess the shape with respect to the kurtosis, these are the graphs:

- Pseudosigma, i.e. dependence of $\ln[(\tilde{x}_{1-p} - \tilde{x}_p)/(-2z_p)]$ on $z_p^2/2$.
- Kurtosis, i.e. dependence of h_p on $z_p^2/2$

Considering the asymmetry of data distribution, this distribution system allows processing only selected types of tasks. On the other hand, the appropriateness of the accepted parameters of skewness and kurtosis can be assessed comprehensively. In particular, the possibility of finding only g or only h distributions is quite interesting.

For the creation of confidence intervals and power analysis, it is often sufficient to find the skewness parameter g as the median of all of the g_p from Eq. (6.191).

6.7 Optimal Combination of Estimates

One of the basic tasks in data analysis is the estimation of unknown parameters. We restrict ourselves to the simple case is scalar parameter a, which can be estimated using M different estimators $\hat{a}_i$ whose mean values are generally equal to $E(\hat{a}_i) = m_i$. These estimates are generally biased and their accuracy can be characterized by the root mean square error.

MSE($\hat{a}_i$), for which is valid

$$\text{MSE}(\hat{a}_i) = E(\hat{a}_i - a)^2 = E(\hat{a}_i - m_i)^2 + E(m_i - a)^2 = \sigma_i^2 + b_i^2 \tag{6.201}$$

where σ_i^2 is the variance of the ith estimate and b_i^2 is the square of the corresponding bias. The question is if an appropriate combination of estimates would lead to a reduction of MSE. Let the combined estimator $\hat{c}$ be a linear combination of estimates $\hat{a}_i$ with weights w_i in the form.

$$c = \sum_i \hat{a}_i\, w_i = \mathbf{w}^T\, \mathbf{a} \tag{6.202}$$

where $\mathbf{w} = (w_1, w_2, ...w_M)$ and $\mathbf{a} = (\hat{a}_1, \hat{a}_2, ...\hat{a}_M)$. For the estimate $\hat{c}$, the mean squared error is then equal to

$$MSE(\hat{c}) = E\left[\mathbf{w}^T\, \mathbf{a} - E\left(\mathbf{w}^T\, \mathbf{a}\right)\right]^2 + E\left[E\left(\mathbf{w}^T\, \mathbf{a}\right) - a\right]^2 = \mathbf{w}^T\, \mathbf{S}\, \mathbf{w} + \left(\mathbf{w}^T\, \mathbf{m} - a\right)^2 \tag{6.203}$$

Here the elements of covariance matrix $\mathbf{S}(M \times M)$ have form

$$S_{ij} = E\left[(\hat{a}_i - m_i)\, \left(\hat{a}_j - m_j\right)\right] \tag{6.204}$$

Task is to analytically determine such a vector $\mathbf{w}^*$ for which MSE ($\hat{c}$) is minimal. Due to the quadratic form of this function, after putting the first derivative equal to zero, it is possible to find the minimum in the form

$$\mathbf{w}^* = \left(\mathbf{m}^T\, \mathbf{m} + \mathbf{S}\right)^{-1} a\, \mathbf{m} \tag{6.205}$$

It is possible to prove that even for the case where the estimates $\hat{a}_i$ are unbiased, the estimate $\hat{c}$ is biased. For example, when $M = 1$, and the case that $m_1 = a_1$ and $b_1 = 0$ it is valid that

$$\text{MSE}(\hat{c}) = E[w_1\, \hat{a}_1 - E(w_1\, \hat{a}_1)]^2 + E(E((w_1\, \hat{a}_1) - a))^2 = w_1^2\, \sigma_1^2 + a^2\, (w_1 - 1)^2 \tag{6.206}$$

For the optimal value of w*, then it is

$$\frac{\mathrm{d}}{\mathrm{d}w_1}\left[w_1^2\, \sigma_1^2 + a^2\, (w_1 - 1)^2\right] = 0 \text{ and then } w_1^* = \frac{a^2}{\sigma_1^2 + a^2} \tag{6.207}$$

It is obvious that for $a \neq 0$ and $\sigma_1^2 \geq 0$ is $w_1^* \leq 1$. This means that $E(\hat{c}) = E(w_1\, \hat{a}_1) < m_1$, and the estimate is biased (underestimated). It can be shown that if normalized weights are used, for which $\sum w_i = 1$ and $w_i \geq 0$, the combined estimator is unbiased. Let the unnormalized weights be $\mathbf{h} = (h_1, h_2 \ldots h_M)$, so the relation holds

$$\mathbf{w} = \left(\mathbf{1}^T\, \mathbf{h}\right)^{-1} \mathbf{h} \tag{6.208}$$

where $\mathbf{1} = (1, 1, \ldots 1)^T$ is a column vector of ones. To determine the optimal vector *h*, the method of Lagrange multipliers is used, when the following function is minimized.

$$L(\lambda, \mathbf{h}) = \mathrm{MSE}(\hat{c}) + \lambda\left(\mathbf{w}^T\,\mathbf{1} - 1\right) \tag{6.209}$$

where λ is the Lagrange multiplier. For the optimal vector of weights, it is then

$$\mathbf{h}^* = \left[\mathbf{S} + (\mathbf{m} - a\,\mathbf{1})\,(\mathbf{m} - a\,\mathbf{1})^T\right]^{-1}\mathbf{1} \tag{6.210}$$

It can be seen that the bias $(m_i - a)$ is in the weight vector. For the optimal weight vector $\mathbf{h}^*$ is $\mathrm{MSE}(\hat{c}) = [\mathbf{1}\,\mathbf{h}^*]^{-1}$. For unbiased estimates where $E(\hat{c}) = a$, when $(m_i - a) = 0$, and $\mathbf{h}^* = \mathbf{S}^{-1}\,\mathbf{1}$, it is therefore true that the combined estimate is also unbiased (a consequence of the normalization of the weights). The corresponding mean squared error is equal to $\mathrm{MSE}(\hat{c}) = [\mathbf{1}^T\,\mathbf{S}^{-1}\,\mathbf{1}]^{-1}$. Let us state that for the case of mutually uncorrelated estimates (when the covariance matrix S is diagonal), the elements of the optimal weight vector are $h_i^* = 1/\sigma_i^2$. The corresponding mean squared error is equal to

$$\mathrm{MSE}(\hat{c}) = \left(\sum 1/\sigma_i^2\right)^{-1} \tag{6.211}$$

The choice of the weight vector therefore affects the bias of the combined estimate. Using the case of two estimators $\hat{a}_1$ with bias b_1 and $\hat{a}_2$ with bias b_2, it is possible to create a bias-reducted combined estimate. Let it be defined as the bias ratio $R = b_1/b_2$. The combined estimate has then form

$$\hat{c} = \frac{\hat{a}_1 - R\,\hat{a}_2}{1 - R} \tag{6.212}$$

and is unbiased. This can be easily demonstrated by modifying of the relationship.

$$E(\hat{c}) = \frac{E(\hat{a}_1)}{1 - R} - \frac{R}{1 - R}E(\hat{a}_2) = \frac{b_1 + a - b_1 - R\,a}{1 - R} = a \tag{6.213}$$

Corresponding variance estimator is

$$D(\hat{c}) = \frac{1}{(1 - R)^2}\left[\sigma_1^2 + R^2\,\sigma_2^2 - 2\,R\,\mathrm{cov}(\hat{a}_1, \hat{a}_2)\right] \tag{6.214}$$

A common case is that the bias can be decomposed into two terms, i.e. $b_j = f_j(N)\,b_j^*(a)\;j = 1, 2$. For similar estimates obtained, e.g. from different subsets of the data, is $b_1^*(a) = b_2^*(a) = b(a)$ and then is $R = f_1(N)/f_2(N)$, where N is the number of points used to compute the estimates. For *Jackknife methods*, an estimate $\hat{a}_1$ from all data $(x_1, x_2, \ldots x_n)$ and an estimate $\hat{a}_2$ determined as an arithmetic mean are combined by default.

$$\hat{a}_2 = \frac{1}{N}\sum \hat{a}_{(-i)} \tag{6.215}$$

where $\hat{a}_{(-i)}$ is the estimate obtained from the data where the i-th element has been omitted Let the bias b_1 can be expressed in the form.

$$b_1 = E(\hat{a}_1 - a) \approx \frac{g_1}{N} + \frac{g_2}{N^2} + \frac{g_3}{N^3} + \cdots \tag{6.216}$$

where the functions $g_1,\ g_2, g_3 \cdots$ are not containing N. For b_2, is then, analogously

$$b_2 = E(\hat{a}_2 - a) \approx \frac{g_1}{N-1} + \frac{g_2}{(N-1)^2} + \frac{g_3}{(N-1)^3} + \cdots \tag{6.217}$$

When it is selected, $R = (N-1)/N$ the combined *Jackknife estimate* $\hat{c}_J \equiv \hat{a}_J$ is equal to $\hat{c}_J = N\,\hat{a}_1 + (N-1)\,\hat{a}_2$. Its bias is already reduced because

$$E(\hat{c}_J - a) \approx -\frac{g_2}{N(N-1)} - \frac{g_3(2N-1)}{N^2(N-1)^2}\cdots \tag{6.218}$$

Quenouille and Tukey [see Johnson (1949)] suggested calculating so-called *pseudo-values*

$$J_i = \hat{a}_1 N - (N-1)\,\hat{a}_{(-i)} \tag{6.219}$$

and from them, it is calculated so-called *Jackknife estimate.*

$$\overline{a}_J = \frac{1}{N}\sum J_i = \hat{a}_1 - (N-1)\,\hat{a}_2 \tag{6.220}$$

It is easy to determine that this estimator corresponds to the choice of the $R = (N-1)/N$ in combined estimator $\hat{c}$. Because it is true

$$\frac{1}{N(N-1)} \approx \frac{1}{N^2}\left(1 + \frac{1}{N} + \frac{1}{N^2} + \cdots\right) \text{ je } E(\overline{a}_J - a) \approx -\frac{g_2}{N^2} \tag{6.221}$$

has a Jackknife estimator first-order bias $\approx 1/N$ removed. Tukey (1958) proved that from mean sample sizes, pseudo-values J_i are approximately normally distributed with mean $\overline{a}_J$ and variance $s_J^2 = \frac{1}{N(N-1)}\sum (J_i - \overline{a}_J)^2$. The random variable $t \approx \frac{(\overline{a}_J - a)}{s_J}$ then has a Student distribution with $(N-1)$ degrees of freedom. Since J_i are mutually correlated, it is appropriate to use a corrected expression.

$$t \approx \frac{(\overline{a}_J - a)}{s_J}\sqrt{\frac{N-1}{2N-1}} \tag{6.222}$$

It is easy to see that if the parameter $a = \mu$ is the mean value, is estimated as the sample arithmetic mean $\hat{a}_1 = \overline{x}$, and is the pseudovalue directly $J_i = x_i$. Then the Jackknife estimate is equal to $\overline{a}_J = \overline{x}$ and the variance $s_J^2 = s^2/N$ is equal to the variance of the sample mean. In this case, it is therefore unnecessary to perform Jackknife estimation.

For *block Jackknife*, the data is divided into m blocks and in each block, the Jackknife estimates $\overline{a}_{Ji}$,s_{Ji}^2 are calculated. Block Jackknife estimates are then arithmetic averages of estimates obtained from all blocks. A special case is when there are only two values in each block (the so-called multihalver) (Tukey, 1987).

The Jackknife procedure can also be used to estimate the variance and bias of the known function $g(a)$ of the distribution parameter a (commonly the mean value μ) (Paar, 1983).

If estimates $\hat{a} = \hat{a}_1$ and $\hat{a}_{(-i)}$ for $i = 1 \ldots N$ are available, a Jackknife estimate of the bias of the function $g(\hat{a})$ can be determined.

$$b_J = (N-1)\sum_{i=1}^{N}\left(g\left(\hat{a}_{(-i)}\right) - g(\hat{a})\right)/N \tag{6.223}$$

Jackknife estimate of the variance of a function $g(\hat{a})$ is in form

$$s_{gJ}^2 = (N-1)\sum_{i=1}^{N}\left(g\left(\hat{a}_{(-i)}\right) - \frac{1}{N}\sum_{i=1}^{N} g\left(\hat{a}_{(-i)}\right)\right)^2 /N \tag{6.224}$$

which is a multiple of the sample variance of the function $g\left(\hat{a}_{(-i)}\right)$. Similarly, the bias and variance of the function $g(\hat{a})$ can also be determined by using of Bootstrap method (Paar, 1983).

6.8 Bootstrap Methods

Bootstrap methods are a typical representative of computer-intensive methods, when the computer is part in the metrology chain. These methods are advantageous for processing data where it can be assumed a priori that the conditions for classical statistical analysis will not be acceptable. Bootstrap methods essentially use a combination of estimates to determine the variance and possibly bias of complex population parameters. In Germany, the name "die Muenchhausen Methode" is used instead of Bootstrap, after Baron von Muenchhausen, who was the hero of many fantasy stories. In one of them, he saved his life by pulling himself out of the swamp by his own hair. The German name came from the translation of one of the works on Bootstrap into German, when it was recommended to "pull yourself out of the statistical swamp by your mathematical hair".

Parameter a estimate $\hat{a}$ in complex situations where their statistical characteristics are not known are traditionally among relatively difficult problems. For the case of an unknown distribution of sample $\mathbf{x} = (x_1 \ldots x_N)$ and arbitrary parameter a, the Bootstrap technique can be advantageously used, which allows both finding the distribution of sample statistics and constructing a confidence interval. The basic idea of the Bootstrap method is simple (Davison & Hinkley, 1997; Efron, 1992; Wehrens et al., 2000). It consists of the generation of an M-tuple of simulated samples $v_1 \ldots v_M$, referred to as Bootstrap samples. Their distribution corresponds to the distribution of the original sample $\boldsymbol{x}$ characterized by the probability density function $f(x)$. From these samples, the M-tuple of Bootstrap estimates $\hat{a}_{Bi} = \hat{a}(\mathbf{v}_i)$ of the parameter a is determined. From this M-tuple of values, confidence intervals can be calculated by using a huge number of methods (Meloun & Militký, 2012).

The basic prerequisite for the success of the entire procedure is the generation of Bootstrap samples. For this purpose, it will be necessary either know or choose the probability density function $f(x)$. The standard non-parametric Bootstrap technique is based on a non-parametric estimate $\hat{f}(x)$ of the sample probability density function in the form

$$\hat{f}(x) = \frac{1}{N}\delta(x - x_i) \tag{6.225}$$

where the Dirac function $\delta(x-x_i) = 1$ for $(x = x_i)$ and everywhere else is $\delta(x-x_i) = 0$. This distribution assigns a probability of $1/N$ to each sample value. Simulated samples are then implemented as random selections from the original sample $\boldsymbol{x}$ with return (i.e. element of the original sample can occur repeatedly in the simulated sample or not be presented). Some results of non-parametric Bootstrap are shown in Fig. 6.11)

It is fact, the probability that arbitrary i-th original sample element will not be appeared in n independent samples (Bootstrap samples) is equal to $P(i) = (1 - 1/n)^n \approx e^{-1} \approx 0,37$. So, for example, if the original sample contains outlier, a state will occur where this outlier will often fall out from Bootstrap samples. For the case when the number of outliers is exactly K, the probability that they will not be in at least one Bootstrap sample is approximately $P_K = e^{-K}$. The number of Bootstrap samples when at least one is not containing any K-tuple of data is equal to $n_K \approx e^K$. For $K = 5$ it is $n_K = 148$ and for $K = 6$ it is $n_K = 1016$. For the case that M is greater than n_K, at least one selection is without outliers (pure). Principle of Bootstrap is shown in Fig. 6.10

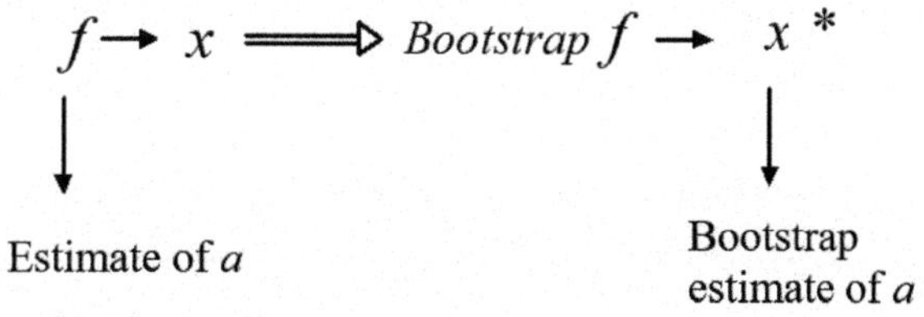

Fig. 6.10 Non-parametric bootstrap

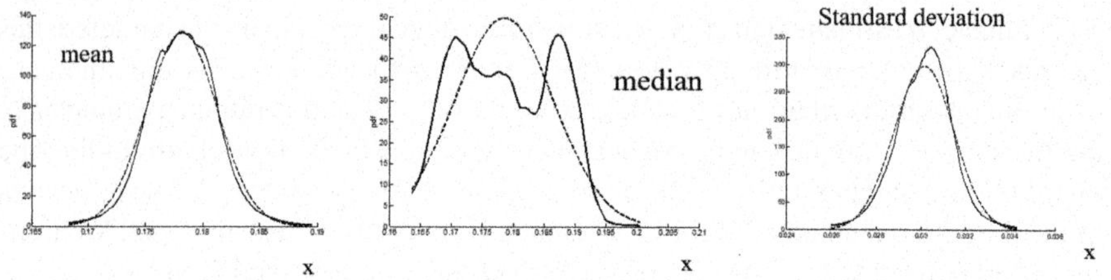

a) mean b) median c) standard deviation

Fig. 6.11 Distribution of selected parameter estimates (arithmetic mean, median, and standard deviation) from Bootstrap samples (solid line) and probability density function of normal distribution with parameters calculated by non-parametric Bootstrap (dashed)

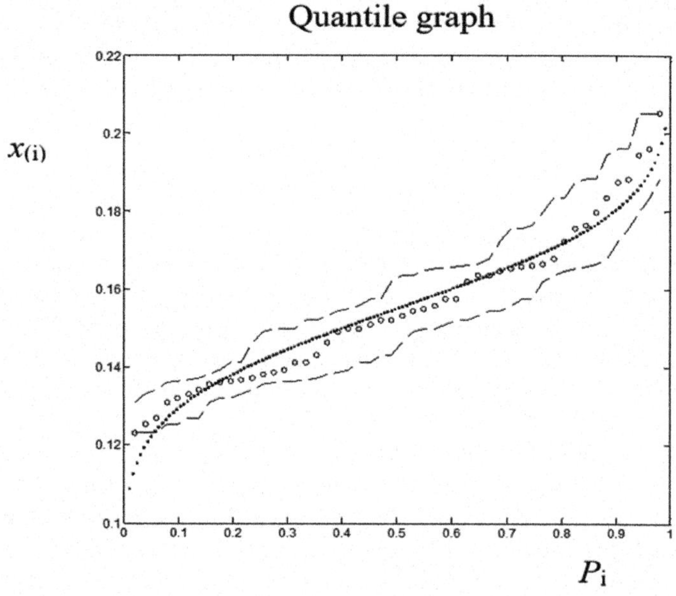

Fig. 6.12 Quantile plot for example of the strength characteristics of cotton yarns produced using ring spinning (circles) with uncertainty limits—non-parametric bootstrap (dashed) and quantile function of a normal distribution (dotted)

In Fig. 6.11, the empirical Bootstrap probability densities of estimates of the arithmetic mean (a), standard deviation (b), and median (c) from an example of the strength of cotton yarns of ring and compact mixed sample ($N = 100$) (parameter estimates are in Table 6.3) are shown. The solid line is the non-parametric density estimate and the dashed line is the probability density function of the normal distribution with estimates of the mean and variance from non-parametric Bootstrap.

It can be seen from Fig. 6.11 that the bimodality of the cotton strength mixed (ring and compact technology) sample was manifested only on the non-parametric Bootstrap estimates of the median.

Another option is to construct a suitable parametric model $f(x)$, *estimate* its parameters, and generate simulated samples using standard procedures. This approach has a number of problems related to possible inhomogeneity, outliers, heteroscedasticity, and autocorrelation.

In more complex tasks such as *regression*, the generation of Bootstrap samples is performed in different ways. The classical regression Bootstrap consists in selecting individual points based on their corresponding residuals, which form the "basic one-dimensional selection". It can be shown that the mean values of the regression parameters estimate from the Bootstrap samples generated in this way is equal to the least square regression parameters estimate for original data. The problem here is mainly the correlation of the residuals and their non-constant dispersion. Especially, for *Bumping variants*, the *paired Bootstrap* is used, where individual points are ranked based on a random selection with a return from a vector of indices. The problem here is mainly possible bad conditionality or collinearity. Therefore, only those samples that have a sufficiently large and positive smallest Eigen number of the $\mathbf{X}^T\mathbf{X}$ matrix are often accepted.

The parametric regression Bootstrap is based on the estimation of the residual variance σ_R^2 and the addition of random errors, e.g. from the normal distribution $N(0, \sigma_R^2)$ to the individual values $f\left(x_i, \hat{a}_{Bj}\right)$.

The non-parametric regression Bootstrap adds residuals generated from the least squares residuals by random selection with return to the values $f\left(x_i, \hat{a}_{Bj}\right)$. Here $\hat{\mathbf{a}}$ are the estimates determined by the classic least squares minimization.

There are still other variants, such as *external Bootstrap*, abstract residuals (*wild Bootstrap*), etc. It is understandable that the method of generation of Bootstrap samples will often be decisive with regard to achieving stability and robustness of Bootstrap estimates. The standard Bagging procedure calculates the combined estimate $\hat{a}_B$ as a simple arithmetic mean of the Bootstrap estimates.

$$\hat{a}_B = \frac{1}{M}\sum_{j=1}^{M}\hat{a}_{Bj} \tag{6.226}$$

Usually is the mean squared error $\mathrm{MSE}(\hat{a}_B) \leq \mathrm{MSE}(\hat{a})$, where $\hat{a}$ is the estimate from the original sample. Additionally, the estimates $\hat{a}_B$ are more stable and the variance of the combined estimate is easily determined as the variance of the Bootstrap estimates.

$$s_B^2 = \frac{1}{M}\sum_{i=1}^{M}\left(\hat{a}_{Bj} - \hat{a}_B\right)^2 \tag{6.227}$$

However, the problem is that the structure of the estimates is not preserved. For example, the regression subset determined by this procedure is no longer an optimal subset, etc. Therefore, the *Bumping technique* is often chosen, where instead of a combined estimate, a Bootstrap estimate $\hat{a}_{Bj}$, for which the required estimate quality criterion is optimal is taken directly. For example, for the mean squared error MSE criterion, it is valid, that

$$\hat{a}_B = \min \text{MSE}\left(\hat{a}_{Bj}\right) \tag{6.228}$$

Bootstrap estimates then also include the estimate $\hat{a}$ from the original sample. The advantage is that the criterion for Bumping does not have to be the same as the criterion used for determining the estimates, and moreover, this estimate is "naturally" robust. This method is suitable for more complex tasks, such as robust regression.

In the case where the purpose is "robustness", even more variants of Bootstrap are performed, when the number of samples is reached more quickly, when at least one does not contain an arbitrary K-tuple of data n_K. In the *double Bumping technique*, the pairs of samples $\mathbf{v}_j$ and $\mathbf{v}_{j+1}$ are constructed. The sample $\mathbf{v}_j$ is classic Bootstrap and the sample $\mathbf{v}_{j+1}$ is Bootstrap from the first sample $\mathbf{v}_j$ and not from the original data. The sample $\mathbf{v}_j$ contains only 0.63 N points, and the sample $\mathbf{v}_{j+1}$ contains $0.63\,N\,0.63 = 0.4N$ points only (both from original data).

To estimate the sensitivity of the variance $D(\hat{a}_B)$, or bias $b(\hat{a}_B)$ on the presence of individual observations, the *Jackknife after Bootstrap method* is used. For example, the variance sensitivity is expressed as $Cr_j = N\left[D(\hat{a}_B) - D_{-j}(\hat{a}_B)\right]$, where D_{-j} is the variance from samples not containing the j-th element of the original data.

When implementing Bootstrap methods, there are always both statistical errors (the sampling distribution $f(x)$ is unknown) and simulation errors (associated with the choice of the number of samples M). Bootstrap methods can also reduce bias by using a modified relationship

$$\hat{a}_{BV} = 2\hat{a} - \hat{a}_B \tag{6.229}$$

For sufficiently large M is $\hat{a}_{BV} \approx \hat{a}^2 - \hat{\sigma}^2/N$. For smooth and approximately linear statistics, the variance of the bias estimate is

$$D(b(\hat{a}_B)) = \frac{\sigma^2}{N} + \frac{N-1}{N\,M} \tag{6.230}$$

and for the variance of the variance estimate is

$$D(D(\hat{a}_B)) = \frac{2\sigma^4}{N^3} + \frac{2}{M}\left(\frac{\sigma^4}{N^2} + \frac{2\sigma^4}{N^3}\right) \tag{6.231}$$

The first term in these expressions is related to data variability and the second to variability due to simulation. It is therefore advisable to choose M greater than about $40N$ (usually a minimum of 400–600 Bootstrap samples is recommended). The larger M is, the better the estimates are up to a certain limit. Therefore, M around 800 is usually recommended for the construction of point estimates and M around 1000 (Chernick, 2008) for the construction of confidence intervals. Confidence intervals can be calculated from the M-tuple of values $\hat{a}_{Bj}$ using a variety of methods.

A. **Calculation from Asymptotic Normality**

It is the simplest procedure based on the idea that M is large enough and estimates $\hat{a}_{Bj}$ $i = 1 \ldots M$ can be treated as sample from a normal distribution. For the so-called Bootstrap estimation $\hat{a}_B$ of the mean value of the parameter a, the Eq. (6.226) is valid and for the corresponding variance, Eq. (6.227) is used. For 100($1 - \alpha$)% the confidence interval of the parameter a the known relation is used

$$\hat{a}_B - u_{1-\alpha/2}\, s_B/\sqrt{M} \leq a \leq \hat{a}_B + u_{1-\alpha/2}\, s_B/\sqrt{M} \tag{6.232}$$

where $u_{1-\alpha/2}$ is the quantile of the standard normal distribution.

B. **Percentile Interval**

This procedure is based on a non-parametric estimate of the limits of the confidence interval (more precisely, the probability interval) based on the order statistics $\hat{a}_{(i)}$, where $\hat{a}_{(i)} \leq \hat{a}_{(i+1)}$, Order statistics are the d% quantile of the estimate $\hat{a}$ distribution

$$d = \frac{i}{M+1} \tag{6.233}$$

The lower bound of the 100($1 - \alpha$) % confidence interval is then

$$\mathrm{LC} = \hat{a}_{(k_1)} \text{ where } k_1 = \mathrm{int}[\alpha(M+1)/2] \tag{6.234}$$

a pro horní mez platí

$$\mathrm{UC} = \hat{a}_{(k_2)} \text{ where } k_2 = \mathrm{int}[(1 - \alpha/2)\,(M+1)] \tag{6.235}$$

Here int(x) is the integer part of the number x. The non-parametric estimate of the distribution function of Bootstrap estimates $\hat{a}_{Bj}$ has the form

$$\hat{F}_B(x) = \frac{\sum_{j=1}^{M} I\left(\hat{a}_{Bj} \leq x\right)}{M} \tag{6.236}$$

Here $I(x)$ is an indicator function taking the value 1 if $\hat{a}_{Bj} \leq x$ and the value 0 elsewhere. Using it, it is possible to express the ($1 - 2\alpha$) confidence interval of parameter a in the form (Efron, 1987)

$$\hat{F}_B^{-1}(\alpha) \leq a \leq \hat{F}_B^{-1}(1-\alpha) \tag{6.237}$$

It is also possible to correct this interval with regard to reducing the bias.

Let the quantity Z_α be defined by relation $\Phi(Z_\alpha) = 1 - \alpha$, where $\Phi(x)$ is the distribution function of the standard normal distribution $N(0,1)$ and Z_0 is defined as $Z_0 = \hat{F}_B^{-1}(\hat{a}_B)$. Then the $(1 - 2\alpha)$ confidence interval of the parameter a with bias reduction (BC) is in the form

$$\hat{F}_B^{-1}(2Z_0 - Z_\alpha) \leq a \leq \hat{F}_B^{-1}(2Z_0 + Z_\alpha) \tag{6.238}$$

Further improvement can be achieved by using a special acceleration constant q (Efron, 1987), which can be expressed in the form

$$q \approx \frac{1}{6} \sum_{i=1}^{N} (E_i)^3 / \sqrt{\left(\sum_{i=1}^{N} (E_i)^2 \right)^3} \tag{6.239}$$

Here E_i is the influence function at the location x_i, which can be replaced by the Jackniffe characteristic

$$E_i \approx (N - 1)\left(\hat{a} - \hat{a}_{(-i)}\right) \tag{6.240}$$

Using the acceleration constant q, the $(1 - 2\alpha)$ confidence interval of the parameter a corrected for bias reduction (BC_a) is in the form (Efron, 1987)

$$\hat{F}_B^{-1}(\Phi(Z[\alpha])) \leq a \leq \hat{F}_B^{-1}(\Phi(Z[1 - \alpha])) \text{ where } Z[\alpha] = Z_0 + \frac{Z_0 - Z_\alpha}{1 - q(Z_0 - Z_\alpha)} \tag{6.241}$$

Based on a simulation study (Frangos and Schucany, 1990), the BC_a confidence interval is the best. If $q = 0$, the BC_a interval is identical to the BC interval, and for $Z_0 = 0$, the BC interval is identical to the classical percentile confidence interval. Hall's modification of the percentile method consists in replacing the empirical distribution function $\hat{F}_B(x)$ with the distribution function $\hat{F}_{BH}(x)$ defined by the relation (Hall, 1988)

$$\hat{F}_{BH}(x) = \frac{\sum_{j=1}^{M} I\left(\left(\hat{a}_{Bj} - \hat{a}\right) \leq x\right)}{M} \tag{6.242}$$

Using $\hat{F}_{BH}(x)$, it is possible to express the $(1 - 2\alpha)$ confidence interval of parameter a in the form (Hall, 1988)

$$\hat{F}_{BH}^{-1}(\alpha) \leq a \leq \hat{F}_{BH}^{-1}(1 - \alpha) \tag{6.243}$$

C. **Studentized Interval**

This interval is based on simple transformations leading to the Studentized random variable t_i

$$t_i = \frac{(\hat{a}_i - \hat{a}_B)\sqrt{N}}{s_{Bi}} \tag{6.244}$$

where s_{Bi} is the sample standard deviation calculated for the i-th Bootstrap sample $\mathbf{v}_i$. The $100(1-\alpha)\%$ confidence interval then has form

$$a_B - t_D\, s_B/\sqrt{M} \le a \le \hat{a}_B + t_D\, s_B/\sqrt{M} \tag{6.245}$$

where the quantile t_D is from Student distribution with $M-1$ degree of freedom

D. **Smoothed Interval**

Based on the values $\hat{a}_i$, an estimate of the probability density function $\mathrm{fe}(\hat{a})$ can be created, e.g. using a histogram or a kernel estimate. Knowing the function $\mathrm{fe}(\hat{a})$, a confidence interval is easily constructed directly from the definition. For the limits of this interval, it is then true that

$$\alpha/2 = \int_{-\infty}^{\mathrm{LC}} \mathrm{fe}(p)\mathrm{d}p \tag{6.246}$$

and

$$\alpha/2 = \int_{\mathrm{UC}}^{\infty} \mathrm{fe}(p)\mathrm{d}p \tag{6.247}$$

Depending on the type of fe estimation, it can be a numerical or analytical integration task. Bootstrap methods generally provide information on both point estimates and confidence intervals. From the point of view of the implementation of Bootstrap methods on the computer, it is the basis of the generation of simulated selections. This operation can be performed very simply in the MATLAB language using the vector trick.

To demonstrate the advanced use of the principles of the parametric and non-parametric Bootstrap methods, it is shown how to construct the confidences limits for the empirical quantile graph $Q(P)$, i.e. the dependence $x_{(i)}$ on P_i (see Sect. 6.5).

A. In a data-driven approach, it is assumed that the original sample contains all the necessary relevant information and the construction of simulated Bootstrap samples must therefore be made from the empirical distribution of the original sample (generation of samples with returns). To determine the confidence limits for the empirical quantile graph, the generation of M non-parametric Bootstrap samples estimating order statistics $\hat{a}_{B(j)} = x_{Bi(j)}\ i = 1 \ldots N, j = 1 \ldots M$ and the calculation of the limits (6.234), (6.235) are sufficient. This only requires sorting all Bootstrap samples to determine order statistics according to j. If $M = 1000$

is chosen, is $LC_i = x_{Bi\,(25)}$ and $UC_i = x_{Bi\,(975)}$. The 95% of all possible samples having the same characteristics as the original sample will be then within the limits LC, UC.

B. In the model-based approach, the original sample is assumed to have a normal distribution with parameters equal to the sample mean and sample standard deviation. The construction of simulated samples must therefore be made from a normal distribution with parameters determined from the original sample. To find the confidence limits for the empirical quantile graph, the generation of parametric Bootstrap samples from the normal distribution (very easily using a pseudorandom number generator) is then performed and the limits (6.234), (6.235) are calculated again as for the non-parametric Bootstrap. If $M = 1000$ is chosen, it is again $LC_i = x_{Bi(25)}$ and $UC_i = x_{Bi\,(975)}$. This percentile estimate is preferred here because it is not "encumbered" by distributional assumptions.

Figure 6.12 shows the result (confidence limits) of quantile graph when using the non-parametric Bootstrap (*A*) for example of the strength characteristics of cotton yarns produced using ring spinning (parameter estimates are in Table 6.3).

The areas limited by dashed lines in Fig. 6.12 show the uncertainty due to the finiteness of the sample and natural variability of its elements.

Bootstrap methods are advantageously applicable to a relatively large area of statistical estimation. However, there are cases where they do not work well (see e.g. Chernick, 2008). Examples are distributions that do not have a finite variance (the central limit theorem does not apply to them). Also, distributions with extremely long tails and distribution of extremes do not perform best. Dependent data with an autocorrelation structure or long-range dependence (see Sect. 6.9) can also cause problems (Chernick, 2008). Good parameter estimates are difficult to obtain for small samples, so even Bootstrap estimates are of little use. It is generally recommended to use non-parametric Bootstrap methods for samples with more than 30 or 50 elements, which is quite limiting in many applications. Parametric Bootstrap methods work better even for small samples (Scholz, 2007).

6.9 Dependent Data

Dependent data is common when:

- It is a measurement corresponding to time series.
- Experiments are performed on the same system in different times.
- It is sampling from continuous processes.
- During the measurement, there are changes that are related to "time" or a quantity that is time-dependent (e.g. temperature).

Thus, time dependence is not always an indication of incorrect measurement. For the analysis of the time dependence of the data, it is advantageous to assume that the measurement if after constant intervals and therefore the index *i* corresponds to the time $t = (i - 1)d$, where d is the sampling interval. This assumption is not necessary, but in many cases, it facilitates interpretation. It is also possible to consider the same assumption for analyzing the "noise" component. In order to distinguish this case from the classical one-dimensional samples, the notation $y(\mathrm{i})$, $i = 1 \ldots N$ will be used for such data. The analysis of dependent data is discussed in detail in Sect. 4.8. Here, only the influence of the dependence of the data on the variance of the sample mean and autocorrelation is discussed. The following artificially generated data are used to illustrate individual procedures:

A. *Data A* generated as independent normally distributed random variables ($N = 100$)
B. *Data B* generated as the autoregressive model of II. Order with normally distributed noise

It is known that the statistical behavior of random processes can be described by the so-called autocovariance function (Maisel, 1971)

$$c(i, h) = c(h) = \mathrm{cov}(y(i)\, y(i, h)) = E(y(0)\, y(h)) \tag{6.248}$$

The last equality is valid for the case of centered data $E(y) = 0$ and so-called weak stationarity (the autocovariance depends on the displacement parameter h and not on the positions i). For the displacement $h = 0$, it results from Eq. (6.248) the variance $\sigma^2 = c(0)$. It is often more advantageous to use the autocorrelation function $R(h)$ defined by the relation (Maisel, 1971)

$$R(h) = \frac{\mathrm{cov}(y(0),\ y(h))}{v} = \frac{c(h)}{c(0)} \tag{6.249}$$

Figure 6.13 shows the autocorrelation functions for both samples, when $h = \text{lag}$.

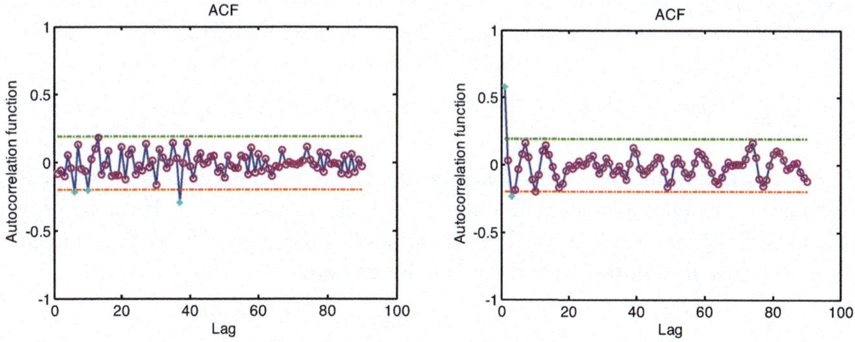

a) Data A

b) Data B

Fig. 6.13 Autocorrelation function for simulated samples

The autocorrelation function for data A lies in the band of insignificance, indicating non-correlation. For data B a significant autocorrelation of the first two orders is evident. Thus, autoregressive model II is identified.

Autocorrelation has a significant effect on the variance of the sample mean $D(y_A)$. For the case of so-called short-range dependence, the sample variance is calculated from the relationship.

$$D(y_A) = \frac{\sigma^2}{N}\left[1 + 2\sum_{k=1}^{N-1}(1 - k/N)\,R(k)\right] = \frac{\sigma^2}{N}G(R) \tag{6.250}$$

where $D(y)$ denotes the variance operator. If the data $y(i)$ are random independent equally distributed quantities, then $G(R) = 1$. For the case of so-called long-range dependence, the variance decreases more slowly.

$$D(y_A) = \frac{\sigma^2}{N^\beta}G(R) \tag{6.251}$$

where the parameter $0 \leq \beta \leq 1$ characterizes the given process. Another complication in the case of autocorrelated data is that the sample variance s^2 is dependent on $R(h)$. For the case of so-called short-range dependence, it is true that

$$E\left(s^2\right) = \sigma^2\left[1 - \frac{2}{N-1}\sum_{k=1}^{N-1}(1 - k/N)\,R(k)\right] \tag{6.252}$$

It is therefore obvious that non-zero autocorrelation will already be reflected in the estimate of the sample variance, which will also be reflected in the confidence intervals of the mean value. In addition, it depends on the type of process. To investigate the structure of these processes, it is advantageous to use the procedure of aggregating data into subsequent groups (blocks) containing L values. Aggregated data $y^{(L)}(i)$ are arithmetic averages of values in blocks of length L (Beran, 2003)

$$y^{(L)}(i) = \frac{1}{L}(y(iL - L - 1 + \cdots + y(iL))L = 1, 2, 3\ldots \tag{6.253}$$

If the original data are weakly stationary, the aggregated data are also weakly stationary with autocovariance function $c^{(L)}(h)$ and variance $\sigma^{2(L)}$. It can be proved (Mandelbrot & Van Ness, 1968) that the variance of the aggregated data is related to the autocovariance of the original data by the relation.

$$\sigma^{2(L)} = \frac{\sigma^2}{L} + \frac{2}{L^2}\sum_{s=1}^{L-1}\sum_{h=1}^{s}c(h) \tag{6.254}$$

An important characteristic is the autocorrelation coefficient of the first order ($h = 1$), for which it is valid

$$R^{(L)}(1) = 2\,\frac{\sigma^{2(2\,L)}}{\sigma^{(2\,L)}} - 1 \tag{6.255}$$

Different types of dependencies can then be easily defined using the characteristics of the aggregated data.

There are three basic types of dependencies (Beran, 2003):

- Random independent equally distributed variables. For this case, ideally all autocovariances are zero $c(h) = 0$ and the data are uncorrelated.
- Short-range dependencies (stationary), when the sum of all autocovariances $c(h)$ $h = 1, 2, \ldots$ converges.
- Long-range dependencies (stationary and non-stationary), when the sum of all autocovariances $c(h)$ $h = 1, 2, \ldots$ diverges.

For short-range (stationary) dependencies, the first-order autocorrelation is zero for sufficiently large L, $R^{(L)}(1) = 0$ for $L \to \infty$. The same is valid for other autocorrelations of all orders h. However, this means that the aggregated data $y^{(L)}(i)$ gradually become less correlated and pass for $L \to \infty$ to the case of independent quantities. Aggregation removes the problem of data dependency here. For sufficiently large L, the variance $\sigma^{2(L)} = \sigma^2/L$ is also the same as in the uncorrelated case.

For long-range (stationary) dependencies, for sufficiently large L the variance is $\sigma^{2(L)} \to \infty$ for $L \to \infty$. This means that the autocorrelation structure does not disappear. For sufficiently large L, it is valid

$$c(h) \approx h^{-\beta} \text{ a } \sigma^{2(L)} \approx L^{-\beta} \tag{6.256}$$

For stationary data, it is $0 < \beta < 1$ and for non-stationary data, it can be β outside this interval. For second-order self-similar processes, $R^{(L)}(h) = R(h)$ and $\sigma^{2(L)} = \sigma^2 L^{-\beta}$. It follows that the correlation structure is preserved and the aggregation here does not lead to uncorrelated data. It can be seen that the characteristic for the long-range dependence is the parameter β. This parameter is related to the so-called Hurt exponent H by relation $H = 1 - 0.5\beta$ (Beran, 2003).

If $H = 0.5$ data are independent. The exponent H is also related to the fractal dimension D_F, because $D_F = 2 - H$ (Mandelbrot & Van Ness, 1968). In general, the l-th central moment of the aggregated data has the form

$$M_l^{(L)} = \frac{1}{N/L} \sum_{k=1}^{N/L} \mathrm{abs}(y^{(L)}(k) - y_A)^l \tag{6.257}$$

The quantity $M_l^{(L)}$ behaves asymptotically as a power function $M_l^{(L)} \approx L^{l(H-1)}$. If it is not a case of long-range dependence, $H = 0.5$. Slope of dependence of $M_l^{(L)}$ on L in logarithmic coordinates is then equal to – 1/2. It must be the case that both N and N/L are large enough.

Usually, variance is chosen and therefore created dependence of $\sigma^{2\,(L)}$ on L in logarithmic coordinates. The presence of short-range dependence data is manifested in these plots by non-linearity for small L. This procedure is very simple, but often does not lead to unambiguous results. In the case of linear dependencies, the slope often differs significantly from theoretical assumptions.

There are a number of more suitable options to estimate β or H. The most commonly used is the power spectral density (PSD) (Beran, 2003)

$$g(\omega) = \frac{1}{2\pi} \int_{h=-\infty}^{\infty} c(h)\, \exp(-i\, h\, \omega)\, \mathrm{d}\omega \quad -\pi < \omega < \pi \tag{6.258}$$

For small frequencies $\omega_k = 2\pi f_k = 2\pi k/N\ k = 1, \ldots m$ is valid

$$g(\omega) \approx \omega^{-(1-\beta)} \quad \omega \to 0 \tag{6.259}$$

The parameter β is again estimated as the slope of the regression line in logarithmic coordinates. A simple estimate of the PSD is the periodogram for which

$$I(\omega) = \frac{1}{N}\left[\left(\sum_{i=0}^{N-1} y(i)\, \cos(\omega\, i)\right)^2 + \left(\sum_{i=0}^{N-1} y(i)\, \sin(\omega\, i)\right)^2\right] \tag{6.260}$$

In Fig. 6.14, the dependences of PSD on ω in logarithmic coordinates for both examples are compared. Discrete Fourier transform DFT was used as PSD estimation. The PSD estimate here is DFT^2/N.

It is evident that for both cases, a relatively high dispersion of data results. For data A, an approximately zero trend can be seen, and for data B, it is a significantly non-linear trend.

It is evident that for both cases, a relatively high dispersion of data results. For data A, an approximately zero trend can be seen, and for data B, it is a significantly non-linear trend.

A simple procedure for verifying the assumption of data independence is the use of a cumulative periodogram, which can be easily calculated from the relation [(Beran, 2003) for Fourier frequencies $f_j = j/N$, $1 \le j \le [N/2]$].

$$\mathrm{CU}(f_i) = \frac{\sum_{j=1}^{i} I(f_j)}{N s^2} \tag{6.261}$$

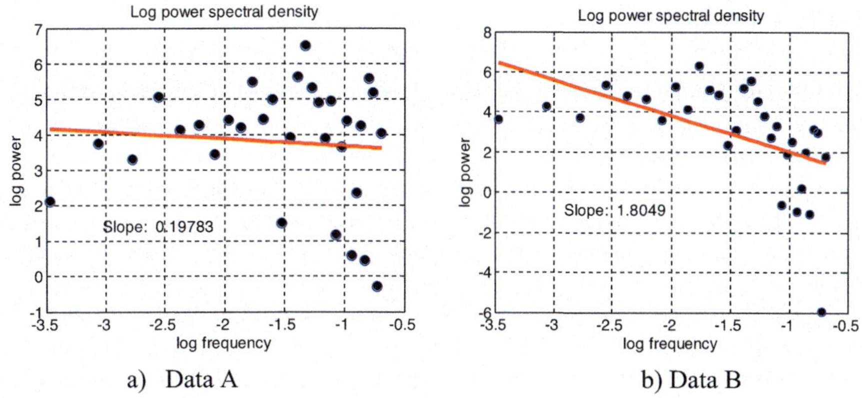

Fig. 6.14 Dependence of ln PSD on ln ω for simulated data

For independent data, plotting $\mathrm{CU}(f_i)$ against f_i yields an approximately linear relationship connecting the points (0, 0) and (0.5, 1). When this assumption is violated, the data systematically deviates from this straight line. The limit lines of the 95% confidence curves are expressed as

$$\mathrm{CU}(f_i) \pm 1.36/\sqrt{(N-2)/2} \tag{6.262}$$

In Fig. 6.15, a cumulative periodogram for simulated samples is shown. It can be seen that the cumulative periodogram for data A lies within the limits for uncorrelated data. For data B, a non-linear trend related to the correlation structure is evident.

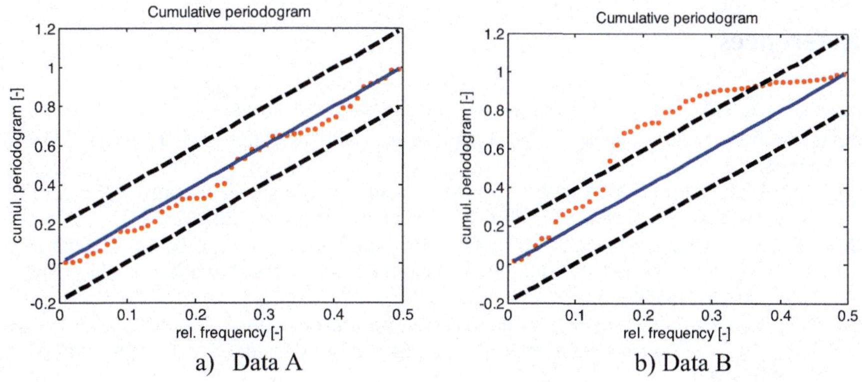

Fig. 6.15 Cumulative periodogram for simulated samples

6.10 Conclusion

It is obvious that the statistical processing of data in practice has a number of specific features that must be considered. It is always advantageous to start with exploratory analysis and comparison, or selection, measurement models and only then choose the next treatment. In line with the concept of "statistical methods mining" (Parzen, 1979), it is often necessary to combine different approaches, such as transformation, robust methods, and computationally intensive methods, to achieve reasonable results.

Especially in the cases of data transformation, it should be kept in mind that it is a data-driven approach and for two different samples from the same distribution, different estimates of the transformation parameter can be obtained. The quality of the transformation expressed by the confidence interval of the parameter λ can be a guide. From the individual procedures of Bootstrap techniques, it is evident that uncertainty can be due to the limitation of the size of the sample (model approach) or due to the knowledge of only one sample (data-oriented approach). This very simple procedure can also be used for other graphs discussed in Sect. 6.6. Information is always obtained about the variability and reliability of individual representations used in the visualization of the distribution, or preliminary data analysis. It is also possible to better assess deviations from ideal shapes. The advantage is that when using computers, this procedure is easy to implement and does not require any strict assumptions other than "belief" in what the data actually are. Statistical processing of data in cases where the data are dependent has a number of specific peculiarities. The formal apparatus of statistics, or adapting the data to the needs of statistical analysis, without a deeper analysis here, can lead to disastrous conclusions.

References

Bartlett, M. S. (1947). The use of transformations. *Biometrics, 3*(1), 39–52.

Beran, J. (2003). *Statistics in musicology*. Chapman and Hall/CRC. https://doi.org/10.1201/9780203496947

Bickel, P. J., & Doksum, K. A. (1981). An analysis of transformations revisited. *Journal of the American Statistical Association, 76*(374), 296. https://doi.org/10.2307/2287831

Boos, D. D., & Hughes-Oliver, J. M. (2000). How large does n have to be for Z and t intervals. *Amer. Statist. 54*, 121–128. http://www.tandfonline.com.sci-hub.club/doi/abs/10.1080/00031305.2000.10474524

Box, G. E. P., & Cox, D. R. (1964). An analysis of transformations. *Journal of the Royal Statistical Society Series b: Statistical Methodology, 26*(2), 211–243. https://doi.org/10.1111/j.2517-6161.1964.tb00553.x

Cameron, M. A. (1984). Choosing a symmetrizing power transformation. *Journal of the American Statistical Association, 79*(385), 107–108. https://doi.org/10.1080/01621459.1984.10477070

Chernick, M. R. (2008). Bootstrap Methods: A Guide for Practitioners and Researchers. J. Wiley, Hoboken

Cleveland, W. S. (1984). Graphical methods for data presentation: Full scale breaks, dot charts, and multibased logging. *The Amer. Statist., 38*, 270–280.

Davison, A. C., & Hinkley, D. V. (1997). *Bootstrap methods and their application*. Cambridge University Press. https://doi.org/10.1017/CBO9780511802843

Doksum, K. A., & Wong, C.-W. (1983). Statistical tests based on transformed data. *Journal of the American Statistical Association, 78*(382), 411. https://doi.org/10.2307/2288649

Draper, N. R., & Cox, D. R. (1969). On distributions and their transformation to normality. *Journal of the Royal Statistical Society Series b: Statistical Methodology, 31*(3), 472–476. https://doi.org/10.1111/j.2517-6161.1969.tb00806.x

Efron, B. (1987). Better bootstrap confidence interval. *J. Amer. Statist. Assoc, 82*, 171–200.

Efron, B. (1992). *Bootstrap methods: Another look at the Jackknife* (pp. 569–593). https://doi.org/10.1007/978-1-4612-4380-9_41

Emerson, J. D., & Stoto, M. A. (1982). Exploratory methods for choosing power transformations. *Journal of the American Statistical Association, 77*(377), 103–108. https://doi.org/10.1080/01621459.1982.10477772

Frangos C. C., Schucany W. R. (1990). Jackknife estimation of the bootstrap acceleration constant. *Comput. Stat. Data Anal, 9*, 271–281.

Hall P. (1988). Theoretical comparison of bootstrap confidence intervals. *Annals Statist., 16*, 927–985.

Harrell, F. E., & Davis, C. E. (1982). A new distribution-free quantile estimator. *Biometrika, 69*(3), 635–640. https://doi.org/10.1093/biomet/69.3.635

Hinkley, D. V., & Runger, G. (1984). The analysis of transformed data. *Journal of the American Statistical Association, 79*(386), 302. https://doi.org/10.2307/2288264

Hoaglin, D. C., Iglewicz, B., & Tukey, J. W. (1986). Performance of some resistant rules for outlier labeling. *Journal of the American Statistical Association, 81*(396), 991–999. https://doi.org/10.1080/01621459.1986.10478363

John, J. A., & Draper, N. R. (1980). An alternative family of transformations. *Applied Statistics, 29*(2), 190. https://doi.org/10.2307/2986305

Johnson, N. L. (1949). Systems of frequency curves generated by methods of translation. *Biometrika, 36*(1/2), 149. https://doi.org/10.2307/2332539

Karian, Z. A., Dudewicz, E. J., & Mcdonald, P. (1996). The extended generalized lambda distribution system for fitting distributions to data: History, completion of theory, tables, applications, the "final word" on moment fits. *Communications in Statistics—Simulation and Computation, 25*(3), 611–642. https://doi.org/10.1080/03610919608813333

Kaskey, G., Kolman, B., Krishnaiah, P. R., & Steinberg, L. (1980). 10 transformations to normality. In *Handbook of statistics* (Vol. 1, pp. 321–341). Elsevier. https://doi.org/10.1016/S0169-7161(80)01012-7

Maisel, L. (1971). *Probability, statistics and random processes*. Simon Schuster.

Mandelbrot, B. B., & Van Ness, J. W. (1968). Fractional Brownian motions, fractional noises and applications. *SIAM Review, 10*(4), 422–437. https://doi.org/10.1137/1010093

Manly, B. F. J. (1976). Exponential data transformations. *The Satistician, 25*(1), 37. https://doi.org/10.2307/2988129

Mardia, K. V. (1980). 9 tests of unvariate and multivariate normality. In *Handbook of statistics* (Vol. 1, pp. 279–320). Elsevier. https://doi.org/10.1016/S0169-7161(80)01011-5

Meloun, M., & Militký, J. (2012). *Interactive statistical data analysis*. Carolinum. (in Czech)

Militký, J., & Meloun, M. (1999). Data treatment in trace analysis. *Conference Microelements, 99*. (in Czech)

Militký, J., & Militká, D. (1985). *Modern mathematic-statistical methods in metallurgy, III. Basic statistical methods*. NHKG. (in Czech)

Neyman, J., & Scott, E. L. (1960). Correction for bias introduced by a transformation of variables. *The Annals of Mathematical Statistics, 31*(3), 643–655. https://doi.org/10.1214/aoms/1177705791

Novitskii, P. V., & Zograf, I. A. (1985). Estimating the errors of means of measurement. *Énergoatomizdat*. (in Russian)

Osborne, J. W. (2002). Notes on the use of data transformations. *Practical Assessment, Research & Evaluation, 8*, 1–7.

Paar W. C. (1983). A note on the jackknife, the bootstrap and the delta method estimators of bias and variance. *Biometrika, 70*, 719–722.

Parzen, E. (1979). *Statistical methods mining and nonparametric quantile domain data analysis.* https://api.semanticscholar.org/CorpusID:116826138

Royston, J. P. (1982). An extension of Shapiro and Wilk's W test for normality to large samples. *Applied Statistics, 31*(2), 115. https://doi.org/10.2307/2347973

Scholz, F. W. (2007). The Bootstrap Small Sample Properties. *Tech Rept -93-051, Boeing Computer Services*

Slifker, J. F., & Shapiro, S. S. (1980). The Johnson system: Selection and parameter estimation. *Technometrics, 22*(2), 239. https://doi.org/10.2307/1268463

Taylor, J. M. G. (1986). The retransformed mean after a fitted power transformation. *Journal of the American Statistical Association, 81*(393), 114–118. https://doi.org/10.1080/01621459.1986.10478246

Thiele, H., Hoaglin, D. C., Mosteller, F., & Tukey, J. W. (1987). Exploring data tables, trends, and shapes. *Biometrical Journal, 29*(2), 249–250. https://doi.org/10.1002/bimj.4710290224

Tukey, J. W. (1958). Bias and confidence in not quite large samples. *Ann Math. Stat. 29*, 614.

Tukey, J. W. (1987). Kinds of bootstraps and kinds of Jackknifes, discussed in terms of a year of weather-related data. Tech. Rep. 292, Department of Statistics, Princeton Univ.

Wehrens, R., Putter, H., & Buydens, L. M. C. (2000). The bootstrap: A tutorial. *Chemometrics and Intelligent Laboratory Systems, 54*(1), 35–52. https://doi.org/10.1016/S0169-7439(00)00102-7

Yeo, I.-K. (2000). A new family of power transformations to improve normality or symmetry. *Biometrika, 87*(4), 954–959. https://doi.org/10.1093/biomet/87.4.954

Chapter 7
Calibration

Things are not always as they appear at first glance.

Abstract Calibration is a very common task connected with the construction of sensors, the development of new instrumental methods, measurements of physical variables that are not directly measurable, and the estimation of unmeasurable characteristics of products. It is composed of calibration model building based on regression analysis and calibration model use, including estimation of uncertainty of results. Standard statistical assumptions are often too restrictive and special methods based on more correct criteria are necessary to be used. The aim of this work is a description of the main types of calibration, including linear and nonlinear models. The basic types of linear calibration models are compared, and their quality is characterized by simulation.

7.1 Introduction

Calibration is a very common practical task that requires the use of statistical methods. Due to the specificity of this task, it is not always possible to use standard statistical assumptions and commonly known solutions described in most statistical manuals and available in statistical programs. Calibration is one of the basic tasks solved in technical practice using regression methods. It is used especially in the construction of sensors of physical quantities, all types of measuring devices, in the development of new instrumental methods in biology, chemistry, etc., and more and more frequently in the estimation of fundamentally unmeasurable characteristics of products (feelings of pleasantness, sensations of taste, flavor, degree of satisfaction) based on measurable properties and characteristics. In all these cases, the desired (target) quantity, usually of a physical nature, is obtained by substituting another (easily measurable) quantity, i.e. the measurement result, into the model function, i.e. the calibration model. In many cases, this model is known and its parameters only need to be estimated. From a metrological point of view, this is a typical problem of so-called indirect measurements, when the measurement result is not identical to the measured quantity (Meloun & Militký, 2012). The basic problem of calibration, affecting in

J. Militký et al., *Textile Material Research—Statistical Tools of Experimental Data Analysis*, https://doi.org/10.1007/978-981-97-6607-9_7

particular, the criterion for estimating the parameters of calibration models and determining the uncertainty of the prediction, is the specification of the statistical nature of the target quantity and the measurement result. The aim of this work is to show some problems occurring when solving classic linear calibration tasks. It is shown what types of tasks occur during calibration and how to proceed in solving them. A regression spline is proposed as a unified calibration model. Its flexibility is ensured by a suitable choice of nodal points. Different types and models of calibration are demonstrated using the example of a calibration line.

7.2 Calibration Task

In calibration tasks, the standard purpose is to determine the values of difficult-to-measure (usually physical with defined units) or completely unmeasurable quantities y (searched quantity, target quantity **T**-target). This quantity is a function (commonly known from physics or chemistry) of an easily measurable quantity x (signal, **M**-measurement). Examples of **T** are concentration, temperature, humidity, pressure, and a number of other physical and chemical quantities. Electric voltage or electric current or light absorbance is standardly used as **M**. A calibrated system is mostly identical to a calibrated device. Here, calibration will actually enable its practical use, i.e. determining the estimate of the quantity **T** from the measured signal **M**. Functional dependence, i.e. model function

$$y = f(x, \mathbf{a}) \tag{7.1}$$

it is either known in advance (from the physical principle of measurement) or is evaluated by regression methods. For example, when measuring temperature with a mercury thermometer, y is the temperature x of the displacement of the column of mercury in the capillary and $f(x, \mathbf{a})$ is a straight line with a slope related to the coefficient of volumetric thermal expansion of mercury.

Especially when calibrating instruments, calibration models in the form of straight lines or low-degree polynomials are commonly used so that they can be easily used without complex approximate procedures (it is a classical determination of inversion values of the calibration model). It is always necessary to determine both the shape of the calibration model and the estimates of the model parameters, which enable the use of the calibration model.

Calibration generally consists of two phases:

- creation of a calibration model
- use of the calibration model.

In the *calibration model creation phase*, the purpose is to find the model $f(x, \mathbf{a})$ using appropriate regression methods. It is usually based on a set of prepared samples that have defined values y_i, $i = 1 \ldots n$. The y_i values are set using other instruments of sufficient accuracy based on physical phenomena (e.g. boiling points, melting points) or using different types of standards. For these samples, the values x_i, $i = 1 \ldots n$ are measured by calibrated system (see Fig. 7.1).

The values of y_i are typically set with a higher precision (which is often below the measurement error level of the calibrated system), so they can often be considered deterministic. The measurement results x_i are subjected to measurement errors, and in accordance with classical regression, an inverse function $x = f^{-1}(y)$ is actually created (see Fig. 7.2) as final result of this phase.

In the case of directly unmeasurable characteristics, set values y_i (usually in ordinal scale) are determined for a set of samples (products) from evaluation by respondents. The corresponding measurable properties are the components of the vector $\boldsymbol{x}_i$. Instead of the standard calibration model, where the set values of y_i are perceived as deterministic, the standard model defined by Eq. (7.1) is used, because the results of respondents' evaluations usually have a significantly higher variability than the results of measurable properties (it is so called inverse calibration).

Example 7.1 *Calibrating the instrument for hand prediction*

The hand of textiles summarizes the sensations caused by human contact with textiles. It is a basic psycho-physical integral characteristic (combination of perceptions) that

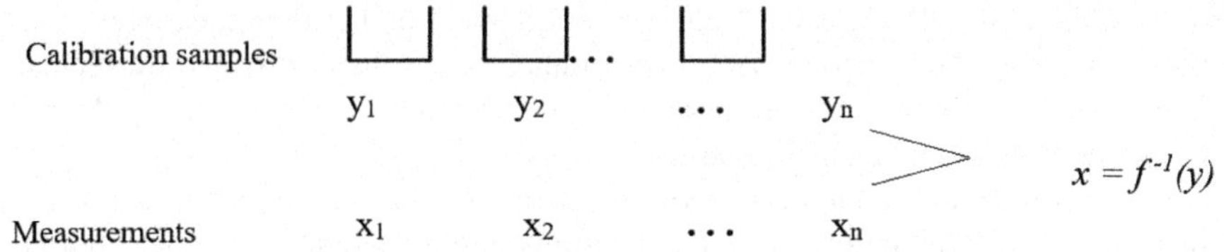

Fig. 7.1 Inputs and outputs of a calibration model

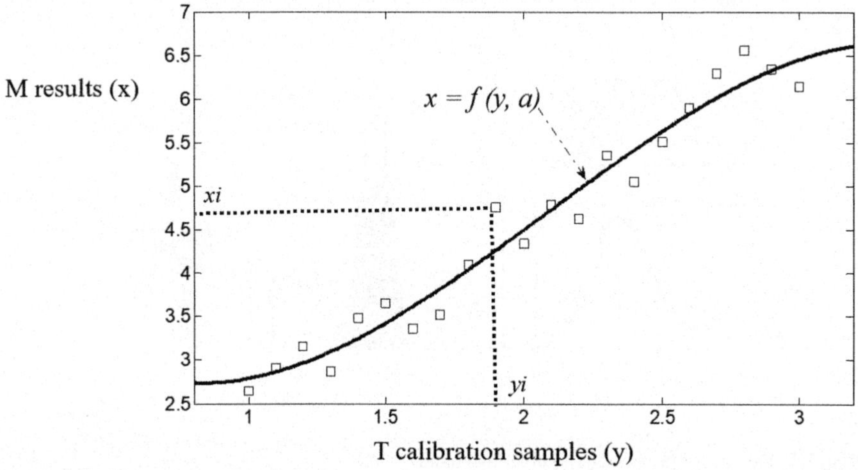

Fig. 7.2 Standard calibration model

is directly unmeasurable (Lundgren, 1969). One of the possibilities for predicting subjective hand is the use of a special device based on stretching a textile sample of a circular shape through a nozzle of a defined shape and size (usually a ring) (Pan et al., 1988). Schematically, the function of this device is shown in Fig. 7.3. It is clear that the device provides characteristics based on the extremes of the curve of the force required for stretching depending on the distance of the stretched sample from the original state. These characteristics are related to hand only indirectly, but they are related to quantities such as the stiffness of the fabric (C1), the smoothness of its surface (C2), and its voluminosity (C3). Therefore, it is not possible to construct a physical model for predicting subjective hand, but it is possible to find a regression (calibration model) based on experiments, allowing to estimate the subjective hand value from measurements on the device (for simplicity, only the maximum force P_t).

Firstly, a representative group of textile samples is selected. The subjective hand of this group of samples is then determined based on the evaluation of the panel of respondents (usually the median in an ordinal scale) (Pan et al., 1988). Result is the set of values y_i. The P_{ti} values determined on the device are used as the measured quantity x_i (in general, other physical and mechanical properties of textiles in tension, pressure, bending, shear, etc., indirectly related to hand can be selected). Therefore, the calibration model is the dependence $y_i = f(P_{ti}, \mathbf{a})$, since the errors of the determination of subjective hand are significantly higher than the errors of the measurement of P_t. Unknown vector of parameters **a** is then estimated by standard least squares regression. This empirical model can be easily generalized to the case of a more of measured characteristics x_i.

In the *calibration model use phase*, the purpose is to determine an estimate of the target physical quantity $\hat{y}$ from the measurement of the output x of the calibrated system (instrument) for the unknown sample. So, in the case of classic calibration, it is about finding the inverse function $\hat{y} = f(x)$ to the calibration model form $x = f^{-1}(y)$ created in first phase. This situation is shown schematically in Fig. 7.4.

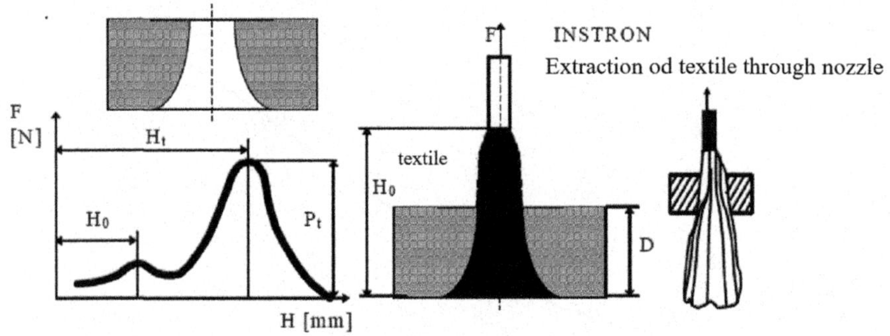

$$F = A + B * h + C * h^2 \qquad 0 \le h \le H_T \quad (A, B, C \text{ coefficients})$$

Fig. 7.3 Device for prediction of subjective hand

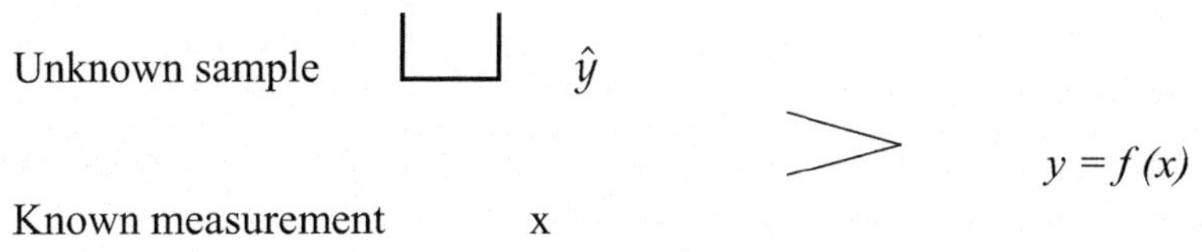

Fig. 7.4 Use of calibration model

The categorization of calibration tasks according to various aspects determining the method of statistical processing is presented in work (Rosenblatt & Spiegelman, 1981).

- Absolute calibration. It is the most common case in technical practice. When constructing a calibration model, a relationship is created between a measurable quantity (M or x), commonly called a signal (electrical voltage and current, electrical resistance, pH, absorbance, etc.), and a physical quantity (T or y) that determines a state or property of the system (content, concentration, volume, temperature, time, etc.). The latter may be more difficult to measure or unmeasurable for unknown sample and must be estimated from calibration model in the second stage of calibration.
- Comparative calibration. It is a procedure in which one instrument is calibrated against another and it is arbitrary which one is used as a standard. Usually, calibration is done for a "new" instrument, and the already calibrated (usually more accurate) instrument is then taken as the standard. The measurement errors on both devices are not negligible, and when building a calibration regression model, it is generally necessary to use regression methods for both variables with random errors.

The complexity of solving the calibration task is mainly related to the selected calibration model. For linear models, it is easy to express confidence bands around the regression model and other statistical characteristics enabling the determination of the accuracy of the estimate of the quantity y from the specified measurement result x. Usually, polynomial models are considered, in which individual components correspond to powers of the measured property.

7.3 Types of Calibrations

The types of calibration are distinguished according to the errors assumed in the quantities x and y in the phase of creating the calibration model (Martens & Naes, 1993):

(a) C-calibration, when the set values of y are considered to be deterministic or measured with a negligible error compared to the measurement of x. Thus, the quantity x is understood as random because it is composed of a deterministic part (of the calibration model f^{-1}) and a random measurement error ε_x. An additive measurement model is commonly assumed

$$x = f^{-1}(y, \mathbf{a}) + \varepsilon_x D(x) = \sigma_x^2 \tag{7.2}$$

Here, $D(x)$ is the variance of the measured quantity, which for the model is expressed by Eq. (7.2) identical to the variance of measurement errors $D(\varepsilon_x) = \sigma_x^2$. If this variance is constant and the measurement errors are mutually independent, it is a so-called homoscedastic model of the action of errors.

For the i-th calibration sample (Fig. 7.1) is $x_i = f^{-1}(y_i, \mathbf{a}) + e_{xi}$ and the parameter estimates can be determined by the method of least squares, which here corresponds to the minimum of the sum of the squares of the deviations in the x values

$$S_x = \sum_{i=1}^{n} e_{xi}^2 \tag{7.3}$$

For linear calibration models, these estimates are the best unbiased linear parameter estimates (Meloun & Militký, 2012). For this type of calibration, it is necessary to solve the inverse problem in the prediction. $\hat{y}_{ce}$.

(b) I-calibration, where the measured quantities x are considered deterministic or measured with a negligible error compared to the setting of the y values. The adjusted variable y is therefore understood as random, composed of a deterministic part (of the calibration model f) and a random measurement error ε_y. Again, an additive measurement model is commonly assumed

$$y = f(x, \mathbf{a}) + \varepsilon_y \quad D(y) = \sigma_y^2 \tag{7.4}$$

Here, $D(y)$ is the variance of the set variable, which is expressed for the model by Eq. (7.4) identical to the variance of the errors $D(\varepsilon_y) = \sigma_y^2$ created by measuring a set of calibration sample values. If this variance is constant and these errors are independent of each other, it is again a homoscedastic model of the effect of errors. For the i-th calibration sample (Fig. 7.1) is $y_i = f(x_i, \mathbf{a}) + e_{yi}$ and the parameter estimates $\mathbf{a}$ can be determined by the method of least squares, which here corresponds to the minimum of the sum of the squares of the errors in y values

$$S_y = \sum_{i=1}^{n} e_{yi}^2 \tag{7.5}$$

Direct prediction $\hat{y}_{\text{in}}$ is used for this type of calibration.

(c) O-calibration, when it is considered that both variables are random and generally have different variances. This is the model where both variables are subjected by random errors that are mutually uncorrelated (see Sect. 7.7.5). The so-called Berkson model of the action of errors (Berkson, 1950), when the values x_i are with errors ε_x and the correlation between the errors ε_x and the actual values $\delta = x + \varepsilon_x$ is considered has similar solution. For these models it is valid, that

$$y = f(x + \varepsilon_x, \mathbf{a}) + \varepsilon_y \quad P = \sigma_y^2/\sigma_x^2 \tag{7.6}$$

The differences between these two models for both variables subjected by random errors are in the estimation of the parameters and their variability. To solve this calibration problem for a straight line, knowledge of the variance ratio P is sufficient.

For $P = 1$, it is actually a task of minimizing the perpendicular distances between the calibration model and the experimental samples points. Direct prediction $\hat{y}_{ok}$ is used for this type of calibration.

The deviations of the i-th point (x_i, y_i) from the calibration model (here, for simplicity, a straight line), which are used in the minimization of the squared deviations for individual types of calibration, are shown in Fig. 7.5.

Thus, when using the standard least squares method, C-calibration minimizes the deviations of e_{xi} measurements x_i from the calibration function; I-calibration minimizes the deviations e_{yi} of the calibration samples y_i from the calibration function; O-calibration (for $P = 1$) minimizes the "perpendicular distances" e_{ci} of the points x_i, y_i from the calibration function (for P different from one is "weighted" by the corresponding variances). From the simple geometry shown in Fig. 7.5, it can be seen that the deviations are given by the relations

$$e_x = x - a_0 - a_1 y \tag{7.7}$$

and

$$e_y = y - b_0 - b_1 y \tag{7.8}$$

Thus, for the variances of both kinds of deviations, it is easy to deduce that

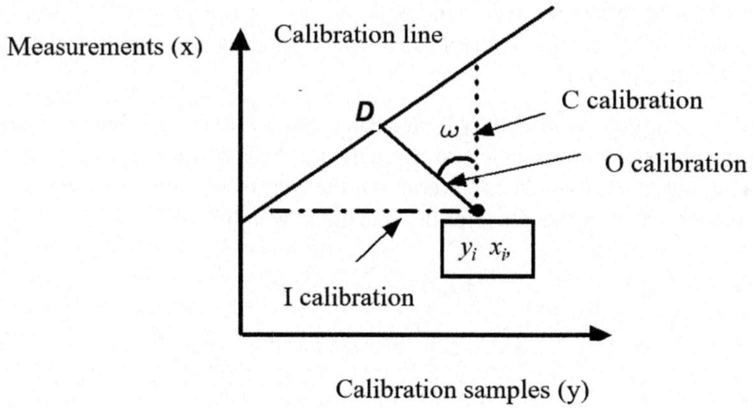

Fig. 7.5 Deviations for different types of calibration

$$D(e_x) = D(x) - a_1^2 D(y) = \sigma_x^2 - a_1^2\sigma_y^2 = w_x^{-1} \tag{7.9}$$

$$D(e_y) = D(y) - b_1^2 D(x) = \sigma_y^2 - b_1^2\sigma_x^2 = w_y^{-1} \tag{7.10}$$

To estimate the parameters of the calibration model for both variables subjected by random errors, it is possible to use the minimization of the weighted sum of the squares of the deviations of the measurement S_{wx} with the weights $w_x = 1/\left(\sigma_x^2\left(1 - a_1^2 P\right)\right)$ or, better, the minimization of the weighted sum of the squares of the deviations of the adjustment S_{wy} with the weights $w_y = 1/\left(\sigma_y^2\left(1 - b_1^2/P\right)\right)$. For the case of using S_{wx} with weights w defined by Eq. (7.9), it is therefore valid

$$S_{wx} = \sum_{i=1}^{n} w_x e_{xi}^2 = \frac{1}{w_x}\sum_{i=1}^{n} e_{xi}^2 = \frac{1}{\sigma_x^2\left(1 - a_1^2 P\right)}\sum_{i=1}^{n} e_{xi}^2 \tag{7.11}$$

Similarly, a criterion for minimizing the weighted sum of squares of the S_{wy} deviations can be expressed as

$$S_{wx} = \sum_{i=1}^{n} w_y e_{yi}^2 = \frac{1}{w_y}\sum_{i=1}^{n} e_{yi}^2 = \frac{1}{\sigma_y^2(1 - b_1^2/P)}\sum_{i=1}^{n} e_{yi}^2. \tag{7.12}$$

For the case of perpendicular distances, $P = 1$, and for the homoscedastic case, i.e. $\sigma_x^2 =$ konst. or $\sigma_y^2 =$ konst. respectively, the minimization of the perpendicular distances leads to the minimization of the sum of the squares of the respective deviations weighted by a constant dependent only on the magnitude of the respective slope.

When estimating intercept a_0, or b_0, the weight term $1/\left(1 - a_1^2\right)$ or $1/\left(1 - b_1^2\right)$ will not be used. When estimating the slope a_1, or b_1, however, it will be necessary to solve a non-linear equation with one unknown. This can be converted to the solution of a quadratic equation.

(d) CR-calibration, proposed by Leng (Pan et al., 1988) (as combined regression), is a generalization of the classical method of least squares leading to the minimization of the "weighted" average of the sum of squares of the horizontal e_y and vertical e_{xi} deviations, for the specified "weight".

$$0 \le g \le 1$$

$$S_{cr} = g\sum_{i=1}^{n} e_{xi}^2 + = (1 - g)\sum_{i=1}^{n} e_{yi}^2 \tag{7.13}$$

For estimating the intercept a_0, the weight term g does not have any effect again. When estimating the slope a_1, however, it is necessary to solve the quartic equation (Pan et al., 1988)

$$g a_2^4 s_x^2 - g a_2^3 \mathrm{cov}_{xy} + (1-g) a_2 \mathrm{cov}_{xy} - (1-g) s_x^2 = 0 \tag{7.14}$$

Here, s_x^2 and s_y^2 are the sample variances and covxy is the sample covariance. Again, the limiting case of CR regression is I-calibration ($g = 1$) and C-calibration ($g = 0$). The perpendicular distance e_c of the points from the calibration line is also given by the relations

$$e_c = e_x \cos(\omega) \quad e_c = e_y \cos(90 - \omega) = e_y \sin(\omega). \tag{7.15}$$

For the angle ω it is then valid, that $\omega = \mathrm{arctg}\left(\frac{e_y}{e_x}\right)$.

It is also possible to use a mixed estimate (CI), when the product of the lengths of deviations e_{xi} and e_{yi} is minimized, i.e.

$$S_{oc} = \sum_{i=1}^{n} \sqrt{e_{yi}^2 e_{xi}^2} \tag{7.16}$$

In the case of classic C-calibration, it is always necessary to solve the inverse problem, i.e. to find a function $f^{-1}(x)$ The inverse calibration task occurs only with C-calibration in the phase of using the calibration model, when the purpose is to find an estimate of the target quantity $\hat{y}$ and the corresponding uncertainty (confidence interval) for the known signal x^*. It is therefore necessary to determine the inverse function of the calibration model $f^{-1}(y)$. This problem can be mathematically transformed into the problem of finding the root of a non-linear equation $f^{-1}(y, a) - x^* = 0$ with respect to the variable y. If this task cannot be solved analytically, a number of numerical procedures are used. Relatively effective is the iterative Newton method, which requires knowledge of the first derivative of the function $f(y)$. The problem is to determine the corresponding variance if $f(y)$ is non-linear (Meloun & Militký, 2012). For strongly non-linear models, other procedures are chosen (Bootstrap-type simulation or estimates based on the likelihood ratio (Huet et al., 1996)).

For the case of small residuals (this is acceptable for most calibration models) and slightly non-linear models, confidence intervals of the estimated target quantity y^* are constructed based on the assumption of its normality. Details are described in the book (Meloun & Militký, 2012). Estimates of the variance of the prediction y^* are also given there, for the case of inverse calibration, they correspond to the estimate of the variance of the prediction for non-linear regression models. For I-calibration and O-calibration, there is no need to solve the inverse problem. In all cases, however, the signal x_0 of the unknown sample is a measured variable (random), so even the prediction $\hat{y}_0$ of the desired characteristic y is a random variable and its variability must be estimated. Individual procedures are comprehensively described in the book (Meloun & Militký, 2012).

In sequel, methods of estimating the parameters of calibration models in connection with various types of disturbances are discussed. The solution for the calibration line is shown in more detail. For cases where a theoretical assignment of the type Eq. (7.1) is not found, the use of regression splines is suggested.

7.4 Models of Error Effect

Till yet considerations assumed that random errors (deviances) are additive and have a constant variance. This is acceptable in practice only if the values of the random variables appear within a range of approximately one order. If the calibration model is constructed for a range of several orders of magnitude (e.g. temperature), the assumption of constant variance cannot be accepted well, and the additive model of the errors action can be replaced by a more suitable model. In these cases, the variance of the random variable is generally no longer identical to the errors variance, which changes the criterion for parameter estimation. Squares of deviations are also only a special case, when they are random variables with a normal distribution. For specific purposes, more robust criteria can be chosen or, for example, criteria limiting the size of the maximum deviation.

It is useful to illustrate different models of the effect of errors on the task of *I*-calibration. For case of *C*-calibration, it is necessary to use the inverse calibration model and change the order of variables, or their type (random vs. deterministic).

Let there be data for calibration $\{x_i, y_i\}$, $i = 1, \ldots, n$, where the variable y is considered as a random variable with "setting" errors of the samples and the variable x as a deterministic variable (or significantly more accurate, so that its variability neglects.). The calibration model $f(x, \mathbf{a})$ is known and a criterion for estimating its parameters needs to be found. Explained, or the set (dependent) variable y is therefore subjected by errors (inaccuracy of sample setting, measurement of values of the calibrated quantity for set samples, variability of the material used, inaccuracies of the calibration model, neglect of changes in the environment, and conditions of use of the calibration model, etc.).

The total error ε of the dependent variable is then a combination of all partial errors. It is usually assumed that the total error of the samples adjustment has zero mean, $E(e_i) = 0$, $i = 1, \ldots, n$. If $E(e_i) =$ const. but not 0, the absolute term is usually missing in the calibration model. The mode of action of errors is characterized by the measurement model (additive, multiplicative, mixed, etc.). The general calibration model, including both the *calibration model* itself and the *measurement model*, can then be expressed in the form

$$y_i = Z_i(X_i, \varepsilon_i, \mathbf{a}) \quad i = 1, \ldots, n \tag{7.17}$$

The specific shape of the function $Z_i(.)$ depends on the assumed mechanism of error action and the form of the calibration function. If the y_i values represent the results of experiments, an additive measurement model is usually assumed for which it is

$$Z_i = f(x_i, \mathbf{a}) + \varepsilon_i \tag{7.18}$$

In many cases, the measured values y_i, $i = 1, \ldots n$ must be positive or at least have a natural origin. Measurement error variances tend to be non-constant. Relative measurement errors are constant for most measuring devices. This is suitable, for example, for the multiplicative measurement model, for which it is

$$Z_i = f(x_i, \mathbf{a}) \exp(\varepsilon_i) \tag{7.19}$$

In practice, there is also a mixed measurement model, which is described by the relation

$$Z_i = f(x_i, \mathbf{a}) \exp(v_i) + \varepsilon_i \tag{7.20}$$

where the errors v_i and ε_i are assumed to be mutually independent.

If the error model is not known, it is possible to start from a general class of models including both additive and multiplicative models. One such general class of models can be expressed using the Box-Cox transformation (Meloun & Militký, 2012), when

$$y^{(\lambda)} = \frac{x^\lambda - 1}{\lambda} \quad \text{for } \lambda \neq 0 \quad \text{or} \quad y^{(\lambda)} = \ln(x) \quad \text{for } \lambda = 0 \tag{7.21}$$

The Box-Cox transformation includes both additive and multiplicative measurement models as special cases. The general measurement model then has the form

$$y_i^{(\lambda)} = f^{(\lambda)}(x_i, \mathbf{a}) + \varepsilon_i \tag{7.22}$$

When identifying the calibration model, the optimal λ is then searched, followed by estimates $\widehat{mathbfa}$ of the model parameters **a**. To find the regression criterion and statistical analysis, the distribution of the random variables y_i must be determined. This distribution is closely related to the error distribution ε_i, which is given by the join probability density function $p_\varepsilon(\boldsymbol{\varepsilon})$. The latter is also a function of certain distributional parameters such as variance, etc.

It is commonly assumed that unimodal, symmetric error distribution $\boldsymbol{\varepsilon}$ has mean value $E(\boldsymbol{\varepsilon}) = 0$. It is often assumed, although sometimes incorrectly, that the measurement errors of ε are also independent. The join probability density $p_\varepsilon(\varepsilon)$ is then given by the product of the marginal densities $p_\varepsilon(\varepsilon_i)$. Thus

$$p_{\boldsymbol{\varepsilon}}(\varepsilon) = \prod_{i=1}^{n} p_\varepsilon(\varepsilon_i) \tag{7.23}$$

A group of distributions, including the normal, uniform, and Laplace distributions, can be expressed in form

$$p_\varepsilon(\varepsilon_i) = Q_N \exp\left(\frac{-|\varepsilon_i|^p}{\vartheta}\right) \tag{7.24}$$

where Q_N is the normalization constant, p determines the type of distribution, and ϑ is a parameter proportional to the variance. If $p = 1$, it is a Laplace distribution, for $p = 2$, it is a normal distribution, and for $p \to \infty$, it is a rectangular distribution. The disadvantage of the distribution of $p(\varepsilon_i)$, defined by relation (7.23), is the fact that for $p < 2$, the neighborhood of the origin is not locally quadratic. Therefore, alternative systems of probability densities are also used, such as the generalized Student distribution, etc. (Meloun & Militký, 2012). For the case of independent random errors $\boldsymbol{\varepsilon}$ and the additive measurement model defined by Eq. (7.18), it is valid

$$p(y_i) = p_\varepsilon(y_i - f(x_i, \mathbf{a})) \tag{7.25}$$

Therefore, it follows that the additive measurement model does not cause any distortions of the distribution of the measured quantities with respect to the distribution of errors.

For the multiplicative measurement model defined by Eq. (7.19) is valid under the assumption of only positive values of the measured quantities valid, that

$$p(y_i) = \frac{1}{y_i} p(\ln y_i - \ln f(x_i, \mathbf{a})) \tag{7.26}$$

This probability density no longer corresponds to the error probability density $p(\varepsilon)$.

From the overview of error models, it follows that the distribution of the measured values of the quantity y can be derived according to the specific conditions of the experiment and ideas about the effect of various types of errors action. In most cases, the normality of measurement errors, their independence, and an additive model of their action are assumed.

The join probability density for the vector of measured values $y = \{y_1, \ldots, y_n\}^T$ is referred to as the likelihood function $L(\boldsymbol{\alpha})$. This function depends on the vector of parameters $\boldsymbol{\alpha}$, which contains the model parameters $\mathbf{a}$ and the distribution parameter s. Maximum likelihood parameter estimates are determined by maximizing the log likelihood function (the logarithm is a monotonic transformation that does not affect the position of the maximum).

$$\ln L(\boldsymbol{\alpha}) = \ln p(\mathbf{y}) = \sum_{i=1}^{n} \ln p(y_i) \tag{7.27}$$

The second equality in Eq. (7.26) is valid only for the special case of independent measurement errors. Maximum likelihood estimators have asymptotic variances that are equal to the inverse of the expected Fisher information matrix (Meloun & Militký, 2012). For the case when $L(\boldsymbol{\alpha})$ is a function of only one parameter, the so-called Fisher information is equal

$$I(\alpha) = E\left\{\left[\frac{\partial \ln L(\alpha)}{\partial \alpha}\right]^2\right\} \tag{7.28}$$

If $L(\alpha)$ changes rapidly with changes of parameter α, the measurement contains a lot of "information" about that parameter. It can be proven that

$$E\left\{\left[\frac{\partial \ln L(\alpha)}{\partial \alpha}\right]^2\right\} = E\left\{\frac{\partial^2 \ln L(\alpha)}{\partial \alpha^2}\right\} \tag{7.29}$$

The quantity $S = \frac{\partial \ln L(\alpha)}{\partial \alpha}$ is called the score function. If the estimate $\hat{\alpha}$ is unbiased, its variance is $\sigma^2(\hat{\alpha}) \geq 1/I(\alpha)$.

For the case of a vector of parameters $\boldsymbol{\alpha}$, the Fisher information analogy is the Fisher information matrix $I(\boldsymbol{\alpha})$, for which it is valid

$$I(\boldsymbol{\alpha}) = E\left\{\frac{\partial \ln L(\boldsymbol{\alpha})}{\partial \boldsymbol{\alpha}}\left[\frac{\partial \ln L(\boldsymbol{\alpha})}{\partial \boldsymbol{\alpha}}\right]^T\right\} \tag{7.30}$$

This matrix is assumed to be positive definite for all $\boldsymbol{\alpha}$. A number of important results can be derived for maximum likelihood estimates:

1. Estimates $\hat{\boldsymbol{\alpha}}$ are for $n \to \infty$ (i.e. asymptotically) unbiased. However, for finite numbers of measurements n, the estimates are biased and the magnitude of the bias $\mathbf{b} = E(\hat{\boldsymbol{\alpha}}) - \boldsymbol{\alpha}$ depends on the degree of non-linearity of the regression model (Meloun & Militký, 2012).
2. The estimators are $\hat{\boldsymbol{\alpha}}$ asymptotically efficient and the variances of the estimators $D(\hat{\alpha})$ are minimal in the class of all unbiased estimators. The covariance matrix of the estimates lies on the lower bound of the Cramer-Rao inequalities (Cheng & Van Ness, 1994). Even this property is not generally valid for finite sample sizes.
3. The random vector $\sqrt{n}(\hat{\alpha} - \alpha)$ has an asymptotically normal distribution with zero mean and variance equal to the inverse of the Fisher information matrix (Meloun & Militký, 2012). If the error distribution is approximately normal, the normality of the estimates is also valid for the finite samples.

For sufficiently large sample size, the above-mentioned properties of estimates $\hat{\alpha}$ can be used. For the finite number of measurements, the bias of the estimates causes significant difficulties.

Knowing the join probability density function $p(\mathbf{y})$, it is possible to find the maximum likelihood estimates or a criterion for their determination, the so-called regression criterion.

In practice, the most common case is when the measurement errors ε are independent with zero mean, constant variance, and normal distribution $N(\mathbf{0}, \sigma^2\boldsymbol{E})$, where E is diagonal matrix with ones on the main diagonal. The additive measurement model defined by relation (7.15) is also assumed. The distribution of the measured quantity y_i is then also normal $N(f_i, s^2)$. So, it is true

$$p(y_i) = \frac{1}{\sqrt{2\pi\sigma^2}} \exp\left(\frac{-(y_i - f_i)^2}{2\sigma^2}\right) \tag{7.31}$$

where $S(\mathbf{a}) = \sum_{i=1}^{n} (y_i - f_i)^2$. Let the vector of estimated parameters be $\boldsymbol{\alpha}^T = (\mathbf{a}^\mathbf{T}, \sigma^2)$. Logarithms of likelihood function $\ln L(\boldsymbol{\alpha})$ are then in the form

$$\ln L(\boldsymbol{\alpha}) = \sum_{i=1}^{n} \ln p(y_i) = -\frac{n}{2} \ln\left(2\pi\sigma^2\right) - \frac{1}{2\sigma^2} S(\mathbf{a}) \tag{7.32}$$

Symbol S ($\mathbf{a}$) is equal to sum of squared deviations

$$s(a) = \sum_{i=1}^{n} (y_i - f_i)^2 \tag{7.33}$$

By the analytical maximization of $\ln L(\boldsymbol{\alpha})$ according to the parameter s^2, it results in $\hat{\sigma}^2 = \frac{s(\mathbf{a})}{n}$. Estimate $\hat{\sigma}^2$, i.e. the residual variance, is biased for small samples. Its unbiased version is

$$\hat{\sigma}^2 = \frac{S(\hat{\mathbf{a}})}{n - m} \tag{7.34}$$

Substituting from Eq. (7.33) into Eq. (7.34), the so-called concentrated likelihood function.

$\ln L(\mathbf{a})$ is determined, for which it is

$$\ln L(a) = -\frac{n}{2}(1 + \ln(2\pi)) - 0.5 \ln S(a) \tag{7.35}$$

The maximum of $\ln L(\mathbf{a})$ apparently corresponds to the minimum of the criterion $S(\mathbf{a})$, which is the criterion of the method of least squares (LSM). Therefore, the method of maximum likelihood is identical to the method of least squares of deviations for this case.

The search for the local extremum of the likelihood function generally leads to the task of non-linear optimization. Using the properties of the maximum likelihood method, it is possible also easily construct confidence intervals and test statistical hypotheses (Meloun & Militký, 2012). When solving calibration tasks, classic unweighted MNČ is used almost exclusively. However, this may lead to unusable results for some situations. It can be shown that (Meloun & Militký, 2012):

(a) for the calibration task that corresponds to the *multiplicative measurement model*, the minimization of the sum of the squares of the deviations in logarithms must be used.
(b) for the calibration task, when the "adjustment" is under conditions of constant relative error (coefficient of variation), it is necessary to use weighted least squares of deviations with weights $w_i = 1/f(x_i, \mathbf{a}) \sim 1/y_i$.

(c) for the calibration task, when the "adjustment" is under the conditions of full accumulation of errors (autocorrelated errors), the minimization of the sum of squares of the deviations in the first differences should be used.
(d) for the calibration task, when the "adjustment" is under conditions where large deviations can occur with a higher probability, it is appropriate to use the minimization of the sum of absolute deviations.
(e) for the calibration task, when it is required that the deviations from the calibration model be approximately the same, the criterion of minimizing the maximum deviation is chosen.

This brief, incomplete list shows that the least squares criterion is far from universal, and during calibration, it is also necessary to examine possible types of errors and their distribution. The advantage of using the maximum likelihood method is that it also allows the determination of variances of individual estimates and further statistical analysis. Details are given in the book (Meloun & Militký, 2012).

7.5 Inverse Calibration

The inverse calibration task only occurs with *C*-calibration. In the stage of use of the calibration model, the purpose is to find an estimate of the target quantity y and the corresponding uncertainty (confidence interval) from the known signal x^*. It is therefore formally a matter of determining the inverse function for the calibration model $f^{-1}(y)$

$$\hat{y} = f\left(x^*, \mathbf{a}\right) = g\left(x^*, \mathbf{a}\right) \tag{7.36}$$

This problem can be mathematically transformed into the problem of finding the root of a non-linear equation

$$f^{-1}(\hat{y}, \mathbf{a}) - x^* = 0 \tag{7.37}$$

with respect to the variable $\hat{y}$. If this task cannot be solved analytically, a whole range of procedures is used. Newton's iterative method, which requires knowledge of the first derivative of the function $f(y)$, is relatively effective. The problem is to determine the corresponding variance if $f(y)$ is non-linear (Meloun & Militký, 2012). For strongly non-linear models, other procedures are chosen [Bootstrap or likelihood ratio estimates (Huet et al., 1996)].

For the case of small deviations (this is acceptable for most calibration models) and low non-linearity of models, confidence intervals of the estimated target quantity y^* are constructed based on the normality assumption (Meloun & Militký, 2012). Estimates of the variance of the prediction y^* for the case of inverse calibration correspond to the estimate of the variance of the prediction for non-linear regression models (Meloun & Militký, 2012).

7.6 Regression Spline

Empirical models are often used for non-linear calibration. Polynomial models in particular are considered sufficiently flexible. Their advantage is that they are linear regression models, so their parameters can be easily estimated. The inverse problem leads to the solution of a polynomial equation, which can be realized analytically for low degrees of the polynomial. The disadvantage of polynomials, especially of higher degrees, is their tendency to oscillate outside experimental points. Moreover, the number of extrema and inflection points is predefined, which is unacceptable for non-associative physical models.

So-called *local polynomial splines* ensure the elimination of these problems while maintaining the simplicity of manipulation and parameter estimation. The calibration model here is formed by a system of local polynomials continuous in the entire definition interval in the specified number of derivatives. For these models, in addition to experimental points, a sequence of so-called knots (nodal points), i.e. $t_j, j = 1, \ldots k$, must also be defined. Knots form the boundaries of intervals where individual local functions are defined. In each interval I_j bounded by knots t_{j-1}, t_j, the calibration model is expressed as $g_j(x)$. The quality of the calibration here depends on the number and positions of individual knots t_j, the shape of the functions $g_j(x)$, and the class C^m to which the calibration model $g(x)$ belongs. Denote that the function $g(x)$ from the class C^m is continuous in the first m derivatives.

A special type of local polynomials is *regression splines*. Splines $S_{m+1}(x)$ are functions of class C^m that are defined as local polynomials of maximum degree $(m + 1)$. Properties of spline functions are defined in work (De Boor, 1998). For calibration purposes, quadratic splines $S_1(x)$ that are continuous and smooth (i.e. continuous in the first derivative) work well. The spline $S_1(x)$ can be simply defined using truncated polynomials

$$S_1(x) = b_1 + b_2 x + b_3 x + \sum_{j=1}^{k} b_{j+3} \left(x - t_j\right)_+^2 \tag{7.38}$$

It is valid, that

$$(x)_+^2 = x^2 \quad \text{for} \quad x > 0$$
$$(x)_+^2 = 0 \quad \text{for} \quad x \leq 0$$

For known values t_j, $S_1(x)$ is a linear regression model with respect to the parameters $\boldsymbol{b}$.

If the additive measurement model is used and the errors ε_i are mutually independent random variables with constant variance, the parameters $b_j, j = 1, \ldots k + 3$ can be estimated by the method of linear least squares. Considering the special form of the quadratic spline expressed through truncated polynomials, the problem of linear least squares can be converted into the solution of a system of linear (normal) equations

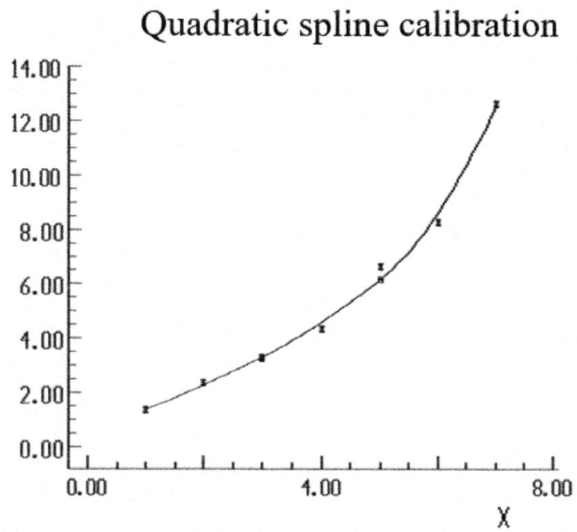

Fig. 7.6 Quadratic spline regression variant *a*—two knots

(Meloun & Militký, 2012). From a numerical point of view, this system of equations is poorly conditioned, and therefore, it is advisable to use special algorithms (e.g. singular value decomposition—SVD). To improve numerical stability, it is beneficial to perform a linear transformation of the x-coordinate range to the interval (1, 2). The flexibility of regression splines is primarily ensured by the appropriate selection of knots (nodal points) t_i. Three alternatives are available in the CALIBRATION module of the ADSTAT system (Meloun & Militký, 2012):

(a) constant length of the interval I_j, containing various number of points;
(b) selection of knots positions such that there is the same number of data in each interval I_j;
(c) user-selected positions and number of knots.

An example of the effect of a different nodal point selection strategy on the spline calibration result is shown in Figs. 7.6 and 7.7.

7.7 Calibration Straight Line

The calibration model in the form of a straight line is one of the most frequently used in practical applications. Both classic C and inverse I calibration can be used, or a whole series of modifications. It is therefore necessary to choose suitable criteria for choosing the type of calibration, or comparing their suitability. The problem is that there are a variety of approaches to choosing the type of calibration that lead to different types of conclusions. By default, the criterion of least squares of deviations is chosen, which, however, depends on the considered type of calibration (see Fig. 7.5). In many cases, when all the data are close to the calibration line, the differences in

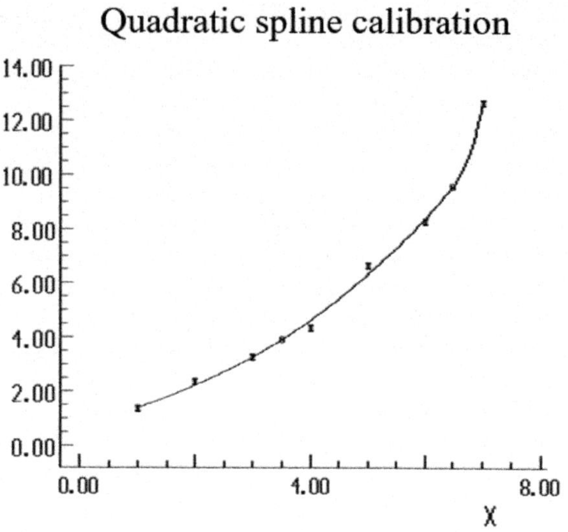

Fig. 7.7 Quadratic spline regression variant *b*—three points

the point estimates of the parameters are not very large, but the interval estimates are different. The initial data creates the sample (x_i, y_i), $i = 1, \ldots, n$. The entire further analysis depends critically on the assumptions about calibration sample setup errors and measurement errors.

For calculations in both phases of creating and using calibration models, only knowledge of sample means, variances, and covariance is usually sufficient

$$\overline{x}, \overline{y}, s_y^2, s_x^2, \text{cov}_{xy} = \frac{1}{n-1}\sum_{i=1}^{n}(y_i - \overline{y})(x_i - \overline{x}) \tag{7.39}$$

In sequel, simple procedures for estimating the value y of the quantity T are shown, as well as a comparison of the properties of individual types of calibration. The properties of the individual estimators are demonstrated on three types of simulated data. The validity of the calibration model $y = 1 + 2 \times$ is assumed, and based on samples of size $n = 21$, estimates of slope and intercept are calculated. Additive measurement models and normally distributed deviations with variance $d^2 = 4$ are selected.

I. **IS data** correspond to the situation where the *random variables are y* (set values). The x_i values are generated from $x_1 = 10$ to $x_{21} = 20$ with a step of 0.5. The y values are generated according to the relationship $y_i = 1 + 2x_i + N_i(0, d^2)$, where $N_i(0, d^2)$ is a random number generated from a normal distribution with zero mean and variance d^2.

II. **CS data** correspond to the situation where the *random variables are x* (measured values). The values of y_i are generated from $y_1 = 21$ to $y_{21} = 41$ with a step of 1. The values of x are generated according to the inverse relation $x_i = -1/2 + 1/2y_i + N_i(0, d^2)$, where $N_i(0, d^2)$ is a random number drawn from a normal distribution with zero mean and variance d^2.

III. **OS data** correspond to the situation where the *random variables are both x and y* (both error-subjected variables). The x_i values are generated from $x_1 = 10 + N_1(0, d^2)$ to $x_{21} = 20 + N_{21}(0, d^2)$ with a step of $0.5 + N_i(0, d^2)$. The y values are generated according to the relation $y_i = 1 + 2x_i + N_i(0, d^2)$, where $N_i(0, d^2)$ is a random number generated from a normal distribution with zero mean and variance d^2.

For all three types of data, a total of 5000 samples were simulated for the case of variance $d^2 = 4$. The results are presented in the form of histograms and tables containing average estimates of the slopes.

7.7.1 C-Calibration Line

It is based on the assumption that the measurement model is valid (i.e. the setting of the calibration samples y is accurate)

$$x = a_0 + a_1 y + \varepsilon_x,\ \varepsilon_x = N\left(0, \sigma_x^2\right) \tag{7.40}$$

For the least squares estimation of the linear calibration model parameters, it can be easily deduced that the model dependence must pass through the center of gravity of the points with coordinates $(\overline{x}, \overline{y})$, which means that the relation $\hat{a}_0 = \overline{x} - \hat{a}_1\overline{y}$ is used for the calculation of intercept. By minimizing the criterion S_x (Eq. 7.3) according to a_1, it is then easily determined that

$$\hat{a}_1 = \text{cov}_{xy}/s_y^2 \tag{7.41}$$

Thus, the calibration model has the form

$$\hat{x} = \overline{x} + \frac{\text{cov}_{xy}}{s_y^2}(y - \overline{y}) \tag{7.42}$$

In Fig. 7.8, the histograms of estimates of slopes $\hat{a}_1$ for the simulated samples **IS**, **CS**, and **OS** are shown.

The average values of slope estimate $\hat{a}_1$ for simulated samples **CS**, **IS**, and **OS** are shown in Table 7.1.

The last column shows the average percentage deviations d_p of $\hat{a}_1$ from the specified slope value (here 0.5) for all three types of simulated data. It can be seen that the distribution of estimates is always approximately symmetrical. For data generated according to the **OS** scheme, the degree of symmetry is the best. The lowest symmetry is for data generated according to the **CS** scheme, where, however, it is the best according to approaching to the specified value. The value of d_p is small, but it should be considered that it will be necessary to use the inversion of $1/\hat{a}_1$ in the phase of using the model (i.e. the results from the third line).

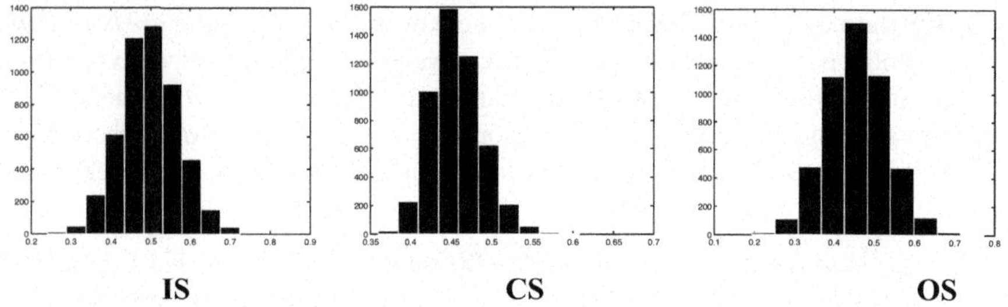

Fig. 7.8 Histograms of slope $\hat{a}_1$ estimates for simulated samples

Table. 7.1 Slope estimates obtained by different methods for simulated **CS**, **IS**, and **OS**

Estimate	CS	IS	OS	d_p [%]
$\hat{a}_1$	0.49866	0.456686	0.456307	5.89
$1/\hat{a}_1$	2.04946	2.19962	2.25579	8.41
$\hat{b}_1$	1.44295	2.00012	1.45098	18.44
$\hat{b}_{1g}$	1.71596	2.09723	1.80293	9.64
$\hat{b}_{1P}$	1.98442	2.18937	2.18101	6.43
$\hat{b}_{1k}$	1.88839	2.16177	2.05429	5.46

The estimate of the inverse function and the prediction of the searched quantity T for the given signal size x^* are in the form

$$\hat{y} = \frac{1}{a_1}x^* - a_0/a_1, \hat{y}_{cl} = \overline{y} + \frac{s_y^2}{\mathrm{cov}_{xy}}\left(x^* - \overline{x}\right) \tag{7.43}$$

The average values of slope estimate $1/\hat{a}_1$ for simulated samples **CS, IS,** and **OS** are shown in Table 7.1.

In Fig. 7.9, the histograms of estimates of slopes $1/\hat{a}_1$ for the simulated samples **IS**, **CS**, and **OS** are shown.

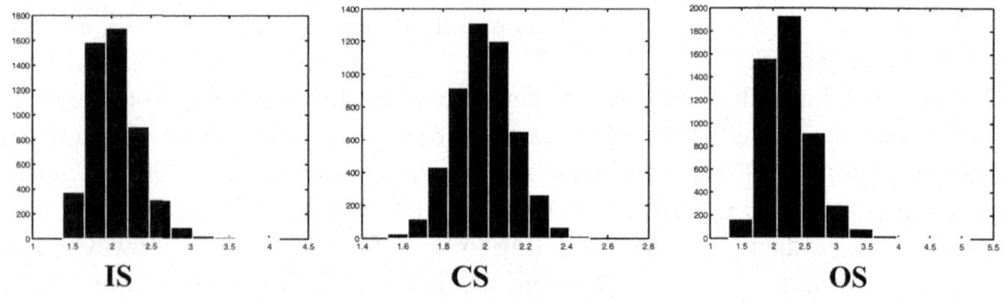

Fig. 7.9 Histograms of slope $1/\hat{a}_1$ estimates for simulated samples

It can be seen that the distribution of estimates $1/\hat{a}_1$ is approximately symmetric only for data generated according to the **CS** scheme, and the average value of the estimate is closest to the specified one (here and further 2), which is to be expected. The value of d_p is low, so the estimates will be less sensitive to the type of error action. The estimate $\hat{y}_{\text{cl}}$ is biased because in general

$$E(\hat{y}_{cl}) \neq \frac{E(x - \hat{a}_0)}{E(\hat{a}_1)} = y_{cl} \tag{7.44}$$

On the other hand, it is a maximum likelihood estimate for which a confidence interval can be easily constructed (Meloun & Militký, 2012). This interval is finite only if $\hat{a}_1$ is sufficiently large. Based on the assumption that the errors ε_x are small enough, asymptotic approximations for the bias B and the mean squared error MSE of the classical estimator $\hat{y}_{\text{cl}}$ were derived (Srivastava & Singh, 1989).

7.7.2 I-*Calibration Line*

This technique was proposed by Krutchoff (1969) for the case where x is deterministic, i.e. the measurements are accurate and errors arise in setting up the calibration samples. It assumes of validity of the measurement model

$$y = b_0 + b_1 x + \varepsilon_y, \quad \varepsilon_y \approx N\left(0, \sigma_y^2\right) \tag{7.45}$$

Again, this model must pass through the center of gravity of the points with coordinates $(\overline{x}, \overline{y})$, which means that for the estimate of intercept b_0, the relation $\hat{b}_0 = \overline{x} - \hat{b}_1 \overline{y}$ is valid. By minimizing the criterion S_y (Eq. 7.5) according to b_1, it is then easily determined that the estimate of the slope is equal to

$$\hat{b}_1 = \frac{\text{cov}_{xy}}{s_x^2} \tag{7.46}$$

Calibration model has then form

$$\hat{y}_{in} = \overline{y} + \frac{\text{cov}_{xy}}{s_x^2}(x - \overline{x}) \tag{7.47}$$

The value $\hat{y}_{\text{in}}$ is directly a prediction of the target quantity for the magnitude of the signal x.

The average values of slope estimate $\hat{b}_1$ for simulated samples **CS**, **IS**, and **OS** are shown in Table 7.1.

In Fig. 7.10, the histograms of estimates of slope $\hat{b}_1$ for the simulated samples **IS**, **CS**, and **OS** are shown.

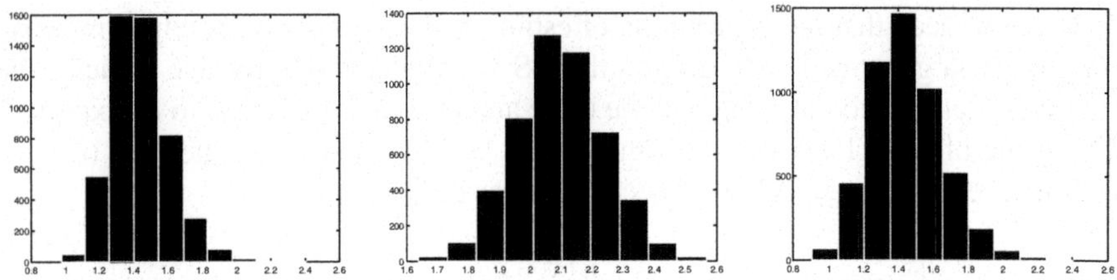

IS **CS** **OS**

Fig. 7.10 Histograms of slope $\hat{b}_1$ estimates for simulated samples

It is visible that the distribution of estimates is approximately symmetric for all types of generated data and is best for **CS** samples. The mean value of the estimate $\hat{b}_1$ is closest to the given one (here 2) for the **IS** data case, which is to be expected. The value of d_p is clearly the highest, so the estimates are highly sensitive to the type of error action.

Based on the assumption that the errors ε_y are small enough, asymptotic approximations for the bias B and the mean square error MSE of the inverse estimate $\hat{y}_{\text{in}}$ were derived (Srivastava & Singh, 1989). For both models, it is possible to estimate the respective error variance (σ_x^2 or σ_y^2) from the residual variance

$$\hat{\sigma}_c^2 = \frac{\text{RSC}}{n-2} \tag{7.48}$$

where RSC is the residual sum of squared deviations. There are works that show the advantages of classical estimates. On the other hand, the inverse estimator is a Bayesian estimator and it is valid, that

$$\hat{y}_{\text{in}} = \left[\frac{F}{F+(n-2)}\right]\hat{y}_{\text{cl}} \tag{7.49}$$

where

$$F = \frac{\hat{a}_1^2 \sum_{i=1}^{n} (y_i - \overline{y})}{\hat{\sigma}_c^2} \tag{7.50}$$

If F is sufficiently larger than the Fisher Snedecor distribution quantile $F_\alpha(1, n - 2)$ distribution, the classical estimator y_{cl} is sufficiently accurate. If this is not the case and F is slightly larger than $F_\alpha(1, n-2)$, the classical estimator is imprecise and it is better to use the inverse y_{in} estimator. If $\hat{\sigma}_c^2$ is small (i.e. the data lie almost on the calibration line) and F is large (the calibration line has a significant slope), the differences between the two types of calibration are not practically significant. It is therefore useful to compare the deviations between these two estimates. It can easily be shown that

$$\frac{\hat{y}_{\mathrm{in}} - \overline{y}}{\hat{y}_{\mathrm{cl}} - \overline{y}} = \frac{\mathrm{cov}_{xy}^2}{s_x^2 s_y^2} = R^2 \tag{7.51}$$

where R^2 is the coefficient of determination, i.e. the square of the correlation coefficient. From this expression, it is clear that:

- if R^2 is equal to one, both estimates are equal,
- because it must be true that R^2 is less than or equal to one, the inverse estimator is always closer to the mean $\overline{y}$ than the classical estimator.

In work (Chow & Shao, 1990), a relation was derived for determining the probability that R^2 will differ from one by a small prescribed value q. This probability P_q is expressed by the equation

$$P_q = P(F_N < (n-2)(1-q)/q) \tag{7.52}$$

Here, F_N denotes the non-central Fisher Snedecor distribution quantile with 1 and $n - 2$ degrees of freedom and the non-centrality parameter $s_y^2 a_1^2 / \sigma_y^2$. When choosing $P_q = 0.05$ and $q = 1\%$, the two estimates practically do not differ if

$$A = \frac{s_y^2 a_1^2}{n \sigma_y^2} \leq 152 \tag{7.53}$$

In practical calculations, it is possible to replace individual parameters with their estimates. A more precise procedure is described in (Chow & Shao, 1990). Interestingly, the asymptotic bias of the inverse estimator is $(n - 3)$ times higher than the asymptotic bias of the classical estimate. For extremely small samples $2 < n \leq 8$, the mean squared error of the estimate $\hat{y}_{\mathrm{in}}$ is always smaller than that of the estimate of $\hat{y}_{\mathrm{cl}}$.

7.7.3 *Comparison of* C- *and* I-*Calibration Lines*

By using Eq. (7.47), it is possible to express the relationship between estimates $\hat{y}_{\mathrm{in}}$ and $\hat{y}_{\mathrm{cl}}$ in the form

$$|\hat{y}_{\mathrm{in}} - \overline{y}| \leq |\hat{y}_{\mathrm{cl}} - \overline{y}| \tag{7.54}$$

This inequality implies that:

(a) $\hat{y}_{\mathrm{in}}$ is closer to the center than $\hat{y}_{\mathrm{ce}}$,
(b) for $\hat{\sigma}_c^2 \to 0$ is $\hat{y}_{\mathrm{in}} = \hat{y}_{\mathrm{ce}}$,
(c) **I**-calibration better describes the behavior of the data in the range of the center $(\overline{x}, \overline{y})$ and *C*-calibration on the tails of the data range.

Considering the minimum root mean square error MSE $= E(y - \hat{y})$, the **I**-calibration is better in the interval

$$\left\{\left(\overline{y} - \sqrt{2\,s_y^2 + s_x^2/a_1}, \overline{y} + \sqrt{2\,s_y^2 + s_x^2/a_1}\right)\right\}$$

than **C**-calibration. In practical calculations, parameters a_1 and σ_x^2 are replaced again by estimates $\hat{a}_1$ and $\hat{\sigma}_c^2$ determined by the above relations.

7.7.4 Combined Calibration

This calibration uses an appropriate weighted combination of $\hat{y}_{\text{cl}}$, $\hat{y}_{\text{in}}$, and $\overline{y}$. One of the first combined estimators is based on the "shift" the classical estimator $\hat{y}_{\text{cl}}$ toward the center of gravity.

$$\hat{y}_p c \hat{y}_{\text{cl}} + (1 - c)\overline{y} \tag{7.55}$$

It can be shown that there exists a value of c for which this estimate is better than the estimate of $\hat{y}_{\text{cl}}$. Another combined estimate proposed in (Srivastava & Singh, 1989) has the form.

$$\hat{y}_{\text{op}} = v\hat{y}_{\text{cl}} + (1 - v)\hat{y}_{\text{in}} \tag{7.56}$$

The choice of $v = \frac{1}{n-2}$ results in an asymptotically unbiased estimate with a small MSE. It is always better than the classical estimator $\hat{y}_{\text{cl}}$ and also better than the inverse estimator $\hat{y}_{\text{in}}$ if the true values are far from the mean $\overline{y}$. Around the mean are the best inverse estimates $\hat{y}_{\text{in}}$.

7.7.5 Calibration for Both Variables Subjected by Random Errors

This type of calibration is closely related to the regression task for the case where both variables are subjected by random errors (Kendall & Stuart, 1973). In general, it is assumed here that the data contain variables with additive random errors, and thus

$$\begin{aligned} x_i &= \delta_i + \varepsilon_{ix} = k_0 + k_1\left(y_i - \varepsilon_{iy}\right) + \varepsilon_{ix} \quad \varepsilon_x = N\left(0, \sigma_x^2\right) \\ y_i &= \eta_i + \varepsilon_{iy}, \quad \varepsilon_y = N\left(0, \sigma_y^2\right) \end{aligned} \tag{7.57}$$

The real (unknown) values of the variables are connected by the relation $\delta_i = k_0 + k_1 \eta_i$, where η_i is the actual unknown deterministic (adjusted, see Fig. 7.1) values. The errors ε_x and ε_y are assumed to be mutually independent, normally distributed, and ε_x is uncorrelated with the quantity δ. Models of this type are referred to as *functional*. Also for these models of errors and least squares criterion, the calibration line passes through the center of gravity of the points with coordinates $(\overline{x}, \overline{y})$, which means that only the slope $\hat{k}_1$ needs to be estimated and the intercept is calculated from the equation $\hat{k}_0 = \overline{x} - \hat{k}_1 \overline{y}$.

One of the first mentions of the so-called orthogonal regression, when perpendicular distances from the regression line are minimized, was published by Adcock in the Journal of Analytical Chemistry as early as 1878 (Adcock, 1878). It is actually a case where both variables are loaded with random errors with equal variances $\sigma_x^2 = \sigma_y^2$. A generalization to the case when $\sigma_x^2 \neq \sigma_y^2$ and variance ratio $P = \sigma_y^2/\sigma_x^2$ is known was published by Kummel (1879) a year later. Orthogonal regression as a special example of minimizing the length of the data projection onto the calibration line was described by Deming (1931). In this case, the generalizations to unequal variances can easily be made.

Wald (1940) proposed a procedure that does not depend on assumptions about the structure of errors in individual variables. It is based on the division of data arranged in ascending order according to the quantity y (instead of y_i, the order statistic $y_{(i)}$ is used) into two groups G1 and G2. Here, G1 contains the first half of data arranged in this way (for an even number of data n, there are data from 1 to $m = n/2$) and G2 contains the second half of data arranged in this way (from $m + 1$ to n). The estimate of the slope of the calibration line b_{1o} is in this case equal to

$$b_{1o} = \frac{\sum_{i=1}^{m} y_{(i)} - \sum_{i=m+1}^{n} y_{(i)}}{\sum_{i=1}^{m} x_{(i)} - \sum_{i=m+1}^{n} x_{(i)}} \tag{7.58}$$

The problem is that the arrangement (ordinal statistics) is correct only when the theoretical values of z_i and not the random variable y_i are known. The properties of slope estimator b_{1o} are described in detail in the work of Gupta and Amanullah (1970). Bartlett (Roy, 1994) proposed dividing ordered data into three groups, and Neyman and Scott (1951) proposed another method of grouping.

It is also simple to use the slope b_{1g} valid for the case of both variables with the same variance in the form of geometric mean of the a_1 and $1/b_1$ slopes for the **C**- and **I**-calibration models. So, it is true that

$$b_{1g} = \sqrt{\frac{a_1}{b_1}} = \sqrt{\frac{\text{cov}_{xy}}{s_y^2} \Big/ \frac{\text{cov}_{xy}}{s_x^2}} = \text{sign}\left(\text{cov}_{xy}\right)\sqrt{\frac{s_x^2}{s_y^2}} \tag{7.59}$$

The average values of slope estimate $\hat{b}_{1g}$ for simulated samples **CS**, **IS**, and **OS** are shown in Table 7.1.

In Fig. 7.11, the histograms of estimates of slopes $\hat{b}_{1g}$ for the simulated samples **IS**, **CS**, and **OS** are shown.

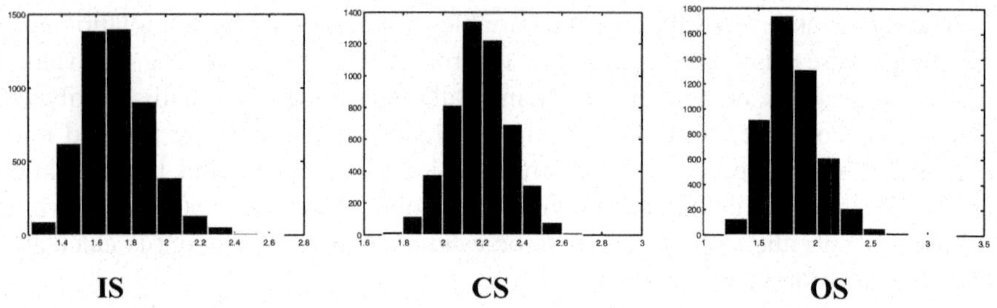

Fig. 7.11 Histograms of slope $\hat{b}_{1g}$ estimates for simulated samples

It can be seen that the distribution of estimates is approximately symmetric for all types of generated data and is best for **CS** samples. The average value of the estimate b_{1g} is closest to the specified (here 2) for the **IS** data case. The value of d_p is relatively low, so the estimates are not very sensitive to the type of error action.

Tessier (1948) showed that the estimate b_{Ig} corresponds to the minimum of the sum of the products of horizontal and vertical deviations from the calibration line, i.e. the minimum of the S_{OC} criterion (Eq. 7.16). To ensure its unbiasedness, it must be true that the variance ratio is equal to one, i.e. $P = 1$ [see Jolicoeur (1975)]. Creasy (1956) showed that the estimate of b_{1g} can have up to infinite variance, in the case that the variance of the data is so large that it is impossible to distinguish whether the original line or a line perpendicular to it describes the data better.

For a known variance ratio $P = \sigma_y^2/\sigma_x^2$, a target quantity prediction $\hat{y}_0$ for the magnitude of the signal x can be determined directly from the relation

$$\hat{y}_0 = \overline{y} - \left[\theta + s_y(\mathrm{cov}_{xy})\sqrt{\theta^2 + P}\right](x - \overline{x}), \theta = \frac{s_y^2 - Ps_x^2}{2\mathrm{cov}_{xy}} \tag{7.60}$$

The slope of this calibration line for a known variance ratio is

$$\hat{b}_{1P} = \left[\theta + \mathrm{sign}(\mathrm{cov}_{xy})\sqrt{\theta^2 + P}\right] \tag{7.61}$$

where sign(x) is the sign function. For $x < 0$, sign(x) = − 1 and for $x > 0$, sign(x) = 1.

The average values of slope estimate $\hat{b}_{1P}$ for simulated samples **CS**, **IS**, and **OS** are shown in Table. 7.1.

In Fig. 7.12, the histograms of estimates of slopes $\hat{b}_{1P}$ for the simulated samples **IS**, **CS**, and **OS** are shown.

It is visible that the distribution of estimates is approximately symmetric for all types of generated data and is best for **CS** data. The mean value of the estimate $\hat{b}_{1P}$ is closest to the given one (here 2) for the **OS** data case, which is to be expected. The value of d_p is relatively low, so the estimates are not very sensitive to the type of error action.

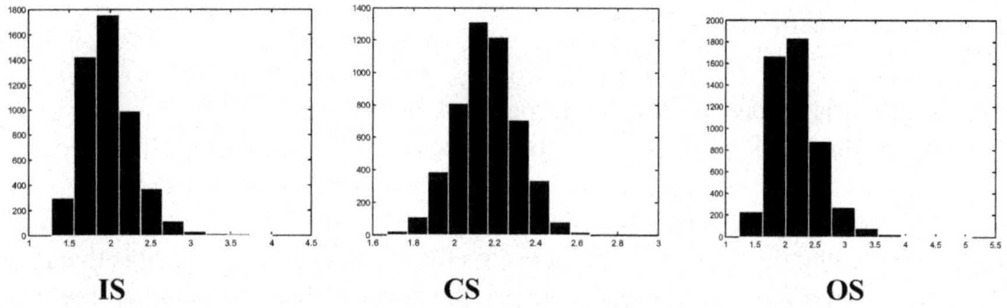

Fig. 7.12 Histograms of slope $\hat{b}_{1P}$ estimates for simulated samples

Details of this calibration and the construction of confidence intervals for the target quantity can be found in the book (Meloun & Militký, 2012).

An interesting variant is the use of the so-called method of least sines (MLS), which corresponds to the minimization of the sum of values $\sin^2(\varphi_i) = \left(\frac{d_i}{c_i}\right)^2$ (Han, 2011). Here, φ_i is the angle between the calibration line and the line connecting the centroid of the data $(\overline{x}, \overline{y})$ to the point (x_i, y_i). This angle is shown in Fig. 7.13. The quantity d_i is the perpendicular distance of the i-th point from the regression line and c_i is the Euclidean distance between the i-th point and the data centroid, i.e. $c_i = \sqrt{(x_i - \overline{x})^2 + (y_i - \overline{y})^2}$.

In work (Han, 2011), it is shown that MLS estimates correspond to the components of the eigenvector, which corresponds to the smallest eigenvalue of the covariance matrix weighted by the weights c_i in the method of principal components.

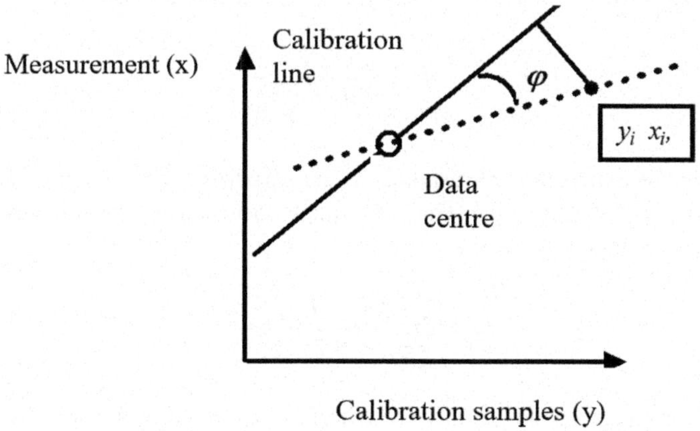

Fig. 7.13 Definition of angle φ

7.7.6 O-Calibration Line

O-calibration is a special case where perpendicular distances from the regression line are minimized. It corresponds to the choice of $P = 1$ in Eq. (7.6). It can be seen from Fig. 7.3 that the line perpendicular to the calibration model $x = a_0 + a_1 y$ and passing through the given point (x_i, y_i) has the form $x = x_i - \frac{1}{a_1}(y - y_i)$. To determine the perpendicular distance of the point (x_i, y_i) from the calibration line, it is therefore necessary to find the intersection D between this line and the perpendicular line from the system of two equations

$$x = a_0 + a_1 y \text{ and } x = x_i - \frac{1}{a_1}(y - y_i) \tag{7.62}$$

The solution x_d for the x coordinate has the form

$$x_d = \frac{x_i + a_1(y_i - a_0)}{1 + a_1^2} \tag{7.63}$$

After substituting into the calibration line, the y_d value of the y coordinate is then found.

$$y_d = \frac{a_1 x_i + a_1^2 y + a_0}{1 + a_1^2} \tag{7.64}$$

The Euclidean distance $d_i = \sqrt{e_{ci}^2}$ between the points $D = (x_d, y_d)$ and (x_i, y_i) can then be expressed on the basis of the Pythagorean theorem in the form

$$d_i = \frac{|a_1 x_i + a_0|}{\sqrt{1 + a_1^2}} \tag{7.65}$$

This distance therefore corresponds to the perpendicular projection of the point (x_i, y_i) onto the calibration line. The least squares criterion of perpendicular distances has for this case the form

$$S_o = \sum_{=1}^{n} \frac{(x_i - a_1 y_i + a_0)^2}{1 + a_1^2} \tag{7.66}$$

This really corresponds to the S_{wx} criterion expressed by Eq. (7.12). From the nullity of the derivatives S_o according to a_0, it follows that the calibration model passes through the center of gravity and $\hat{a}_0 = \overline{x} - \hat{a}_1 \overline{y}$. From the nullity of the derivative S_o according to a_1, a quadratic equation emerges after adjustments.

$$\alpha a_1^2 + \beta a_1 + \gamma = 0 \tag{7.67}$$

where $\alpha = \bar{z} - \bar{x}\,\bar{y}$, $\beta = \bar{x}_q - \bar{y}_q - \bar{x}^2 + \bar{y}^2$, $\gamma = \bar{x}\,\bar{y} - \bar{z}$.

Used symbols are defined by relations

$$\bar{z} = \left(\sum_{i=1}^{n} x_i y_i\right)/n \quad \bar{x}_q = \left(\sum_{i=1}^{n} x_i^2\right)/n \quad \bar{y}_q = \left(\sum_{i=1}^{n} y_i^2\right)/n \tag{7.68}$$

The solution of the quadratic Eq. (7.68) has the form

$$\hat{a}_1(1, 2) = \frac{-\beta \pm \sqrt{\beta^2 - 4\alpha\gamma}}{2\alpha} \tag{7.69}$$

From these two solutions, the one that corresponds to the minimum S_o is then selected. Jackson and Dunlevy (1988) found slope estimate $\hat{a}_1$ using variances and covariance

$$\hat{a}_1 = \frac{s_x^2 - s_y^2 + \sqrt{\left(s_x^2 - s_y^2\right)^2 + 4\mathrm{cov}_{xy}^2}}{2\,\mathrm{cov}_{xy}} \tag{7.70}$$

The average values of slope estimate $\hat{b}_{1k}$ for simulated samples **CS**, **IS**, and **OS** are shown in Table 7.1.

In Fig. 7.14, the histograms of estimates of slopes $\hat{b}_{1k}$ for the simulated samples **IS**, **CS**, and **OS** are shown.

It can be seen that the distribution of estimates is approximately symmetric for all types of generated data and is best for **CS** sample. The mean value of the estimate $\hat{b}_{1k}$ is closest to the given one (here 2) for the OS data case, which is to be expected. The value of d_p is the lowest, i.e. the estimates are the least sensitive to the type of error action.

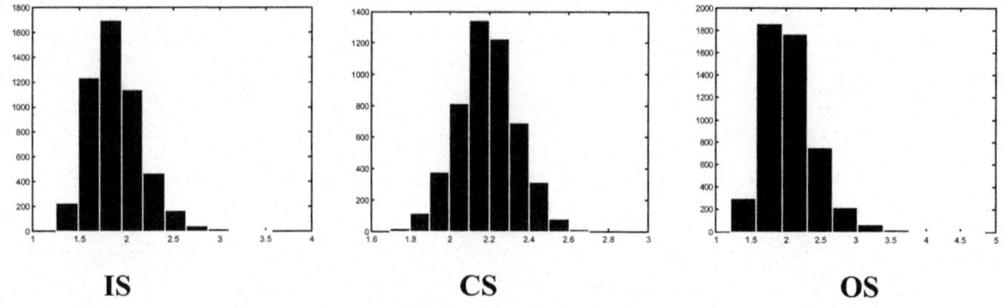

Fig. 7.14 Histograms of slope $\hat{b}_{1k}$ estimates for simulated samples

It can be seen that there are a number of procedures to find an estimate $\hat{a}_1$ of the slope for the O-calibration. One simple procedure is to determine a calibration model with coefficients corresponding to the coefficients of the eigenvector of the largest principal component in the PCA principal component method (Meloun & Militký, 2012).

Example 7.2 A simple example from work (Martens & Naes, 1993) is used to illustrate linear calibration. The data are shown in Table 7.2. The basic statistical characteristics are

$$\overline{x} = 8.6346 \quad s_x^2 = 72.327 \quad \overline{y} = 6.4808 \quad s_y^2 = 70.308 \quad \text{cov}_{xy} = 66.284$$

C-Calibration

By linear least squares, it was found that $x = 2.524 + 0.9427y$, $\hat{y}_{ce} = -2.68 + 1.06x$. The correlation coefficient was $R = 0.9051$ and the residual variance $\hat{\sigma}_c^2 = 0.89427$. Calibration limits were also calculated (Meloun & Militký, 2012):

Limit of detection: 5.718.
Critical limit: 2.859.
Determination limit: 10,320.

Figure 7.15 shows the calibration line and its 95% confidence interval.

In Fig. 7.16, the course of the concentrated probability function $\ln L(\mathbf{a})$ determined from Eq. (7.53) is shown.

I-Calibration

By linear least squares, it was found that $\hat{y}_{\text{in}} = -1.432 + 0.916x$.

Table. 7.2 Data for linear calibration

Sample number	Target y	Measurement x
1	2	5
2	4	6.5
3	4.5	6.25
4	5	6.5
5	5	8
6	6	6.5
7	6.5	8
8	6.5	9
9	7.75	10.5
10	8	11
11	9	12.5
12	9,5	10.5
13	10.5	12

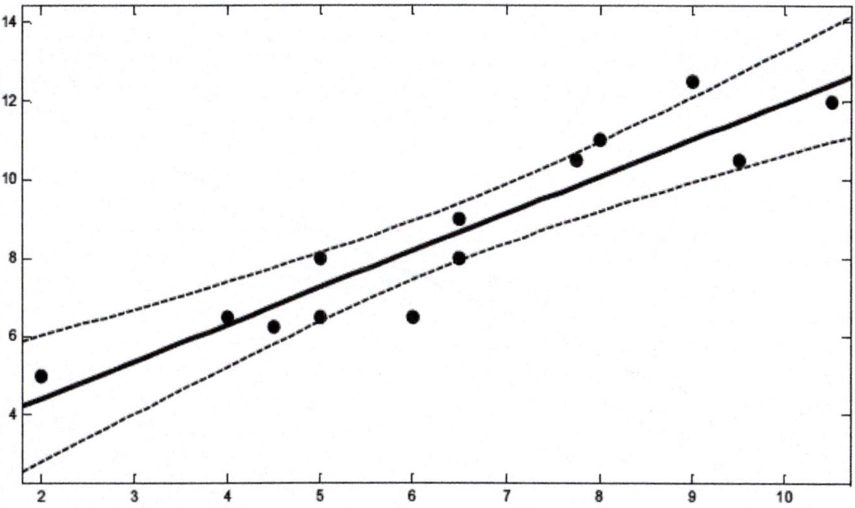

Fig. 7.15 C-calibration line

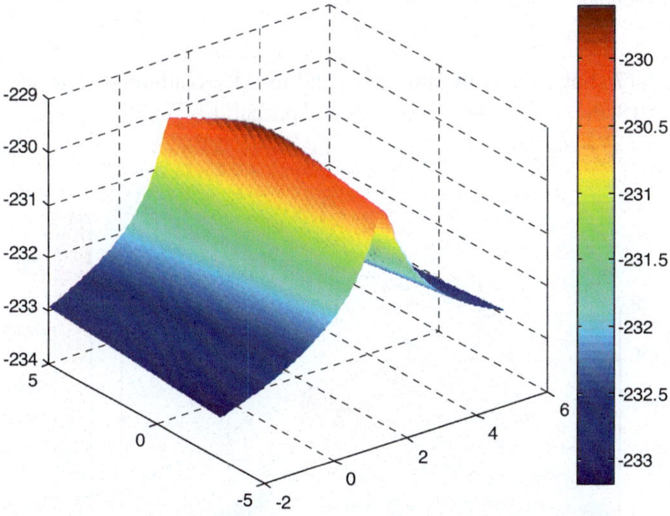

Fig. 7.16 Concentrated likelihood function for *C*-calibration

Calibration limits were also calculated (Meloun & Militký, 2012):

Limit of detection: 4.53.
Critical limit: 1.54.
Determination limit: 13.53.

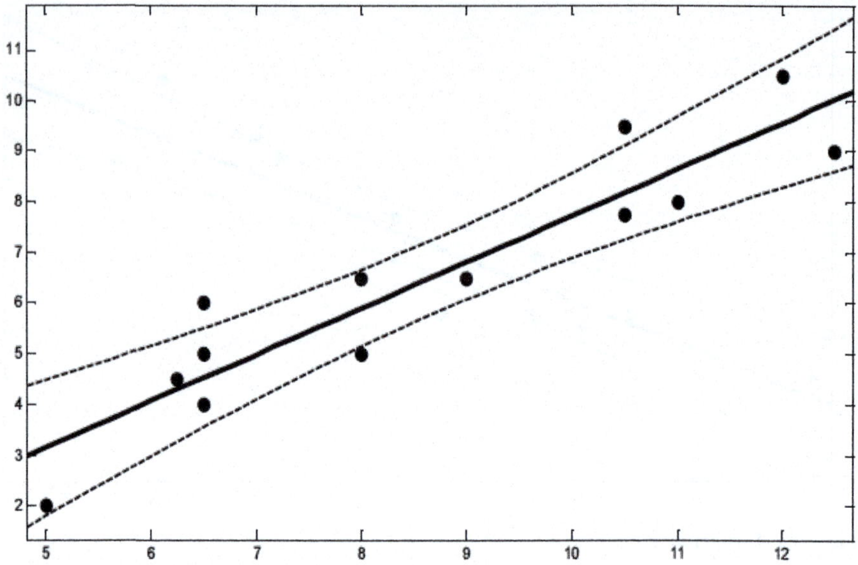

Fig. 7.17 I-calibration line

Figure 7.17 shows the calibration line and its 95% confidence interval.

In Fig. 7.18, the course of the concentrated probability function $\ln L(\mathbf{a})$ determined from Eq. (7.53) is shown.

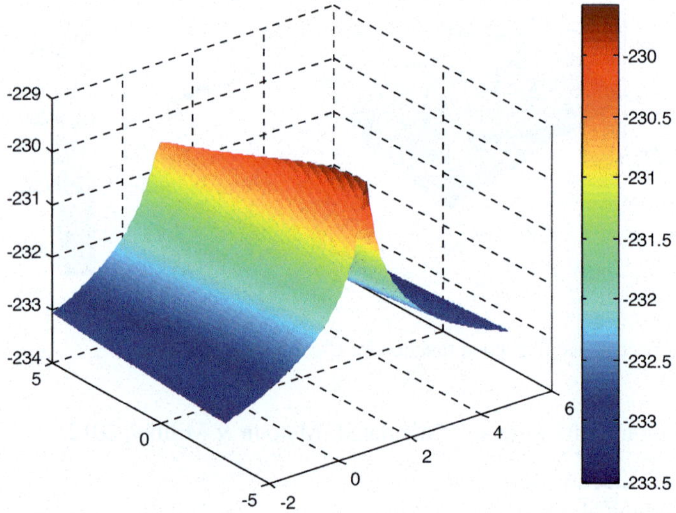

Fig. 7.18 Concentrated likelihood function for I-calibration

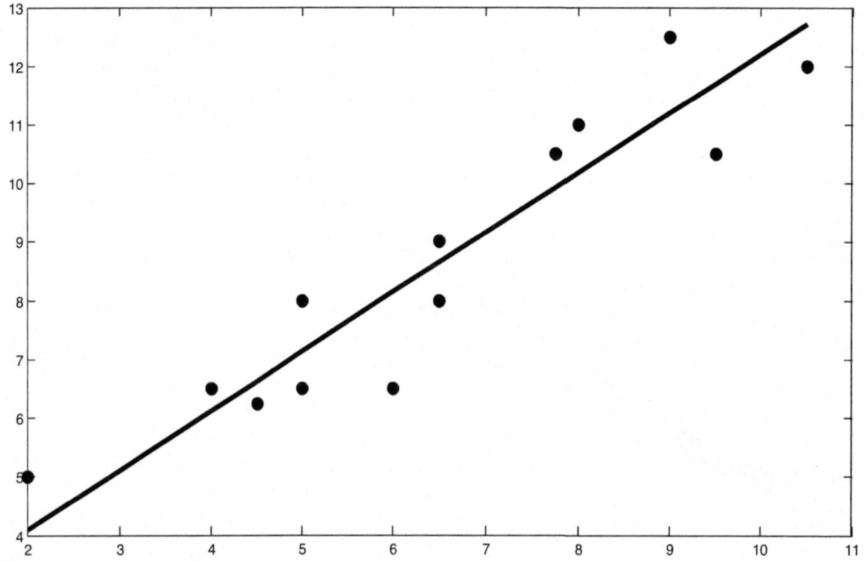

Fig. 7.19 O-calibration line

From Eq. (7.52), the coefficient of determination $R^2 = 0.864$ was determined and from Eq. (7.53), parameter $A = 5.375$ was calculated. Thus, the classical estimate does not differ significantly from the inverse one. The interval in which $\hat{y}_{\text{in}}$ is better than $\hat{y}_{\text{cl}}$ in terms of MSE is equal to $\langle -5.4173, 18.378 \rangle$. It can therefore be summarized that $\hat{y}_{\text{in}}$ is better than $\hat{y}_{\text{cl}}$ in a wide interval, but the difference between the two estimates is practically insignificant.

O-Calibration

By orthogonal linear least squares, it was found that $\hat{y}_{\text{in}} = 2.054 + 1.0153x$. Figure 7.19 shows the calibration line.

A comparison of all three calibrations is shown in Fig. 7.20.

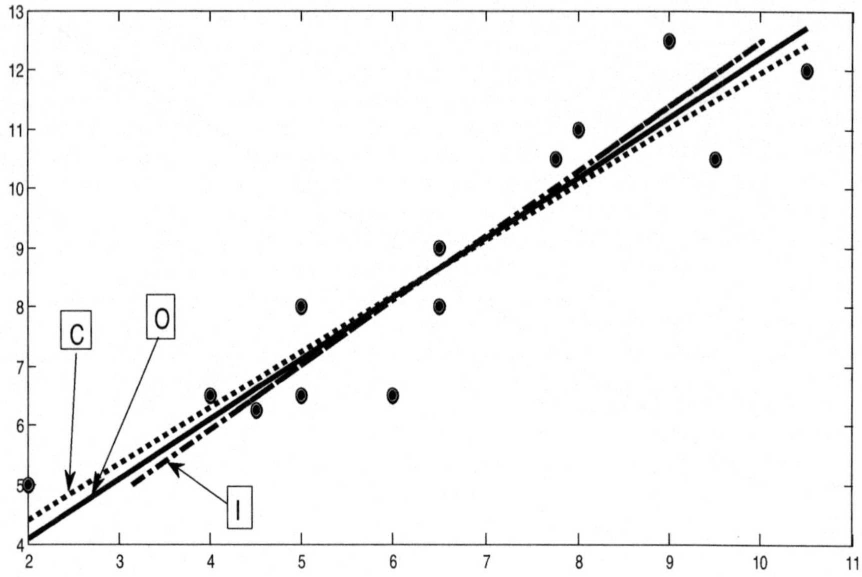

Fig. 7.20 Comparison of individual calibrations

7.8 Conclusion

As can be seen, the choice of calibration type depends on the nature of the variables and their variability.

<u>Variable $x(M)$</u>:

- usually quite precisely and simply experimentally determined (electric quantity, absorbance),
- error ε_x includes random component and bias due to unconsidered variables (temperature, ambient humidity, etc.),
- variance σ_x^2 can be non-constant.

<u>Variable $y(T)$</u>:

- usually physical variable determined often from external information (other devices, standards, etc.),
- error ε_y includes measurement errors by other methods,
- variance σ_y^2 is usually an increasing function of y.

The variance ratio $P = \sigma_y^2/\sigma_x^2$ can vary within the limits $(0, \infty)$. The choice between I and C-calibration can be made either on the basis of the variance ratio or using other criteria. Also, in linear calibration, the problem of choosing a measurement model must be solved in general.

References

Adcock, R. J. (1878). A problem in least squares. *The Analyst, 5*(2), 53. https://doi.org/10.2307/2635758

Berkson, J. (1950). Are there two regressions? *Journal of the American Statistical Association, 45*(250), 164. https://doi.org/10.2307/2280676

Cheng, C.-L., & Van Ness, J. W. (1994). On estimating linear relationships when both variables are subject to errors. *Journal of the Royal Statistical Society Series B: Statistical Methodology, 56*(1), 167–183. https://doi.org/10.1111/j.2517-6161.1994.tb01969.x

Chow, S.-C., & Shao, J. (1990). On the difference between the classical and inverse methods of calibration. *Applied Statistics, 39*(2), 219. https://doi.org/10.2307/2347761

Creasy, M. A. (1956). Confidence limits for the gradient in the linear functional relationship. *Journal of the Royal Statistical Society Series B: Statistical Methodology, 18*(1), 65–69. https://doi.org/10.1111/j.2517-6161.1956.tb00211.x

De Boor, C. (1998). *A practical guide to splines.* Springer Verlag.

Deming, W. E. (1931). XI. The application of least squares. *The London, Edinburgh, and Dublin Philosophical Magazine and Journal of Science, 11*(68), 146–158. https://doi.org/10.1080/14786443109461671

Gupta, Y. P., & Amanullah. (1970). A note on the moments of the Wald's estimator. *Statistica Neerlandica, 24*(3), 109–123. https://doi.org/10.1111/j.1467-9574.1970.tb00113.x

Han, H. (2011). *Least Sine Squares and Robust Compound Regression Analysis*, Ph.D. Thesis. Stony Brook University.

Huet, S., Bouvier, A., Gruet, M.-A., & Jolivet, E. (1996). *Statistical tools for nonlinear regression.* Springer. https://doi.org/10.1007/978-1-4757-2523-0

Jackson, J. D., & Dunlevy, J. A. (1988). Orthogonal least squares and the interchangeability of alternative proxy variables in the social sciences. *The Statistician, 37*(1), 7. https://doi.org/10.2307/2348374

Jolicoeur, P. (1975). Linear regressions in fishery research: Some comments. *Journal of the Fisheries Research Board of Canada, 32*(8), 1491–1494. https://doi.org/10.1139/f75-171

Kendall, M. G., & Stuart, A. (1973). *The advanced theory of statistics* (Vol. 2). Charles Griffin and Co.

Krutchkoff, R. G. (1969). Classical and inverse regression methods of calibration in extrapolation. *Technometrics, 11*(3), 605. https://doi.org/10.2307/1267027

Kummell, C. H. (1879). Reduction of observation equations which contain more than one observed quantity. *The Analyst, 6*(4), 97. https://doi.org/10.2307/2635646

Lundgren, H. P. (1969). New concepts in evaluating fabric handle. *Textile Chemist and Colorist, 1*(35).

Martens, H., & Naes, T. (1993). *Multivariate calibration.* Wiley.

Meloun, M., & Militký, J. (2012). *Interactive statistical data analysis.* Carolinum. (in Czech)

Neyman, J., & Scott, E. L. (1951). On certain methods of estimating the linear structural relation. *The Annals of Mathematical Statistics, 22*(3), 352–361. https://doi.org/10.1214/aoms/1177729581

Pan, N., Yen, K. C., Zhao, S. J., & Yang, S. R. (1988). A new approach to the objective evaluation of fabric handle from mechanical properties. *Textile Research Journal, 58*(8), 438–444. https://doi.org/10.1177/004051758805800802

Rosenblatt, J. R., & Spiegelman, C. H. (1981). [A Bayesian analysis of the linear calibration problem]: Discussion. *Technometrics, 23*(4), 329. https://doi.org/10.2307/1268218

Roy, T. (1994). Fitting a straight line when both variables are subject to error: Pharmaceutical applications. *Journal of Pharmaceutical and Biomedical Analysis, 12*(10), 1265–1269. https://doi.org/10.1016/0731-7085(94)00057-3

Srivastava, V. K., & Singh, N. (1989). Small-disturbance asymptotic theory for linear-calibration estimators. *Technometrics, 31*(3), 373. https://doi.org/10.2307/3556147

Teissier, G. (1948). La Relation D'Allometrie sa Signification Statistique et Biologique. *Biometrics, 4*(1), 14. https://doi.org/10.2307/3001695

Wald, A. (1940). The fitting of straight lines if both variables are subject to error. *The Annals of Mathematical Statistics, 11*(3), 284–300. http://www.jstor.org/stable/2235677

GPSR Compliance

The European Union's (EU) General Product Safety Regulation (GPSR) is a set of rules that requires consumer products to be safe and our obligations to ensure this.

If you have any concerns about our products, you can contact us on ProductSafety@springernature.com

In case Publisher is established outside the EU, the EU authorized representative is:

Springer Nature Customer Service Center GmbH
Europaplatz 3
69115 Heidelberg, Germany

The manufacturer's authorised representative in the EU is Springer Nature Customer Service Centre GmbH, Europaplatz 3, 69115 Heidelberg, Germany. If you have any concerns regarding our products, please contact ProductSafety@springernature.com

Printed and bound by CPI Group (UK) Ltd, Croydon, CR0 4YY
01/10/2026
02225907-0003